T0213291

Data Visualisation with R

Thomas Rahlf

Data Visualisation with R

111 Examples

Second Edition

 Springer

Thomas Rahlf
Rheinische Friedrich-Wilhelms-Universität
Bonn
Bonn, Germany

ISBN 978-3-030-28446-6   ISBN 978-3-030-28444-2   (electronic)
https://doi.org/10.1007/978-3-030-28444-2

This Springer imprint is published by the registered company Springer Nature Switzerland AG.
The registered company address is: Gewerbestrasse 11, 6330 Cham, Switzerland

Preface to the Second English Edition

Due to the continued interest in the book, Springer Verlag has decided to publish an English translation of the second edition. Again, many thanks to Ralf Gerstner from Springer, and also to Tracey Duffy for translating all of the additions.

In 2017, Ulrike Grömpling reviewed the English translation of the first edition in the *Journal of Statistical Software*. Inspired by the book, and following the release of the second edition, she has published an R package "prepplot" which greatly simplifies the configuration of figure regions for base R graphics. I highly recommend trying this one out.

Meanwhile, the third edition of Paul Murrell's standard work *R Graphics* has been published. When I read in his foreword that my book—among others—was one inspiration for restructuring part of his book, I felt very honored.

Bonn, Germany   Thomas Rahlf

Preface to the Second German Edition

The feedback on the first edition of this book (published by Open Source Press in 2014 and now out of print), and on the English edition that has been published in the meantime, was so pleasing that I was happy to accept the offer made by Springer Verlag to publish a new, updated edition of the book. The concept of explaining complete examples and the restriction to Base Graphics have been retained. Compared to the first edition of the book, this new, updated edition contains 11 additional, new examples, hence the new subtitle "111 Examples". There are two main additions: firstly, a section on visualising network relationships has been added to the chapter on categorical data. In addition to examples of classic network diagrams, an adapted heat map, and a multiple bar chart, this section also contains a chord diagram and a riverplot. Although the chord diagram and riverplot may appear unusual at first glance, they can now be found in publications of respected scientific journals such as Science, Nature or Cell. Examples on the use of georeferenced grid formats and on cartograms have been added to the chapter on maps. Three examples for the integration of data created with R in interactive JavaScript illustrations have also been added to the book. R now provides multiple concepts and packages that can be used to create JavaScript visualisations more or less directly. Ultimately, such packages form a type of container in R. In each case, a specific syntax developed by the authors of these packages translates the scripts written in this form into the notation required for the underlying JavaScript library. This means that we have to rely on the scope of the language of the R package and on the quality and flexibility of the translation routines. I am not sure if this is the right path. In this book, I have taken a different path. In three examples in Chap. 12, the data are prepared with R such that they are integrated in existing, only slightly adapted JavaScript code. This is done using Highcharts and Mapael, two JavaScript libraries that can be used to create very aesthetic illustrations "out of the box" with minimal change effort.

The primary objective of this book is still to explain how to create presentation graphics. For the exploratory visualisation within the scope of data analysis, I refer to the book Graphical Data Analysis with R (CRC Press, 2015) by Antony Unwin. The first people I would like to thank are Agnes Herrmann and Iris Ruhmann at

Springer, who have enabled me to create this updated edition of the book. For helpful comments, suggestions, and exchange of ideas for this edition, I would also like to thank Alberto Cairo, Martin S. Fischer, Sebastian Jeworutzki, Nikola Sander, Antony Unwin, January Weiner, and Stefan Fichtel. January Weiner has kindly included a comment in his riverplot package that is helpful for example 6.4.4.

Bonn, Germany   Thomas Rahlf

Preface to the English Edition

This book is a translation of the German book "Datendesign mit R" that was published 2014 by Open Source Press. Due to the encouraging strong interest in the German edition Springer Verlag offered to publish an English translation. First of all I would like to thank Ralf Gerstner from Springer for this and for his helpful suggestions for improvement, as well as Annika Brun for translating most of the text, Colin Marsh for copy editing, and Katja Diederichs for converting all scripts from German to English. Last year I benefited a lot from a communication with Antony Unwin. His book "Graphical Data Analysis with R" can be seen as complementary to my own: while this one focusses on presentation of graphics, you will benefit from his book if you are interested in exploring data graphically.

Bonn, Germany   Thomas Rahlf

Preface to the German Edition

Some 20 years ago, when I reviewed a score of books on statistical graphics and graphic-based data analysis, things were completely different: there were proprietary formats and operating systems, their character sets were incompatible, and graphic and statistical software was expensive. Since the turn of the century, the situation has changed fundamentally: the Internet has come of age, open-source projects have attracted more and more followers, and a handful of enthusiasts provided version 1.0 of the free statistical programming language R. Many developers were inspired to collaborate on this project. R reached version 3 in 2013, and in addition to the basic software, more than 7000 freely available extension packs are currently available. Companies and organisations such as Google, Facebook or the CIA are using R for their data analysis. Its graphic capabilities are again and again emphasised as its strong point. Pretty much all technologies relevant for data visualisation are quickly integrated into R. Through numerous functions, detailed designs of every imaginable figure, creation of maps and much more are made possible. All it takes is to know how—and that is where this book wants to contribute.

What This Book Wants to Be—and What It Doesn't Want to Be

This book is not an introduction that systematically explains all the graphic tools R has to offer. Rather, its aim is to use 100 complete script examples to introduce the reader to the basics of designing presentation graphics, and to show how bar and column charts, population pyramids, Lorenz curves, box plots, scatter plots, time series, radial polygons, Gantt charts, heat maps, bump charts, mosaic and balloon charts, and a series of different thematic map types can be created using

R's Base Graphics System. Every example uses real data and includes step-by-step explanations of the figures and their programming. The selection is based on my personal experiences—it is likely that readers will find one or another illustration lacking, and consider some too detailed. However, a large scope should be covered. This book is aimed at:

- R experts: You can most likely skip Part I. For you, the examples are particularly useful, especially the code.
- Readers that have heard of R and maybe even tried R before and are not daunted by programming; you will profit from both parts.
- Beginners: for you, the finished graphics pictured here will be most helpful. You will see what R can do. Or, in other words: you will realise that there is such a tool as R, and that it can be used to create graphics you have wanted to create for a long time, but merely never knew how. The code will be too complicated for you, but you may be able to commission others to do your graphics programming in R.

Windows, Mac, and Linux

All of the scripts and working steps will yield identical results when executed in Windows, Mac OS X or Linux. All of the examples were created in Mac OS X and then tested in Ubuntu 12.04 and an evaluation copy of Windows 8.1.

Acknowledgements

The following people deserve my thanks for hints, comments, feedback, data, discussions or help: Gregor Aisch, Insa Bechert, Evelyn Brislinger, Giuseppe Casalicchio, Arnulf Christl, Katja Diederichs, Günter Faes, Mira Hassan, Mark Heckmann, Daniel Hienert, Bruno Hopp, Uwe Ligges, Lorenz Matzat, Meinhard Moschner, Stefan Müller, Paul Murrell, David Phillips, Duncan Temple Lang, Martijn Tennekes, Patrick R. Schmid, Thomas Schraitle, Valentin Schröder, Torsten Steiner, Michael Terwey, Katrin Weller, Bernd Weiss, Nils Windisch, Benjamin Zapilko, and Lisa Zhang. This manuscript particularly benefited from discussions with an infographic designer and a data journalist. Stefan Fichtel looked over every figure and provided critical feedback. For selected figures, he designed his own suggestions; this has been an invaluable help. We did not always agree, and I have disregarded his advice here or there. Therefore, any remaining errors and shortcomings are mine. Björn Schwentker went to the trouble of proof reading large parts of the manuscript. I am very grateful for his valuable notes which have surely made some parts of the text clearer and more readable. Finally, I want to thank Markus Wirtz for tackling the experiment of ultimately printing everything into a book.

On the Internet

The figures are conceived for different final output options. The format of the book implies that some details have become very small, e.g. in maps and radial column charts. Particularly for such cases, please refer to the book's website, on which all figures are available in high resolution or as vector graphics in PDF format:

http://www.datavisualisation-r.com

Bonn, Germany   Thomas Rahlf

Contents

1	**Data for Everybody**		1
	1.1	Data Visualisation Between Science and Journalism	1
	1.2	Why R?	3
	1.3	The Concept of Data Design	3

Part I Basics and Techniques

2	**Structure and Technical Requirements**		7
	2.1	Terms and Elements	7
	2.2	Illustration Grids	7
	2.3	Perception	10
	2.4	Typefaces	13
		2.4.1 Fonts	15
		2.4.2 Free Typefaces	16
	2.5	Symbols	18
		2.5.1 Symbol Fonts	19
		2.5.2 Symbols in SVG Format	21
	2.6	Colour	21
		2.6.1 Colour Models	21
		2.6.2 Colour in Statistical Illustrations	23

3	**Implementation in R**		25
	3.1	Installation	25
	3.2	Basic Concepts in R	26
		3.2.1 Data Structures	27
		3.2.2 Import of Data	30
	3.3	Graphic Concepts in R	37
		3.3.1 The Paper-Pencil-Principle of the Base Graphics System: High-Level and Low-Level Functions	42
		3.3.2 Graphic Parameter Settings	44
		3.3.3 Margin Settings for Figures and Graphics	50
		3.3.4 Multiple Charts: Panels with mfrow and mfcol	51

		3.3.5	More Complex Assembly and Layout	53
		3.3.6	Font Embedding	55
		3.3.7	Output with cairo_pdf	56
		3.3.8	Unicode in Figures	57
		3.3.9	Colour Settings	60
	3.4		R Packages and Functions Used in This Book	61
		3.4.1	Packages	61
		3.4.2	Functions	66
		3.4.3	Schematic Approach	75
4	**Beyond R**			**77**
	4.1		Additions with LaTeX	77
	4.2		Manual Post-processing and Creation of Icon Fonts Using Inkscape	82
		4.2.1	Post-processing	82
		4.2.2	Creation of Icon Fonts	84
5	**Regarding the Examples**			**89**
	5.1		An Attempt at a Systematics	89
	5.2		Getting the Scripts Running	91

Part II   Examples

6	**Categorical Data**			**95**
	6.1		Bar and Column Charts	95
		6.1.1	Bar Chart Simple	96
		6.1.2	Bar Chart for Multiple Response Questions: First Two Response Categories	101
		6.1.3	Bar Chart for Multiple Response Questions: All Response Categories	105
		6.1.4	Bar Chart for Multiple Response Questions: All Response Categories, Variant	108
		6.1.5	Bar Chart for Multiple Response Questions: All Response Categories (Panel)	110
		6.1.6	Bar Chart for Multiple Response Questions: Symbols for Individuals	113
		6.1.7	Bar Chart for Multiple Response Questions: All Response Categories, Grouped	116
		6.1.8	Column Chart with Two-Line Labelling	121
		6.1.9	Column Chart with 45° Labelling	123
		6.1.10	Profile Plot for Multiple Response Questions: Mean Values of the Responses	124

	6.1.11	Dot Chart for Three Variables	127
	6.1.12	Column Chart with Shares	130
6.2	Pie Charts and Radial Diagrams		132
	6.2.1	Simple Pie Chart	133
	6.2.2	Pie Charts, Labels Inside (Panel)	135
	6.2.3	Seat Distribution (Panel)	136
	6.2.4	Spie Chart	139
	6.2.5	Radial Polygons (Panel)	142
	6.2.6	Radial Polygons (Panel): Different Column Arrangement	144
	6.2.7	Radial Polygons Overlay	145
6.3	Chart Tables		146
	6.3.1	Simplified Gantt Chart	147
	6.3.2	Simplified Gantt Chart: Colours by People	151
	6.3.3	Bump Chart	152
	6.3.4	Heat Map	155
	6.3.5	Mosaic Plot (Panel)	158
	6.3.6	Balloon Plot	160
	6.3.7	Tree Map	162
	6.3.8	Tree Maps for Two Levels (Panel)	164
6.4	Network Relationships		168
	6.4.1	Undirected Network	169
	6.4.2	Chord Diagram	174
	6.4.3	Directed Network	178
	6.4.4	Riverplot	182
	6.4.5	Heat Map for Relationships	185
	6.4.6	Multiple Bar Chart	188
7	**Distributions**		**191**
7.1	Histograms and Box Plots		191
	7.1.1	Histograms Overlay	191
	7.1.2	Column Charts Coloured with ColorBrewer (Panel)	193
	7.1.3	Histograms (Panel)	196
	7.1.4	Box Plots for Groups: Sorted in Descending Order	199
	7.1.5	Box Plots for Groups: Sorted in Descending Order, Comparison of Two Polls	203
7.2	(Population) Pyramids		207
	7.2.1	Pyramid with Multiple Colours	208
	7.2.2	Pyramids: Emphasis on the Outer Areas (Panel)	210
	7.2.3	Pyramids: Emphasis on the Inner Areas (Panel)	213
	7.2.4	Pyramids with Added Line (Panel)	216
	7.2.5	Aggregated Pyramids	217
	7.2.6	Bar Charts as Pyramids (Panel)	220
7.3	Inequality		223
	7.3.1	Simple Lorenz Curve	224

	7.3.2	Lorenz Curves Overlay	226
	7.3.3	Lorenz Curves (Panel)	229
	7.3.4	Comparison of Income Proportions with Bar Chart (Quintile)	231
	7.3.5	Comparison of Income Proportions with Bar Chart (Decile)	234
	7.3.6	Comparison of Income Proportion with Panel-Bar Chart (Quintile)	236

8 Time Series **239**
8.1	Short Time Series		239
	8.1.1	Column Chart for Developments	239
	8.1.2	Column Chart with Percentages for Growth Developments	242
	8.1.3	Quarterly Values as Columns	245
	8.1.4	Quarterly Values as Lines with Value Labels	247
	8.1.5	Short Time Series Overlayed	249
8.2	Areas Underneath and Between Time Series		252
	8.2.1	Areas Between Two Time Series	252
	8.2.2	Areas as Corridor with Time Series (Panel)	254
	8.2.3	Forecast Intervals (Panel)	256
	8.2.4	Forecast Intervals Index (Panel)	259
	8.2.5	Time Series with Stacked Areas	262
	8.2.6	Areas Under a Time Series	264
	8.2.7	Time Series with Trend (Panel)	267
8.3	Presentation of Daily, Weekly and Monthly Values		270
	8.3.1	Daily Values with Labels	270
	8.3.2	Daily Values with Labels and Week Symbols (Panel)	272
	8.3.3	Daily Values with Monthly Labels	276
	8.3.4	Time Series from Weekly Values (Panel)	277
	8.3.5	Monthly Values (Panel)	280
	8.3.6	Monthly Values with Monthly Labels	282
	8.3.7	Monthly Values with Monthly Labels (Layout)	285
8.4	Exceptions and Special Cases		288
	8.4.1	Time Series as Scatter Plot (Panel)	288
	8.4.2	Time Series with Missing Values	290
	8.4.3	Seasonal Ranges (Panel)	293
	8.4.4	Seasonal Ranges Stacked	295
	8.4.5	Season Figure (Seasonal Subseries Plot) with Data Table	297
	8.4.6	Temporal Ranges	300

9 Scatter Plots **303**
9.1	Variants		305
	9.1.1	Scatter Plot Variant 1: Four Quadrants Differentiated by Colour	305
	9.1.2	Scatter Plot Variant 2: Outliers Highlighted	308

	9.1.3	Scatter Plot Variant 3: Areas Highlighted	311
	9.1.4	Scatter Plot Variant 4: Superimposed Ellipse	313
	9.1.5	Scatter Plot Variant 5: Connected Points	316
9.2	Exceptions and Special Cases		319
	9.2.1	Scatter Plot with Few Points	319
	9.2.2	Scatter Plot with User-Defined Symbols	321
	9.2.3	Map of Germany as Scatter Plot	324

10 Maps **327**

10.1	Introductory Examples	327
	10.1.1	Maps of Germany: Local Telephone Areas and Postcode Districts	327
	10.1.2	Filtered Postcode Map	329
	10.1.3	Map of Europe NUTS 2006 (Cut-out)	331
10.2	Points, Diagrams, and Symbols in Maps	333
	10.2.1	Map of Germany with Selected Locations and Outline (Panel)	333
	10.2.2	Map of Germany with Selected Locations (Pie Charts) and Outline	336
	10.2.3	Map of Germany with Selected Locations (Columns) and Outline	338
	10.2.4	Map of Germany as Three-Dimensional Scatter Plot	341
	10.2.5	Map of North Rhine-Westphalia with Selected Locations (Symbols) and Outline	345
	10.2.6	Map of Tunisia with Self-Defined Symbols	347
10.3	Choropleth Maps	350
	10.3.1	Choropleth Map of Germany at District-Level	350
	10.3.2	Choropleth Map of Germany at District-Level (Panel)	353
	10.3.3	Choropleth Map of Europe at Country-Level	358
	10.3.4	Choropleth Map of Europe at Country-Level (Panel)	360
	10.3.5	World Choropleth Map: Regions	364
10.4	Exceptions and Special Cases	366
	10.4.1	World Map with Orthodromes	366
	10.4.2	City Maps with OpenStreetMap Data (Panel)	369
	10.4.3	Georeferenced Map in Grid Format	373
	10.4.4	Cartogram (Panel)	380

11 Illustrative Examples **385**

11.1	Table with Symbols of the "Symbol Signs" Type Face	385
11.2	Polar Area Charts with Labels (Panel)	387
11.3	Polar Area Charts Without Labels (Panel)	394
11.4	Polar Area Chart (Poster)	397
11.5	Nighttime Map of Germany as Scatter Plot	401
11.6	Scatter Plot Gapminder	403
11.7	Map of Napoleon's Russian Campain in 1812/13, by Charles Joseph Minard, 1869	408

12   Interactive Visualisation with JavaScript: Highcharts and Mapael .. 413
      12.1   Scatter Plot in Highcharts 414
      12.2   Time Series in Highcharts 423
      12.3   Choropleth Maps with Mapael 429

Appendix 447
      A Data 447
      B Bibliography 449

Chapter 1
Data for Everybody

The type and scope of data, our attitudes towards them, and their availability have fundamentally changed in recent years. Never before has there been more data than today. Never have they been so readily available. And never have there been greater opportunities for analysis, preparation, and presentation. So-called infographics, frequently animated and interactive, are spreading through the Internet. Genuine, standard-setting offerings are based on the work of extensive teams of experts and become research subjects themselves. In this book, all data are visualised using the free statistical software R. Getting started is easier than for many other programming languages, since R is specifically designed for data and statistics, and for their visualisation.

1.1 Data Visualisation Between Science and Journalism

Some scientists, like mathematician Stephen Wolfram, believe that the process of data analysis can largely be automated and, in that context, even speak about a democratisation of science. Others, like Google chief economist Hal Varian on the other hand, believe that this would mean several skills had to be acquired and that these would become future central key qualifications. Over the last few years, a plethora of new websites, books, and other publications have emerged devoted to visualisation of data. Here, the priority is their narrative rather than exploratory visualisation. One of the most well known examples is the mission of Gapminder author and inventor Hans Rosling to find catchy ways to illustrate statistics on global societal developments for a wide audience. In 2012, Time Magazine counted Hans

---

Electronic Supplementary Material The online version of this chapter (https://link.springer.com/chapter/10.1007/978-3-030-28444-2_1) contains supplementary material, which is available to authorized users.

T. Rahlf, *Data Visualisation with R*, https://doi.org/10.1007/978-3-030-28444-2_1

Rosling among the "100 most influential people in the world". Almost-forgotten social scientists who provided didactic visualisations, above all Otto Neurath, have been rediscovered. However, it is not as if the wheel were being reinvented. Data visualisations have always and continuously played an important role in science. Imaging processes are integral parts of many medical analyses, and almost all natural sciences use figurative representations of data to visually communicate their results. In the scope of statistical methodology, some scientists had already conducted basic research on statistical graphics many years ago. Aside from the works of William S. Cleveland, Edward Tufte's book *The Visual Display of Quantitative Information* was a revolutionising breakthrough. The book was published in 1983, and the first edition already saw 16 reprints. In combination with two subsequently published works, *Envisioning Information* and *Visual Explanations*, Edward Tufte found a genuine way to defining the standard of the topic. Also in economics, there is a long tradition of data presentation. Companies have not only collected and analysed data for internal purposes for years, but have also translated them into figures. Publications specifically prepared for external use showcase impressive displays of presentation graphics in annual reports. Finally, official statistics have been successfully dedicating their efforts towards presenting data not only in tabular but also graphic form. Here, a progressive trend towards better visualisation of official data material is evident virtually from year to year on both a national and international level. The massive influx of data pushing us to analyse them has one side-effect: their new, potential availability and openness brings with it a change in attitude towards usage rights and viewing rights. Increasing demands for transparency not only affect government but also company data. Environmental and weather records, expenditure data or those from the health or education sector, state parliament elections, legislative texts, traffic data or time tables for public transport should be free and publicly available. Big Data and Open Data have led to new methods and new approaches. An innovative variant that adopted the name "Data Science" takes this to mean a combination of programming skills, mathematical and statistical knowledge, and content-specific expertise. Drew Conway visualised this combination using a Venn diagram, which also clearly shows the overlaps.[1]

Generally, data science is highly mathematical and elaborate. However, even the journalistic sector shows a strong interest in data. Data journalism, research, and visualisations primarily offered by The New York Times and The Guardian are on the rise.

Additional popularity is enjoyed by individual offerings from "information designers" such as Catherine Mulbrandon, Stephen Few, Robert Kosara, Ben Fry or Nathan Yau, who develop their own data visualisation software, found consulting businesses, offer global workshops or set up blogs with thousands of registered users. If viewed from the viewpoint of "traditional" statistical graphics, some miss the point, and some are not only considered too colourful, too playful or too busy, but also confusing or even misleading. This is where a recent discussion arose that will eventually profit both sides.

---

[1] http://drewconway.com/zia/2013/3/26/the-data-science-venn-diagram.

1.2   Why R?

In scientific articles, R as programming language has become common and well liked. Outside of science, however, its potential for creation of purpose-adapted figures is not widely known. This is not surprising, since it is a known fact that graphic designers or journalists find programming difficult. Of course, it would be wrong to claim that mastering R was so easy that a first appealing graphic could be created within minutes. On the other hand, it offers several advantages that could be extremely valuable for journalists or data designers:

- All graphics can be saved in vector format (PDF, EPS or SVG) and processed with well-known vector graphic programs such as Adobe Illustrator or the free Inkscape, making each graphic element individually customisable.
- In R, the colour or shape of every graphic element can be changed whichever way you like. Text, symbols, arrows or entire drawings can be added, and different diagrams combined.
- The basic shapes of the most important diagram types, such as bar charts, line or pie charts, can frequently be created with a single command for a quick first impression.
- R can also handle maps, and can therefore be used for any geo-visualisations. The necessary maps can be loaded as, e.g., files in the common shape format.
- Since graphics are entirely programmed in a script, every step can be traced, every error found, and changes easily made. This also enables quality control by third parties and disclosure of the graphic source code in the scope of maximal open-data transparency.
- R is free.
- R is open.
- R can be extended with many software modules (packages) to visualise special graphic types or to perform advanced upstream data analyses. A growing international community continuously offers new extensions.

R graphics can also be used as a basis for interactive online graphics; a JavaScript package like, e.g., D3.js can be used to bring chart elements saved as SVG to life. An alternative is the complete JavaScript package Shiny which enables users to write interactive online data applications directly in R.

1.3   The Concept of Data Design

This book follows a 100% approach: every example shows the entire design of a specific figure. The starting point is always the result, the initial questions were always: how does a specific graphic have to look, or how can existing data best be visualised? Irrespective of specific software, the first step was always a sketch

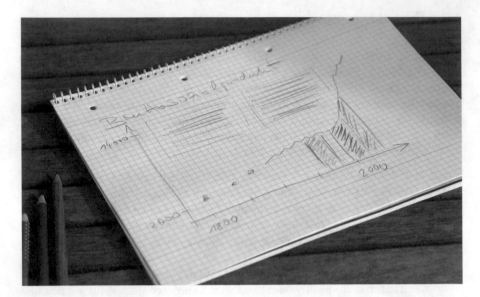

Fig. 1.1 Sketch of a picture

(Fig. 1.1). Only with the next step did the search for appropriate tools (packages and functions) and their use begin.

For the most part, the used data come from social science and official statistics, whereas some originate from business management, macroeconomics, politics, medicine, meteorology or social media. It was my goal to find suitable data for all of the selected presentation types. Obviously, this was sometimes more, sometimes less successful. However, data were not "tweaked", but used in the exact form in which they were provided. This means that scripts are sometimes a little bigger than in an ideal setting with data optimally prepared for the problem. On the other hand, this is much close to reality and may be useful when navigating some of your data pitfalls. All figures are designed as PDF files for preferably lossless and flexible further use. On average, 40 lines of code were required to create the result. It usually took a day to get from the first idea to the final product, but sometimes a week. In my opinion, the time investment is worth it, if you want to convey a message with your data.

Part I
Basics and Techniques

Chapter 2
Structure and Technical Requirements

Before we address the actual implementation of graphics in R, we will take a closer look at the structure of figures. Two examples for the different perception of graphics will be followed by a definition of graphic elements using schematic overviews that we call the "styling pattern", based on the term used in graphic design. Then follow explanations of important "ancillary elements" for figures, the used typefaces and symbols, as well as colour.

2.1 Terms and Elements

A figure can comprise one or more charts or graphs. For our purpose, the latter two terms will therefore be used synonymously. A chart consists of a data area (in R: plot region) and optional axes, axis labels, axis names, point names, legends, headings, and captions. A figure can contain several charts. In this case, every individual chart can have headings and captions, axes, legends, etc.; furthermore, there are titles and subtitles for the entire figure. If one figure contains several charts, then this is called a panel.

2.2 Illustration Grids

A figure principally comprises a title (1), a subtitle (2), a y-axis (3) including label (4) and name (5), the data area (6), a legend (7), an X-axis (8) including label (9) and name (10), and ultimately the sources (11). Figures can also contain further elements such as annotations, lines or symbols (Fig. 2.1).

© Springer Nature Switzerland AG 2019
T. Rahlf, *Data Visualisation with R*, https://doi.org/10.1007/978-3-030-28444-2_2

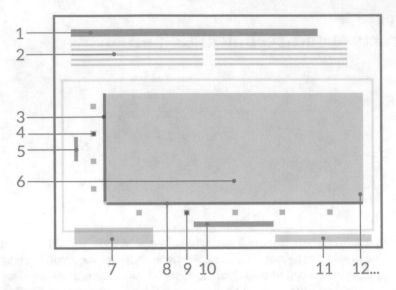

Fig. 2.1 Elements of a figure

First, consideration should be given to the aspect ratio of the figure. If, for example, both parameters of a scatter plot are percentages, and the range of values is to be represented from 0 to 100, then it would be logical to have axes of equal length, i.e. ,a square data area. In other cases, making a decision is not that simple. R gives you the opportunity to exactly specify these parameters during graphic creation (Sects. 3.3.3 and 3.3.7).

Do we need a legend? When? Where? The best-case scenario is one that can work without a legend. Generally speaking, this is the case with time series, because the labels can be written directly on the data: they are connected with lines and therefore clearly defined. However, this is not the case with scatter plots, where the meaning attributed to the colours has to be explained in a legend. In R, the legend() function lets you choose almost any setting for shape and placement of the legend.

If we include several charts in one figure, then we are creating a panel. In this case, certain elements can appear repeatedly (Fig. 2.2).

The arrangement of individual elements can vary, as can the number of charts included in one figure (Fig. 2.3).

In this book, we present examples for figures containing more than 40 graphics. R offers different ways for definition of such panels (Sect. 3.3.4 and 3.3.5).

Clearly, there cannot be universally applicable rules for the creation of an styling pattern. However, the following notes should be kept in mind:

1. It does make a difference whether graphics are free standing or embedded in body text. In the latter case, the heading is different, font sizes have to be adjusted for each element, and an explanatory subheading and explanatory labels and arrows are omitted or used more sparingly.

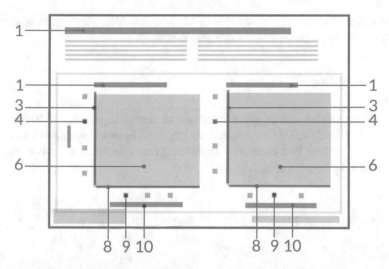

Fig. 2.2 Elements of a figure with two diagrams

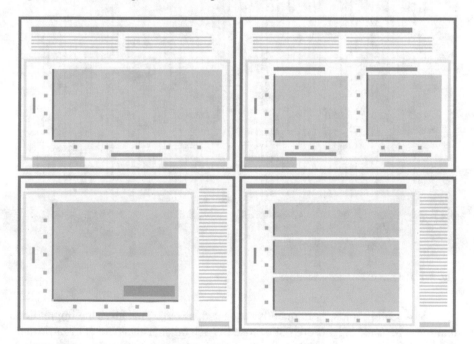

Fig. 2.3 Exemplary layout of individual elements

2. Generally, there is not just one adequate presentation of the data, but several. Whether to use stacked columns, or several column charts in one panel has to be decided for each specific case, depending on the specific data.

3. The source and title of a figure within an essay, book or website can be omitted, if they can be included where the figure is embedded.

2.3 Perception

The most important aspect when it comes to designing figures is the accurate perception of the data. This can be severely impaired by an unfortunate choice of the presentation format. Two examples: In the first example, body heights of selected celebrities are presented (Fig. 2.4)

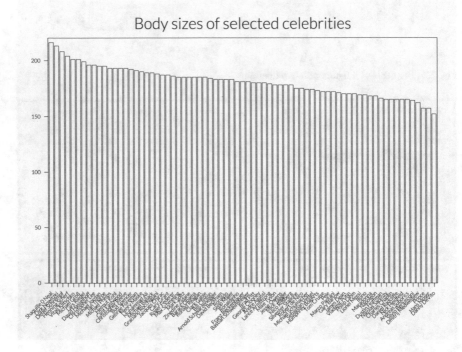

Fig. 2.4 Heights of selected celebrities

y-Axis scaling starts, as frequently requested, at zero. This gives the impression that these persons' heights are quite close to each other. This effect is even further enhanced by the use of (unfilled) bars, whose total volume takes up a large part of the entire area of the figure.

This contradicts our everyday experiences of considerable differences in body height. Searching the Internet reveals a picture showing Danny de Vito next to Christopher Reeve. Most people looking at this picture will likely think that the bar chart does not adequately reflect the differences in height. For these data, the following dot chart, repeatedly recommended by William Cleveland, is the more sensible option (Fig. 2.5).

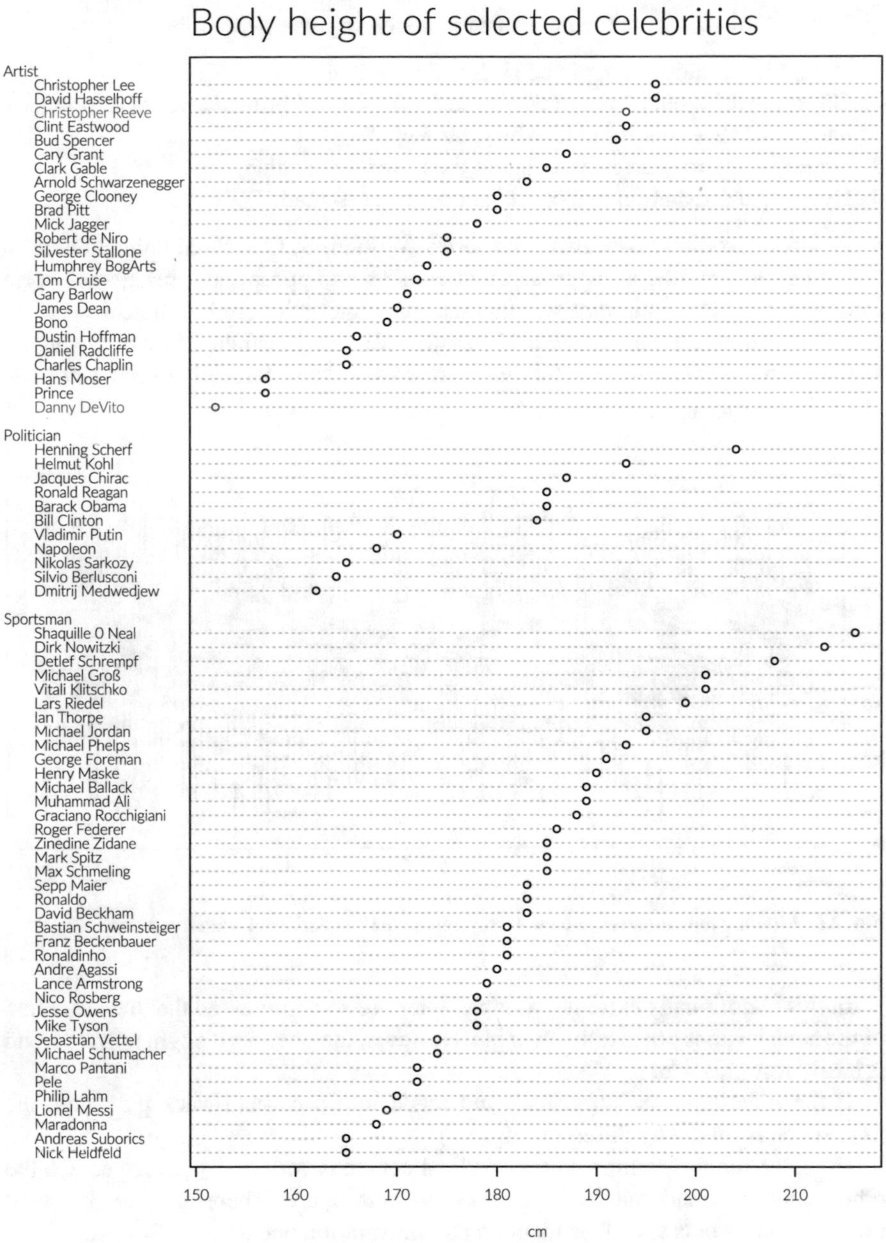

Body height of selected celebrities

Fig. 2.5 Heights of selected celebrities as Dot chart

Four differences from the bar chart markedly improve perception:

1. Height information is depicted as dots rather than bars.
2. Grouping different "types of celebrities" offers an additional level of information and generally enhances clarity by grouping.
3. Scaling does not start at zero, but at the lowest data value.
4. Horizontal orientation makes the names easier to read.

A second example concerns time series. William S. Cleveland coined the term "banking" to describe an approach that ensures an appropriate presentation form for line charts. The main idea was that data characteristics are best perceived when data lines on average approach a 45° angle. We will illustrate this point using an example that depicts the monthly temperatures in New Jersey between 1895 and 2011 (Fig. 2.6).

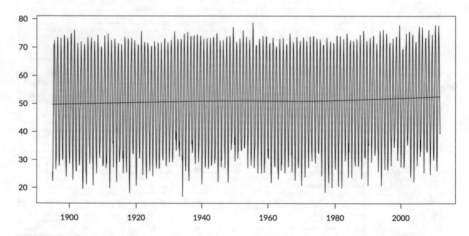

Fig. 2.6 Monthly temperatures in New Jersey between 1895 and 2011 with trend line

In this intentionally extreme example, lines are so compressed that the exact line of the actual data is practically invisible. However, the obvious if slight upward trend is easily recognisable.

If the illustration is "stretched" and made into a cut-and-stack plot, then the impression is markedly different (Fig. 2.7).

Here, the monthly temperatures' cyclical course is very easily perceived. On the other hand, no trend can be discerned from this figure. Therefore, the choice of presentation form is also dependent on the information one desires to convey.

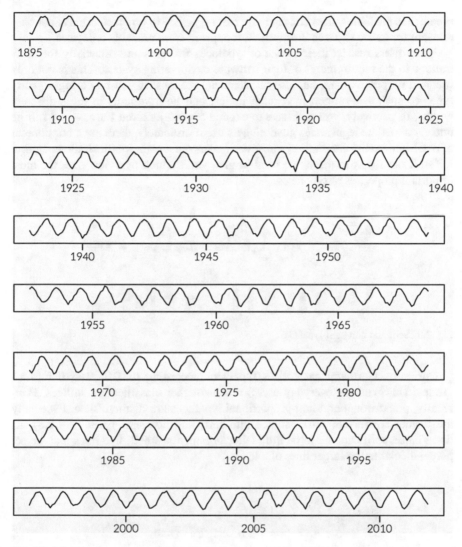

Fig. 2.7 Monthly temperatures in New Jersey between 1895 and 2011 as cut-and-stack plot

2.4 Typefaces

Typefaces make up a not insignificant part of figures. Unfortunately, they are usually neglected. However, use of the correct typeface can contribute much to clarity. An interesting study is available thanks to Sven Neumann from the design department of the HTW Berlin. He studied readability of typefaces in the public domain, and concluded from a survey of more than 100 people that the distance from which a typeface is readable varies considerably from typeface to typeface. This is not only relevant for traffic signs as illustrations, too, profit from readable typefaces.

Many users restrict their choice of typefaces for texts and especially for illustrations to the requirements of their software or operating system. This is not only founded on pragmatism, but also has financial reasons: if you buy a high-quality typeface such as Frutiger, in regular, italics, and bold variants in three different widths, respectively, you will have to expend several hundred Euros—and still be unclear about the legal status governing its use. Fortunately, there are a fair number of free high-quality alternatives whose use makes sense even for illustrations. Before we shift our focus to these, we want to give you a quick overview over the most important properties of typefaces.

Fig. 2.8 Serif and non-serif typefaces

Currently, Germany categorises typefaces according to DIN 16518 into 11 groups. However, for everyday use, a much rougher classification suffices. Principally, proportional and non-proportional typefaces are distinguished. Especially the former are further differentiated into serif and non-serif types (Fig. 2.8). At first glance, serifs appear to be little letter ornaments: small, fine lines that are set perpendicular to the larger lines of a letter.

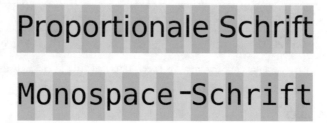

Fig. 2.9 Proportional and non-proportional typefaces. *Source*: de.wikipedia.org, Algos

Such typefaces are usually used for long texts, as serif-type typefaces are proven to be more pleasant to read. Non-serif typefaces on the other hand are used for headings or short texts. A proportional typeface (Fig. 2.9) is characterised by individual letters taking up different spaces in width. A lower 'l' or 'i' takes up less

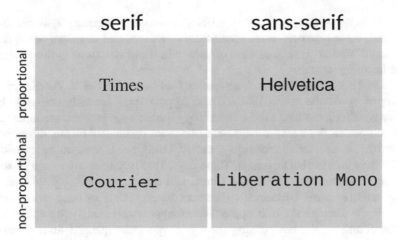

Fig. 2.10 Examples for typefaces

space than an 'm'. In non-proportional typefaces. on the other hand, every letter takes up the same space.

The arguably most well-known non-proportional (serif) typeface is Courier, also known as the typewriter font—whereas the most well-known non-serif proportional typeface is Helvetica by Max Miedinger and Eduard Hoffman. Times, the most well-known proportional serif typeface, had already been designed a quarter of a century earlier, by Stanley Morison and Victor Lardent (Fig. 2.10).

Almost every typeface, irrespective of being proportional or non-proportional, serif or non-serif, is produced in different cuts. Many restrict themselves to regular, bold, and italics variants (Fig. 2.11).

regular **bold** *italics*

Fig. 2.11 Regular, bold, and italics cuts (Linux Libertine)

2.4.1 Fonts

Even though the previously mentioned typefaces have been omnipresent for decades, hardly anyone would know their names were they not used as "fonts" on a computer. While typefaces used to be hardwired into printers, nowadays they come with the respective operating systems or, at least in part, with software applications or web servers.

But even today, there are marked differences between operating systems. On the one hand, some typefaces were specifically developed and licensed, while on the other hand 'similar' typefaces were offered with slight variations to their names to avoid licensing issues.

The first platform-spanning, high-quality font base were the 35 PostScript fonts developed by Adobe in the 1980s. These Type-1 fonts not only comprise high-quality typefaces, but also used a technology usable by Windows, Macintosh, and Unix computers. For a long time, patent litigations dominated further development and led to an alternative technology called TrueType. It used to be considered qualitatively inferior, but no longer. These days, TrueType fonts are used as standard fonts in Windows, Mac OS X, and Linux, and are readily exchangeable between the operating systems. Unfortunately, current developments are once more going in two different directions: while Apple favours Apple Advanced Typography (AAT), Microsoft and Adobe are pushing the OpenType technique. At least OpenType fonts can also be used in Linux and Mac OS X. The technology offers interesting opportunities for visualisations as well, because number variants with different upper and lower line heights can be selected, as long as they are included in the font.

At that time, Acrobat Reader contained 14 original PostScript fonts: Courier (regular, oblique, bold, bold oblique), Helvetica (regular, oblique, bold, bold oblique), Symbol, Times (roman, italic, bold, bold italic), and ITC Zapf Dingbats. In more recent versions, Helvetica and Times were replaced with Arial and Times New Roman. Helvetica was considered a font that did not look particularly good on screens during the pioneering era of personal computers. An alternative was Arial, which became a part of the Windows (3.1) operating system in the early 1990s. In contrast to frequent claims, Arial does not look like Helvetica, but they do share the same metrics. These fonts are therefore exchangeable without altering line or page breaks in a text. A few years ago, a new Unicode variant was added to the typeface, but without bold or italic variants (Fig. 2.12).

High-quality fonts such as Calibri (since 2012 also available in two light variants) only became available in Windows with the advent of Vista, but some of them were considered to be copies of existing fonts (e.g., the Suego typeface as a clone of Frutiger).

2.4.2 Free Typefaces

Aside from typefaces that come with operating systems or those that can be purchased, there are now a series of high-quality "free" typefaces, whose usage is governed by open font licences. Google's website offers a carefully selected choice of such fonts for download. There, you will also find descriptions of the necessary steps. The typefaces are offered as TrueType typefaces and can be installed by double-clicking on the TTF file in Windows, Mac OS X or Linux.

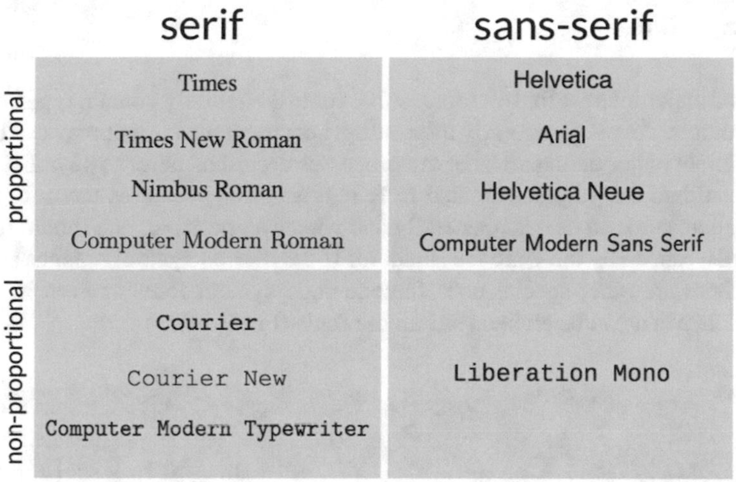

Fig. 2.12 Examples of fonts with "variants"

With a few exceptions, the examples in this book use the typface family "Lato" by Łukasz Dziedzic. This font also has its own website (Fig. 2.13).

Fig. 2.13 The free fonts Lato, Gentium Plus, and Liberation Mono

2.5   Symbols

A fact of major interest for our topic is that fonts do not only contain typefaces, but also symbols. This is particularly interesting, because it means that they can be used in R without major detours. Using the processes described here, any symbol can be embedded into the graphics created in R in a way that preserves them for further downstream work processes. Generally and whenever possible, one should resort to standards, especially the Unicode standard. If the desired symbols cannot be found there, there are other special non-Unicode code symbol fonts or even individual symbol files that can be embedded into the fonts (Fig. 2.14).

Fig. 2.14 Display of the glyphs of the code block "miscellaneous symbols" of the font Symbole (Fontmatrix)

Symbols, pictograms, ideograms or icons have most likely been part of every culture and have been around since the dawn of time. Everyone has some experience with signs whose meaning is immediately obvious or is subject to general agreement: Here is a first-aid station; there is food; this should only be cleaned, not washed. These types of signs again and again became a focus of attention for designers. Famous examples are the Isotype pictures, designed by Gerd Arntz for Otto Neurath, or the sport pictograms by Otl Aicher, designed for the Olympic Games in Munich in 1972, and still an unchallenged classic.

It was a short leap to the idea of embedding symbols in typefaces (fonts), and a large leap to not do this through separate symbol fonts such as the ancestral ITC Zapf Dingbats that operating systems and printers somehow, during the printing process, threw together with the "normal" fonts.

Frequently, symbol fonts would just re-assign letters. If you selected Zapf Dingbats as a font and typed "a", then a flower would appear, if not directly then definitely on the printout; with an exclamation mark, scissors. The underlying mechanisms were plentiful and historically grown rather than systematically developed. After Dingbats came the three Wingdings fonts that emerged from Lucida; a few years later, Webdings. Dingbats and Wingdings share many of the glyphs they contain, but they code them at different locations.

With the advent of Unicode, the starting position underwent a positive reformation. The hardest work, to code all globally existing characters in a binding scheme, had to include pictograms sooner or later. Only in October 2010, in version 6.0, were symbols included into Unicode on a large scale. At that point, various loci of the monumental typeface already included technical symbols, mathematical, typographic symbols or even the old-school Zapf Dingbats, some of whose symbols were already included into their own code block 161 in version 1 of Unicode.

What was actually stipulated was a rather mixed collection of pictures that had always been used here and there. Even without designer training, this fact is obvious when looking at the uniform presentations required by Unicode. Politely put, as stated by Andreas Stötzner, the block tables with the now fixed reference glyphs demonstrate the difficulty in reaching a globally valid graphic design language starting from local references. Or to put it differently: the visual variety [...] points towards the extent of the design task necessary for future implementation of fonts.

The different ways font designers define symbols as simple as the phone—if they even bother to develop a glyph for the character defined in Unicode—is evident in the five telephone symbols currently incorporated into Unicode for a few exemplary selected fonts (Fig. 2.15).

The identical symbols are either simple "take overs" by the font designer, or the operating system uses a so-called fall back font, if the glyph is not available in the selected font.

For less commonly used symbols, differences can be much more drastic. While the glyph for GRAPES from the code block "Miscellaneous Symbols and Pictographs" are reasonably alike in the fonts Segoe UI Symbol and Symbola, they interpret the neighbouring symbol MELON drastically differently.

Looking at the phone symbols in the first line of Fig. 2.15, it becomes obvious that colour has found its way into fonts. However, there are currently no standards; once more, Apple and Microsoft go their own ways.

2.5.1 Symbol Fonts

While the total number of standardised encoded symbols markedly increased with Unicode 6.0, there are still numerous additional symbols that would lend themselves to use within illustrations. On the one hand, there are non-standardised, but still high-quality symbol fonts, on the other hand there are many high-quality symbols that can be embedded into a font for use in illustrations.

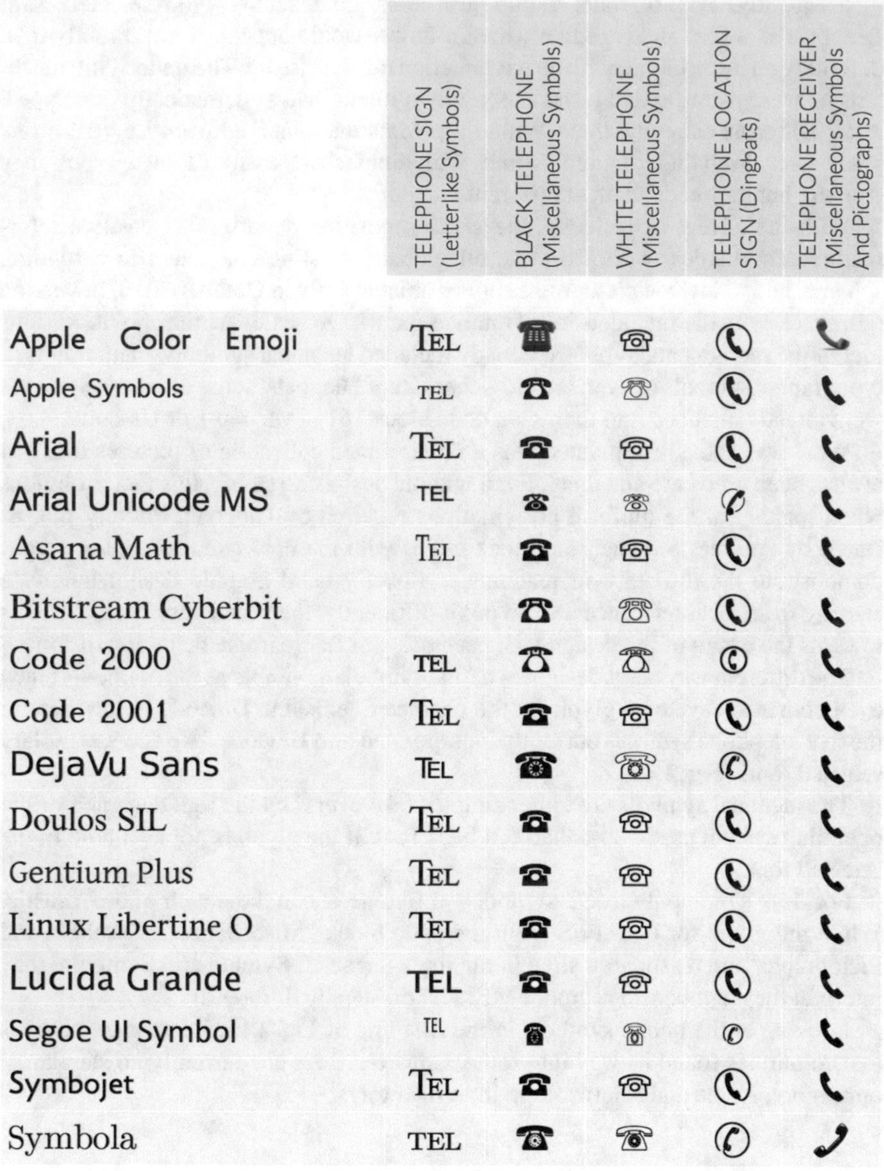

	TELEPHONE SIGN (Letterlike Symbols)	BLACK TELEPHONE (Miscellaneous Symbols)	WHITE TELEPHONE (Miscellaneous Symbols)	TELEPHONE LOCATION SIGN (Dingbats)	TELEPHONE RECEIVER (Miscellaneous Symbols And Pictographs)
Apple Color Emoji	TEL	☎	☏	◖	✆
Apple Symbols	TEL	☎	☏	◖	✆
Arial	TEL	☎	☏	◖	✆
Arial Unicode MS	TEL	☎	☏	∅	✆
Asana Math	TEL	☎	☏	◖	✆
Bitstream Cyberbit		☎	☏	◖	✆
Code 2000	TEL	☎	☏	◖	✆
Code 2001	TEL	☎	☏	◖	✆
DejaVu Sans	TEL	☎	☏	◖	✆
Doulos SIL	TEL	☎	☏	◖	✆
Gentium Plus	TEL	☎	☏	◖	✆
Linux Libertine O	TEL	☎	☏	◖	✆
Lucida Grande	TEL	☎	☏	◖	✆
Segoe UI Symbol	TEL	☎	☏	◖	✆
Symbojet	TEL	☎	☏	◖	✆
Symbola	TEL	☎	☏	◖	✆

Fig. 2.15 Telephone glyphs in different fonts

A successful example for a symbol font outside of Unicode symbols (therefore with a different character assignment) is the "Symbol Sign" font by Sander Baumann. You can download this font, which is also free for commercial use, through, e.g., Fontsquirrel (www.fontsquirrel.com).

2.5.2 Symbols in SVG Format

If the desired glyphs are not available as Unicode symbols or in special symbol fonts with non-standardised symbols, then it makes sense to do a specific search for suitable individual symbols. Symbols in SVG format (scalable vector graphics) offer good prerequisites in this regard, because it only takes a few steps to develop them into fonts for further use in illustrations. Step-by-step instructions will be given in Sect. 4.2.

The "Noun Project" is a very good starting point for a search; here, designers from all over the world offer symbols under a free licence (thenounproject.com). In Sect. 4.2, we will embed two symbols from the stock of the Noun Project into a font file.

Another good address is Fotalia (www.fotalia.com), an international picture agency that also offers a plethora of licence-free symbols.

2.6 Colour

Looking at any book on "graphic design" will usually reveal extensive discourses on the topic of colours. This is not really surprising, since humans can easily distinguish several tens of thousands colours, a trained eye many more. Some women have four instead of three colour receptors, which enables them to distinguish between up to even 100 million colour gradations.

2.6.1 Colour Models

In the world of computers, there are two main colour systems, the RGB and the CMYK colour model. For the RGB colour model, light colours are created through addition by mixing the parts red, green, and blue. Using this principle, colours in active light sources, like displays or computer screens, are created. Every colour has a potential value between 0 and 255. The result is $256 \times 256 \times 256 = 16.7$ million colours. For the CMYK colour model, the colours cyan, magenta, yellow, and black are combined through subtraction. Using this principle, colours on passive media, like paper, are created. The two colour systems differ greatly in certain areas. Metallic colours (gold and silver), for example, or different types of reflection (matte and gloss) cannot be easily displayed on a computer screen. Ink jet printers can print considerably more colours than a mediocre computer screen can display; on the other hand though, the colour space output is much greater with RGB colours than with CMYK colours. This means that some colours look radiant on a website, but pale in print. This affects especially turquoise and orange. If either of those colours is used, one should check if the print yields the desired result.

This can be quite clearly seen in the so-called CIE standard colour chart (Commission Internationale de l'Éclairage) or rather a diagram that compares different colour palettes in a colour space: a shoe sole-like shape represents the detection capability of the human eye, and the contained polygons the colour spectrum of various devices or colour models. Figure 2.16 depicts the RGB and CMYK colour models' colour spaces.

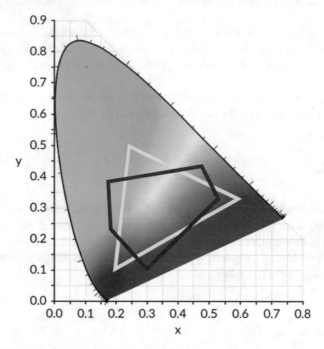

Fig. 2.16 RGB (*yellow*) and CMYK (*blue*) colour spaces

On top of that, pretty much all devices create colours differently, which means that device-specific adjustments are necessary. To do this, device manufacturers provide ICC profiles (International Color Consortium, in Windows: ICM, Image Color Matching Profile) that are compared with the colour profiles of the files to be displayed or printed. In theory, this makes colours on the output medium appear "natural". In reality though, more or less extensive calibration is required in most cases. Windows, Mac OS X, and Linux provide tools to convert RGB into CMYK pictures. In Mac OS X, this can be done with the ColorSync program that is part of the operating system (Fig. 2.17).

For demonstration purposes, we have purposefully selected colours whose RGB and CMYK versions differ for some of the examples in this book; this means you can compare this print version with the version on the book's website.

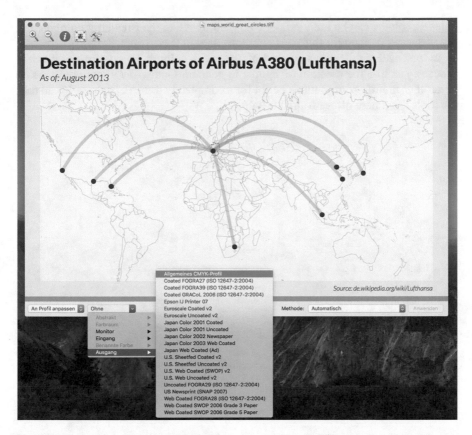

Fig. 2.17 Conversion of an exemplary TIFF file from RGB to CMYK using ColorSync (Mac OS X)

2.6.2 Colour in Statistical Illustrations

On opening a classical book on data visualisation, one is often surprised to discover that precious little consideration has been given to "colour" when compared with other topics. However, it has received increased attention in statistics over the past years, certainly facilitated by the technical development. Important stimuli came from thematic cartography, among other fields. Cynthia A. Brewer in particular has developed a series of tips that have become widely adopted. A significant contributor to the dissemination was one of her websites, which offers a simple selection of prefabricated colour palettes according to her self-developed principles (www.colorbrewer2.org).

Some years ago, Achim Zeileis, Kurt Hornik, and Paul Murrell advanced Cythia Brewer's idea with a focus on statistical graphics. Their propositions are based on the so-called Hue-Chroma-Luminance colour space, which they claim allows easier colour perception than the RGB colour space. Here, in contrast to the Brewer colour palettes, colours can be individually compiled. Just like the Brewer colour palettes, the software intended for this use is implemented in R (see Sect. 3.3.9)

In practical applications, choosing the right colours is usually an iterative process, just like the creation of figures in general. One will try one combination or another, modify compilations or opt for a completely different palette. To do this, the tools listed here offer invaluable help.

Chapter 3
Implementation in R

By now, several thousands of packages, compilations of special functions and extensions, can be downloaded from the official Internet platforms for R. Some packages offer more than a thousand functions. Some functions, such as the par() function so important for our topic, can receive more than 80 arguments. At the beginning, a system like this appears as hard to negotiate as a jungle or a maze.

In this book, we will use a very specific selection of those that have shown to be necessary or helpful for the creation of the examples. It is obvious that we cannot give a complete introduction to R; these days, there are a multitude of good books and websites available to do just that.

3.1 Installation

For Windows, Mac OS X, and Linux, R can be downloaded from the project's website as a pre-compiled installation package. As per standard, R comes with a series of packages; further packages can be installed within R—either via a function or via a menu—as needed. The graphic user platform RStudio is well suited for day-to-day working with R; it is also freely available and provides an identical work environment for Windows, Mac OS X, and Linux (Fig. 3.1).

© Springer Nature Switzerland AG 2019
T. Rahlf, *Data Visualisation with R*, https://doi.org/10.1007/978-3-030-28444-2_3

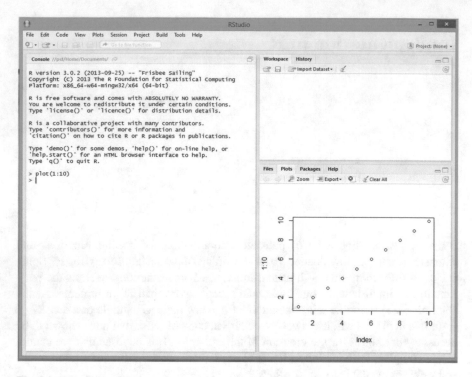

Fig. 3.1 RStudio for Windows

3.2   Basic Concepts in R

R is an interpreter. This means that every input is processed directly and row by row.
If you enter anything in the command prompt, the entry is executed upon completion
of the line (on pressing the Enter key). If you enter $3 + 1$, the result 4 will be
immediately displayed. Additionally, you will be informed that this 4 is an element
of an object at position 1.

```
> 3+1
[1] 4
>
```

Using the command row, you can also gain access to R's outstanding help.
Using the command help() will give you extensive explanations on any function;
for example

```
> help(plot)
```

opens a new window with the help text (see Fig. 3.2).

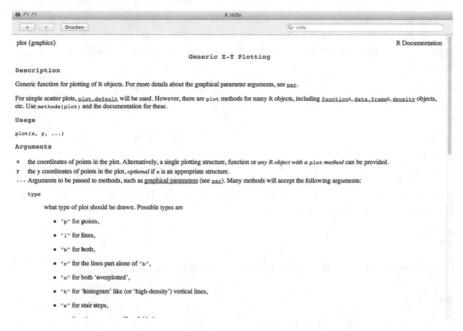

Fig. 3.2 R Help

3.2.1 Data Structures

R knows different data structures. On the one hand, they are vectors, matrices, and arrays. These differ in their dimension. In other words: a vector is a series of numbers or characters, a matrix is a table with rows and columns, and an array is a "multidimensional" table. For three dimensions this could be a series of several tables with the same number of rows and columns. The individual elements of the objects can be uniformly addressed across all data structures using square brackets. The simplest way to create a vector in R is by assignment via the c() function:

```
> x<-c(3 5, 3 ,9, 1 ,7)
> x
[1] 3 5 3 9 1 7
```

To create a matrix instead of a vector, the array() function is used. Here, the dimension has to be defined. The following example will create a table with two rows and three columns:

```
> y<-array(c(3, 5, 3, 9, 1, 7), dim=c(2,3))
> y
     [,1] [,2] [,3]
[1,]   3    3    1
[2,]   5    9    7
```

Higher-dimensional arrays are created the same way; however, we will not be doing that in this book. The output of the previous example also shows how to address elements of a vector or array: by use of the object name followed by square brackets

```
> x[2]
[1] 5
```

and

```
> y[2, 2]
[1] 9
```

The mnemonic "CAR" (columns after rows) can be used to remember the order: rows are always first, columns second. Without a value before or after the comma, the entire row or column is returned.

```
> y[, 2]
[1] 3 9
```

Do not be confused that the values are listed in a row, not a column.

```
> y[1, ]
[1] 3 3 1
```

You can also define output areas:

```
> y[, 2:3]
      [,1] [,2]
[1,]     3    1
[2,]     9    7
```

In arrays, both rows and columns can contain names:

```
> colnames(y)<-c("V1", "V2", "V3")
> rownames(y)<-c("Case 1", "Case 2")
> y
       V1 V2 V3
Case 1  3  3  1
Case 2  5  9  7
```

Data Frames

The elements of vectors or matrices always have to be of the same type, such as numbers only. To save different types in one object, e.g., numbers and letters, the object type "data frame" is available in R. A data frame can be created from, e.g., an array:

```
> z<-as.data.frame(y)
> z
       V1 V2 V3
Case 1  3  3  1
Case 2  5  9  7
```

The result looks exactly like the matrix. However, if you display the structure of the object:

```
> str(y)
 num [1:2, 1:3] 3 5 3 9 1 7
 attr(*, "dimnames")=List of 2
 ..$ : chr [1:2] "Case 1" "Case 2"
 ..$ : chr [1:3] "V1" "V2" "V3"
```

and compare it with the structure of the data frame:

```
> str(z)
'data.frame':  2 obs. of 3 variables:
 $ V1:  num  3 5
 $ V2:  num  3 9
 $ V3:  num  1 7
```

then it becomes obvious that "observations" (obs.) and "variables" (variables) have been defined instead of "dimensions". Variables of data frames can be displayed via their name, if a dollar sign is attached to the object's name, followed by the variable's name:

```
> z$V2
[1] 3 9
```

To create a data frame with different types of data, one of the options is to add a text column to the data frame z:

```
> z$4<-c("Yes", "No")
> z
         V1    V2    V3    V4
Case 1    3     3     1   Yes
Case 2    5     9     7    No
```

However, you can also address individual rows or columns as you would in a matrix:

```
> z[2, ]
         V1    V2    V3    V4
Case 2    5     9     7    No
```

and

```
> z[,2]
[1] 3 9
```

Time Series

A special type of object that is also very useful for illustrations is used for time series:

```
>ts(1:20, frequency = 12, start = c(1950, 1))
     Jan Feb Mar Apr May Jun Jul Aug Sep Oct Nov Dec
1950   1   2   3   4   5   6   7   8   9  10  11  12
1951  13  14  15  16  17  18  19  20
```

Lists

Contrary to common opinion, data frames are not a case apart of a matrix, but rather of an uncommon data type: "lists". These are compilations of vastly different types of elements—in the case of a data frame, they can be used to compile numerical and alphanumerical vectors. Generally speaking, lists allow the compilation of different types of objects. The first element of a list could, e.g., be a string, the second a vector, the third a data frame:

```
> examplelist<-list ("A", x, data)
> examplelist
[[1]]
[1] "A"

[[2]]
[1] 3 5 3 9 1 7

[[3]]
      V1 V2 V3
1 Peter  2  3
2  Paul  3  2
3  Paul  2  2
4 Marie  1  3
```

The object type "lists" is quite useful if several objects are transferred in a graphic function. This is something we will make use of in several examples in this book.

Objects for Maps: SpatialPolygonsDataFrame

A SpatialPolygonsDataFrame is a special form of data frames whose most notable addition to a data frame is a list of polygons that can be used to draw maps.

3.2.2   Import of Data

Frequently, data will not be included in a script, but rather imported from an external file. To do this, R offers two "in-house" options: first, there is a proprietary, binary data format. Any type of R objects can be saved in such a file. Existing files can be saved or opened as follows:

```
> save (data1, data2, data3, file="data/testdata.RData")
> dataframes<-load("data/testdata.RData")
> dataframes
[1] "data1" "data2" "data3"
> data1
  V1 V2 V3
1  1  2  3
2  2  3  2
3  2  2  2
4  3  1  3
```

Be aware that the object dataframe is a string vector with the names of the three data frames. You can still directly retrieve the three data frame objects saved in the testdata.RData file. Another option is the use of the data() function. Depending on the suffix of the file transferred as argument, this will load either a binary R data file (suffix .Rdata or .rda), a script file (suffix .r), or a txt or csv file using the read.table() function.

When importing files in Excel or CSV format (Fig. 3.3), you can specify whether string columns are uploaded as factors (see below) or as string variables. To import an Excel 2007 file, you can either use the read.xls() function from the gdata package, or the read.xlsx function from the xlsx package. gdata requires a Perl installation, xlsx a Java installation. Both packages offer a series of options.

```
> library(gdata)
> data1<-read.xls("data/data1.xlsx")
> data2<-read.xls("data/data2.xlsx")
> data1
   V1 V2 V3
1   1  2  3
2   2  3  2
3   2  2  2
4   3  1  3
> data2
      V1 V2 V3
1 Peter  2  3
2  Paul  3  2
3  Paul  2  2
4 Marie  1  3
```

	A	B	C			A	B	C
1	V1	V2	V3		1	V1	V2	V3
2	1	2	3		2	Peter	2	3
3	2	3	2		3	Paul	3	2
4	2	2	2		4	Paul	2	2
5	3	1	3		5	Marie	1	3

Fig. 3.3 Two examples for Excel data

R knows numeric, integer (special case of numeric), logical, string, date, and so-called factor vectors:

```
> f1<-factor(data2$V1)
> f1
[1] Peter Paul Paul Marie
Levels: Marie Paul Peter
```

A factor is just a categorical variable. You can also assign ⬃
      labels for the factor attributes:
```
> data1$V1<-factor(data1$V1, labels=c("Peter", "Paul", "Marie"))
> data1
```

```
      V1 V2 V3
1 Peter  2  3
2  Paul  3  2
3  Paul  2  2
4 Marie  1  3
```

Labels have to be entered in alphabetical order of the values. With large data frames you can use the sort(unique(x)) function to create such a list for a vector x to be factored.

Vice versa, a factor can also be converted back into a string vector:

```
> as.character(f1)
[1] "Peter" "Paul" "Paul" "Marie"
```

SPSS files (Fig. 3.4) can be imported using either the spss.get() function from the Hmisc package by Frank E. Harrell Jr. or spss.system.file() from the memisc package by Martin Elff, which enables efficient import of a subset of a bigger SPSS file. Equally, you can choose whether string variables should be created from the labels of existing factor variables when importing SPSS files.

Fig. 3.4 SPSS data

Using Hmisc:

```
> library(Hmisc)
> data5<-spss.get("data/ZA4615_AC12.SAV")
> table(data5$V9)
 1 - NOT IMPORTANT    ..   .. ..  ..  .. 7 - VERY IMPORTANT
            48   755    0  0   0   0 2672
```

Or using memisc:

```
> library(memisc)
> data <- spss.system.file("data/ZA4615_AC12.SAV")
> data5 <- subset(data, select = c(v60, v61))
```

```
> codebook(data5$v60)
- - - - - - - - - - - - - - - - - - - - - - - - - -


data5$v60 'CONFIDENCE: COLLEGES, UNIVERSITIES'


- - - - - - - - - - - - - - - - - - - - - - - - - -


Storage mode: double
Measurement: nominal
Missing values: 97-Inf
            Values and labels        N      Percent
  1    'ABSOLUTELY NO CONFIDENCE'    36    1.1    1.0
  2    '..'                          46    1.4    1.3
  3    '..'                         207    6.3    5.9
  4    '..'                         603   18.2   17.3
  5    '..'                        1166   35.3   33.5
  6    '..'                        1001   30.3   28.8
  7    'GREAT TRUST'                247    7.5    7.1
 99 M  'NOT SPECIFIED'              147           5.0
```

Equivalent mechanisms are available for Stata. R also offers packages for connections to Oracle, MySQL and SQLite databases and for connections via an ODBC interface. Contrary to the import of Excel or SPSS data, no files are being imported in this case but requests are sent to a database. The result sets of the request can then be imported. MySQL data are shown in Fig. 3.5.

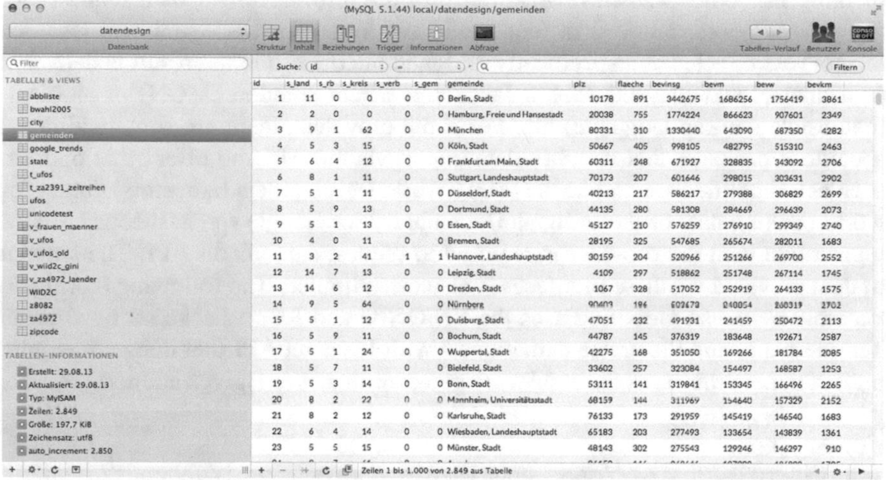

Fig. 3.5 MySQL data (German municipalities)

They are imported as follows (XX has to be replaced with actual data):

```
>library(RMySQL)
>con <- dbConnect(MySQL(),user="XX",password="XX",dbname="XX", ↘
    host="XX")
```

```
>sqlcommand<-"select municipality, postal code, area of ⬎
    municipality limit 2"
>data4<-dbGetQuery(con,sqlcommand)
>data4
      municipality  postal code    area
1      Berlin, City          10178   891
2 Hamburg, Free and Hanseatic City          20038     755
```

If data are being retrieved from a local database, then host has to be replaced with a socket containing the path to the mysql.sock file. In this case:

```
con <- dbConnect(dbDriver("MySQL")
username="XX",
password="XX",
dbname = "XX",
unix.socket="/Applications/XAMPP/xamppfiles/var/mysql/mysql.sock⬎
    ")
```

Please note that MySQL provides the complete scope of Unicode, not just the basic multilingual plane, from version 5.5.3 (since 2010).

Data via APIs

These days, data offered through the Internet are increasingly available in a form that can be retrieved via a so-called API (application programming interface). Its main use is automatic data requests on websites, for example, to integrate parts of these into other websites. However, they are also useful for data requests for analysis purposes. To do this, requests in the form of a URL (uniform resource locator) are directed at a specifically designed interface, and the result is delivered as a distinct text file: either as a JavaScript object notation (JSON) or an XML (extensible markup language) file. For some providers, there are some additional specific R packages that simplify data import via the API and offer other options, such as a simplified search like the WDI (World Development Indicators) library for world banking data or the RGoogleDocs library for Google spreadsheets. Usually, registration with the provider is required to request data via an API. The amount of data provided via these interfaces is also often limited. The following (abridged) example by Bryan Goodrich uses the RJSONIO and RCurl packages by Duncan Temple Lang, and illustrates data request with R via an API that does not require registration. In this case, the results are geo coordinates to a given address.

```
library(RJSONIO)
library(RCurl)
json<-getURL("http://maps.googleapis.com/maps/api/geocode/json?⬎
    sensor=
  false&address=6000+J+Street,+Sacramento,+CA")
x<-fromJSON(json)
x$results[[1]]$geometry$location
```

In case of success, the result for the last line will be returned as

```
 lat      lng
38.56443 -121.42622
```

Please refer to the specified thread for further details.

RDF Data

A still relatively new variant of data access, increasingly used for statistical data,
is the RDF format (Resource Description Framework). RDF data are offered by,
e.g., Eurostat, the World Bank or by the Open Data Portal of the US government
(data.gov). The Request language SPARQL is closely related to SQL and can be
used with RDF. Willem Robert van Hage and Tomi Kauppinen offer the SPARQL
package for R, which enables import of RDF data into R using SPARQL. The
following example shows its use with data derived from data.gov:

```
library(SPARQL)
endpoint <-"http://services.data.gov/sparql"
query <-
"PREFIX dgp1187: <http>//data-gov.tw.rpi.edu/vocab/p/1187/>
SELECT ?ye ?fi ?ac
WHERE {
?s dgp1187>year ?ye .
?s dgp1187>fires ?fi .
?s dgp1187:acres ?ac .
}"
qd <- SPARQL(endpoint,query)
df <- qd$results
sort.df<-df[order(df$ye) ,]
attach(sort.df)
plot(ye, fi, type="b")
```

The result is returned as the number of forest fires in the USA between 1960 and
2008 (Fig. 3.6).

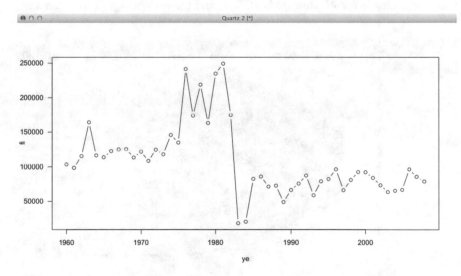

Fig. 3.6 Result of the SPARQL request

Map Data

Geo data, especially national, regional or administrative borders, are frequently
saved in so-called shapefiles. This is a very old data format in which map information and associated attributes are saved in a series of files with the same name,
but different file extensions. London's constituencies, for example, are provided in
the following format:

```
greater_london_const_region.shp
greater_london_const_region.prj
greater_london_const_region.dbf
greater_london_const_region.shx
```

The shp file contains the actual geometrical data. These are either points or lines
or areas in the shape of polygons. The prj file only contains a so-called projection set,
a string with information stating how the originally three-dimensional data are two-
dimensionally represented in the shp file. The dbf file contains a data record in which
"factual data", such as statistics for each geometry, are saved in the form of a data
frame. The shx file finally links factual and geometrical data. Furthermore, other
files with fixed suffixes can contain additional information. One of the considerable
downsides of the dub file type for use with factual data is that it cannot readily store
Unicode data. This is why it is often useful to restrict yourself to using geometrical
data and to link those with data from other files. In this book, we will be using
shapefiles for graphical representation of maps. Even though R can also be used for
extensive processing and examination of map data, and for geo-statistical analyses,
please refer to advanced literature for further information on this. Software to view
shapefiles is especially useful for our purposes, such as the open-source QGIS,
which is available for Windows, Mac OS X, and Linux. With QGIS you can, e.g.,

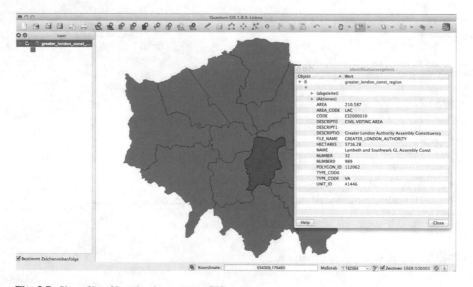

Fig. 3.7 Shapefile of London in quantum GIS

select individual polygons of a shapefile and view the associated information, or get information about the projections used (Fig. 3.7). In R, shapefiles are imported using the readShapeSpatial() function from the maptools package. How to link your own data with the polygons in the correct order is explained in detail in Sect. 10.3.

3.3   Graphic Concepts in R

R's graphic capabilities follow two different approaches (Fig. 3.8).

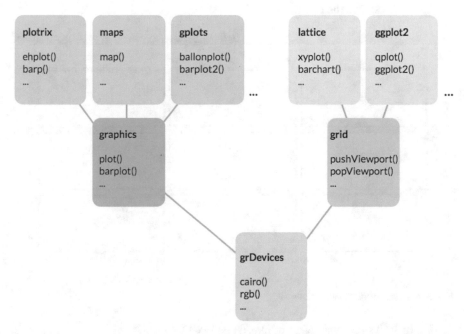

Fig. 3.8 Graphic organisation in R (adapted from Murrell)

All of the examples in this book use "Base Graphics", R's traditional graphic system. In addition, there is Paul Murrell's grid, which offers very flexible options, but is based on a different construction method. Both are based on the same graphic-device functions, but are largely incompatible. There are bridges between these two worlds, but because their methods are fundamentally different, one should opt for one or the other approach. The popular packages lattice and ggplot2 are based on grid. Some packages, such as sp, offer functions (or methods for the extension of functions) for both approaches. To better understand the way graphics operate in R, we first have to look at the results of the call of the central function plot() with different data. data1 and data2 are the test data from the previous section. Once more the test data as a reminder:

```
> data1
  V1 V2 V3
1  1  2  3
2  2  3  2
3  2  2  2
4  3  1  3
> data2
     V1 V2 V3
1 Peter  2  3
2  Paul  3  2
3  Paul  2  2
4 Marie  1  3
```

Figure 3.9 shows the result.

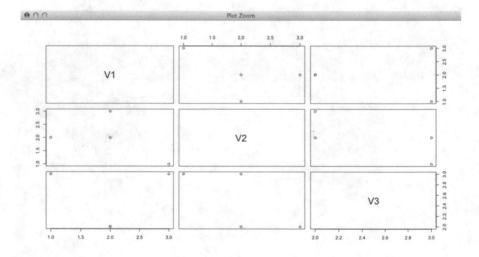

Fig. 3.9   Result of plot(data1, cex=2)

Figure 3.10 shows the result of plot(daten2$V1).

```
library(maptools)
data3<-readShapeSpatial("data/london/greater_london_const_region↘
    .shp")
plot(data3)
```

Here, data3 is of the SpatialPolygonsDataFrame type (Fig. 3.11). For this type, no axes are plotted by default. plot is a so-called generic function. This means that it is actually a group of functions from which a suitable one is selected each time. In fact, there are currently well over 30 functions (or rather, methods) within plot() that deliver the appropriate representation for numerous objects. So if you call plot(x), then R initially searches for a function that is suitable for the object type of x. If none is found, then a plot.default method is selected. Many things work in this way in R. In the first case, we presented R with a matrix, so that the plot() call results in a multivariate illustration. In the second case, it was a single categorical variable from which a bar chart was created. In the last case, the method was provided by the

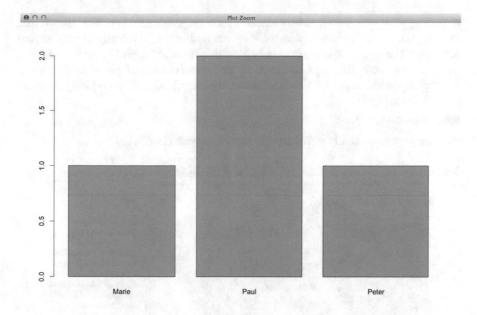

Fig. 3.10   Result of plot(data2$V1)

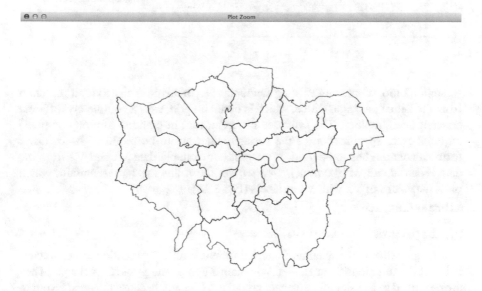

Fig. 3.11   Result of plot(data3)

sp package that was automatically loaded by the maptools package. This means that existing functions can be supplemented by methods through retrospectively loaded packages. The sp package does not only provide a method for plot(), but also, among many others, the native spplot() function for representation of geo data. If we use the spplot() function (which needs a second argument) instead of the plot() function in the last example:

```
spplot(data3, "NAME")
```

then we get a completely different map representation (Fig. 3.12).

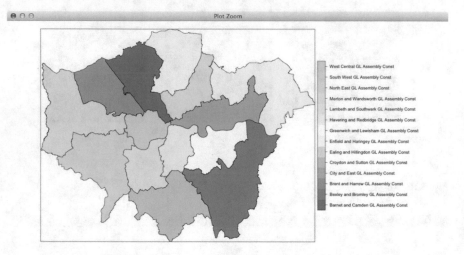

Fig. 3.12   spplot

spplot(), too, is based on existing elements of R, in this case on a xyplot() function from the lattice package. lattice, which is based on grid, uses a completely different concept than graphics. The syntax is different, too. In contrast to plot(), xyplot() requires exact specification of the variables that are to be plotted. To do this, a formula notation has to be used. If we plot our example data V1 and V2 from the data record data2 with xyplot(), then we get the following representation, which plots the values of V2 with V1 as labels (Fig. 3.13):

```
library(lattice)
```

```
xyplot(data2$V2 – data2$V1, cex=2)
```

An again different appearance and a different syntax for creation are characteristics of the functions from Hadley Wickham's ggplot and ggplot2 packages. They are described as a system for "elegant graphics": Elegant, because they endeavour to independently reach appropriate decisions concerning the dimensions of individual elements, scaling, colours, etc. The output is based on the principles of *Grammar of Graphics* by Leland Wilkinson (Fig. 3.14).

```
library(ggplot2)
qplot(data2$V1)
```

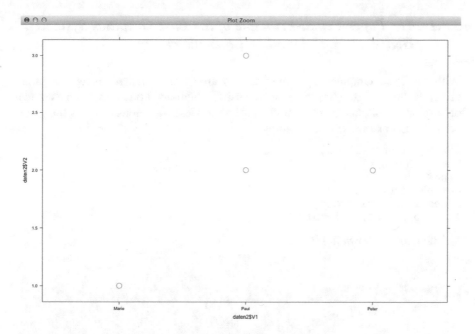

Fig. 3.13 Result of xyplot(data2$V2 ∼ data2$V1, cex=2)

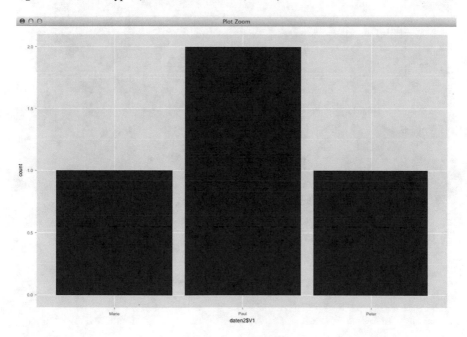

Fig. 3.14 Result of qplot(data2$V1)

3.3.1   The Paper-Pencil-Principle of the Base Graphics System: High-Level and Low-Level Functions

With the Base Graphics System, graphics are always created using the same principle: first, a figure has to be created using a high-level function (such as plot() or barplot()). This can then be enriched using low-level functions (such as lines() or points()). Let us start with an example:

```
par(bg="lightyellow")
bar<-c(1,4,3,4)
line<-bar/2
bp<-barplot(bar)
lines(line, col="red")
lines(bp, line, col="blue")
```

The result is shown in Fig. 3.15.

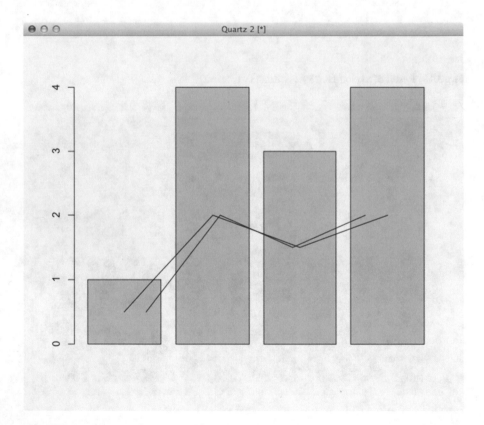

Fig. 3.15 A minimal working example

First, we assign a light yellow as background colour. Then, we define a vector bar that contains four values. After that, a vector line is defined by dividing all bar vector values by 2. The call for the bar plot with barplot() is saved in a variable bp. barplot() is a high-level function that creates a new graphic regardless of whether the call is saved in a variable or made directly. bp<-barplot(bar) and barplot(bar) will therefore yield the same result. In the next line, lines() calls the drawing of a line. lines() is a low-level function that does not create a new graphic, but adds something to an existing graphic. Since the data we transfer to the function is one vector, these values are used as y-values for the graphic, and a running index 1, 2, 3, 4 is automatically added as x-values. col="red" colours the line red. However, as can be seen from the illustration, the line does not quite fit the bars: the first point is markedly right of the centre in the first bar, the second a little right of the centre in the second bar, the third a little left of the centre in the third bar, and the fourth point markedly left of the fourth bar's centre. With the second call, we transfer bp and line. This might sound odd at first, since bp is the bar plot saved in a variable. However, it works because with bp, we saved an object with the property "X-values", among others. R automatically uses this property, if we use the object within the lines() function. And as you can see, the X-positions are now "correct": because here, not only the number of bars, but also their width is considered. We can confirm this, if we use the print(bp) function to output the values:

```
> print(bp)
[,1]
[1, ] 0.7
[2, ] 1.9
[3, ] 3.1
[4, ] 4.3
```

This additive composition of individual elements is the basic principle of the Base Graphic System and affects almost all areas. One of the consequences of this is that the omission of axes does not affect plot size. R will keep the space free, so that the axis() function can be used to add axes at will. This is a fundamental difference from the lattice graphics. In the following example, we first create a layout that displays two graphics next to each other. No axes are drawn in the second one:

```
> par(mfcol=c(1, 2))
> plot(1:10)
> plot(1:10, axes=FALSE)
```

However, the position of the individual elements within the two graphics is identical (Fig. 3.16).

If you now use

```
> axis(1)
> axis(2)
> box(lty="solid")
```

to add axes and a frame to the second graphic, it will look exactly like the one on the left. In this book, we use the following high-level plot functions: barp(), barplot(), boxplot(), dotchart(), dotchart2(), hist(), monthplot(), pie(), plot(), profile.plot(),

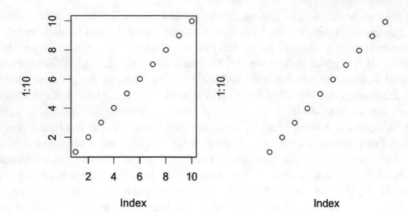

Fig. 3.16 Call of plot() with and without axes

radial.pie(), radial.plot(), and scatterplot3d(). Many high-level functions have a parameter add=TRUE, so they can be added to an existing graphic just like low-level functions (Fig. 3.17).

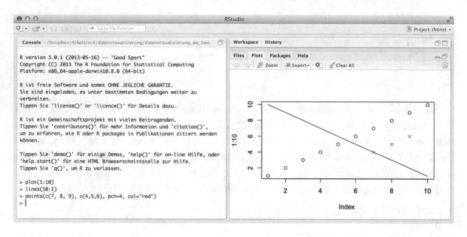

Fig. 3.17 Additive supplementation of a graphic using lines() and points()

3.3.2 Graphic Parameter Settings

Illustration labels are dynamically scaled if the output window is resized, because default settings define them as a factor of the total illustration size. For many applications, these presets are inconvenient. Fortunately, literally every minute detail can be set up via the parameters. Graphics parameters are always transferred to functions as arguments in the form of parameter=value combinations. This can be done in two ways: either through calling of the par() function, which only accepts these graphic parameters as arguments, or as additional arguments in the individual graphic functions. Using the par() function, approximately 80 parameter settings can be adjusted. Which of these can also be used as arguments of a graphic function depends on the method or object type. This means that, e.g.,

```
plot(1:10, lwd=3)
```

and

```
par(lwd=3)
plot(1:10)
```

will yield the same results. The parameter lwd stands for "line width". There are parameters that are only used in specific graphic functions, but cannot be set up via par(). This is why

```
plot(1:10, type="l")
```

creates a line diagram, but

```
par(type="l")
plot(1:10)
```

does not yield the desired result. However, you could also create the above line with lines(1:10) or with point(1:10, type="l"), whereas lines(1:10, type="p") plots a series of points instead of a solid line. This is very systematic, but not necessarily perspicuous. Unfortunately, there are also instances where a parameter has different effects depending on whether it is set using par() or a graphic function. For example, R has different ideas about what "background" means, depending on whether you are using the argument bg (background) in the par() or plot() function. While the

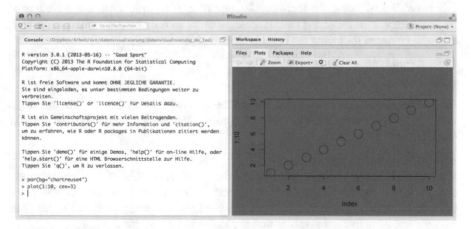

Fig. 3.18 Argument bg in the par function

call within par() expectedly fills the background of the entire illustration window with the selected colour, a call within plot() fills the background of points within the illustration with the selected colour, assuming you have set the value p as the argument for type (which is the default) and you have chosen a number for a so-called open symbol as the value for the plot symbol pch (plotting character). All symbols with a number between 21 and 25 are open. To make the effect more obvious, we have used the cex=3 function to increase the point size by a factor of 3 from the standard in Figs. 3.18 and 3.19.

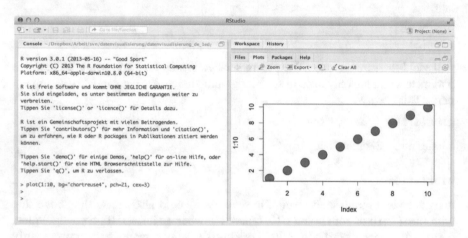

Fig. 3.19 Argument bg in the plot function

It is important to note that a call via par() remains valid until the setting is changed, while a call within a graphic function such as plot() remains valid only for this function. Of the approximately 80 possible arguments, we use around 50 in this book. We will describe them briefly hereafter. A comprehensive overview can be accessed via the help(par) call. If an argument can only be used with par(), then this will be noted following the argument.

Used Arguments of the par() Function

`adj`
Horizontal orientation of text elements in the functions text() (text within figures), mtext() (for margin texts), and title() (for titles). Values between 0 (far left) and 1 (far right) are possible.

`bg (different meanings)`
If used within par(): background colour of the entire figure. Use within par() always creates a new figure. Colours can be specified by name, value or function (see below). If used within plot(), it is not the background of the plot, but rather the background of individual symbols that is plotted in the specified colour.

bty
The box type drawn around the figure. The following values can be specified: o
(default), l, 7, c, u or]. The created box will correspond to the entered character,
with, e.g., c indicating a line on the top, left, and bottom, but not on the right. A 7
means: a line on the right and top, etc. The value n suppresses the box; in that case,
only the axes are drawn. We hardly ever use this option, since the same effect can
be replicated more flexibly with plot(x, type="n"), followed by a call of the axis(n,
...) function.

cex
The factor by which text and symbol sizes are increased relative to default.

cex.axis, cex.lab, cex.main, cex.sub
The factor by which axis labels, axis titles, figure title and figure subtitle are enlarged
relative to default.

col
Colour(s) of the data. The axes and labels are not affected. For multiple variables,
a vector can be specified. Colours can be specified by name, value or function (see
below).

col.axis, col.lab, col.main, col.sub
Colour of axes labels, axes title, figure title, and figure subtitle.

family
Specification of the typefaces to be used.

fg
The figure's foreground colour. If called within plot(), this means all axes and the
box; if called via par(), data colour is also included. Axis labels are unaffected in
either case.

fin(only par())
Circumference of the figure's extent in inches. Generally, this extent is not explicitly
set, but rather results from the margin settings.

las
Angle of axis labels. 0: parallel to the axes (for the y-axis, this means: rotated 90°
counter-clockwise); 1: all horizontal; 2: all perpendicular (for the x-axis, this means:
rotated 90° counter-clockwise); 3: all vertical (for the x- and y-axis: rotated 90°
counter-clockwise)

lend
Line end. 0: round; 1: cropped; 2: square. Generally, the difference is only visible
with very thick lines.

lheight
Line height for text.

lty
Line type. The commonest ones are 0: blank, 1: continuous (default); 2: dashed; 3: dotted; 4: alternating dotted and dashed; 5: long dash, 6: alternating long and short dash.

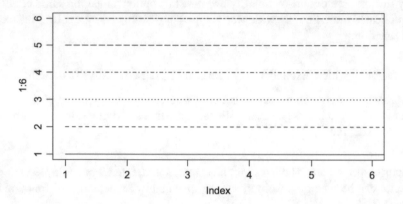

For multiple variables, a vector can be specified. Further options and variants can be found in the help file. The individual line types can be viewed using the following two lines:

```
>plot(1:6, type="n")
>for (i in 1:6) lines(rep(i, 10), lty=i)
```

lwd
Line width. Default: 1. For multiple variables, a vector can be specified.

mai, mar
Margin settings. As value, a vector with the four specifications c(bottom, left, top, right) is required. With mai, margins are specified in inches ("margins in inches"), with mar, as number of lines. We will mostly use absolute values in inch in this book.

mfcol, mfrow (only par())
Number of charts in one figure. The respective result is a matrix with m rows and n columns. With mfcol, counting is done by columns, with mfrow, by rows. Therefore, par(mfcol=c(2,3)) means: two rows, three columns, counting by columns. The order of the graphics is therefore: top left, bottom left, top middle, bottom middle, top right, bottom right. par(mfrow=c(2,3)) also means two rows and three columns, but the order is: top left, top middle, top right, bottom left, bottom middle, bottom right.

mgp
Axis and axis label distance. A vector with three values has to be specified, with the first defining the distance of axis titles, the second the distance of axis labels, and the third the distance of axis lines (each given in "mex units").

new (only par())
Causes plot.new to either create a new figure (confusingly with new=FALSE, which is the default), or to plot the chart in the existing figure. We will be using this argument multiple times.

oma, omd, omi (only par())
The outer margin of the figure (not of the chart/s), specified with a vector c(bottom, left, top, right). The units are: oma: number of text lines; omd: proportion of the figure's total size; omi: inches. We will almost exclusively use inches in this book.

pch
Point character to be used. The numbers 1–25 define a specific symbol, each; additionally, individual characters can be used as symbols (see below; Fig. 3.20). For multiple variables, a vector can be specified. The available symbols can be displayed using

```
plot(1:25, rep(1:25), pch=1.25, axes=FALSE, xlab="", ylab="", ⬊
     ylim=c(0.75, 1.5))
text(1:25, rep(1.25, 25), 1:25)
```

1	2	3	4	5	6	7	8	9	10	11	12	13	14	15	16	17	18	19	20	21	22	23	24	25

Fig. 3.20 Plot symbols in R

21–25 are "open" symbols that are filled with the background colour (bg).

srt
Angle of text insertion with the text() function.

tck
Length of the axes' scale lines. The expression is given as a proportion of the length of the shorter side of the data region. Negative values cause outfacing scale lines. If the data region is 15 cm wide and 10 cm high, the setting would be tck=0.1, that is 10 cm by 0.1 = 1 cm long, inward-facing scale lines. As per standard, R uses a different argument, namely:

tcl
Lengths of the scale lines in relation to the line height of the text. The standard value is −0.5, which means outfacing scale lines measuring half the text height of the labels in length.

xaxp
For non-logarithmic scales: smallest and largest label values and the number of intervals in the form c(x1, x2, n). For logarithmic scales, a code between 1 and 3 that is further explained in the help file. Generally, R's approach leads to very useful spacings. Since R tries to avoid overlaps of the scale line labels at all cost, this can sometimes cause missing labels in positions where one would really like

labels. In such instances, the easiest alternative is manual positioning and labelling of the scale lines using the arguments at and labels of the axis() function.

xaxs
The type of calculation to be used for axes range. Possibilities are r and i. With i, the range is chosen so that the axis labels look "nice". With r, the data area is first stretched by 4% at each end.

xaxt
X-axis type. n means: no axis; every other value causes them to be plotted.

xlog (only par())
Use of a logarithmic x-axis scale. Within plot(), log="x" has to be used.

xpd
Logical argument that allows protrusion of the plotted elements from the data region. With xpd=FALSE, all elements outside the data region are cropped.

yaxp
See xaxp.

yaxs
See xaxs.

yaxt
See xaxt.

ylog
See xlog.

3.3.3   Margin Settings for Figures and Graphics

A central topic for our examples is the precise margin settings. It is important to distinguish between an inner and outer margin. You can specify spacing for both margins either relatively or absolutely. The advantage of relative specification is that the distances are automatically adjusted if the size of the output window is changed. In those of our examples that output into a specific file, we only use absolute specifications. Absolute specifications are given in inches, and the parameters are mai (margins in inches) for the inner and omi (outer margins in inches) for the outer margins. For both, four values are given in the following order: bottom, left, top, right. An inner margin with one inch, respectively, and an outer with half an inch, respectively, is set up like this:

```
par(mai=c(1,1,1,1), omi=c(0.5, 0.5, 0.5, 0.5))
```

Within these defined margins, R positions the figure using the specified distances for the individual elements. With plot(), the axis titles are positioned at the inner border of the inner margin, the figure title created with the plot() parameter main is positioned vertically centred between the upper margin of the data area and the upper inner margin. Margins that are specified this way also define the starting positions for the mtext() function: the margin text created with this function is written on the row defined by line on the outside of the margin, specifically on the side specified by side=1, 2, 3 or 4. The parameter adj=0 causes orientation on the left, adj=1, orientation on the right margin.

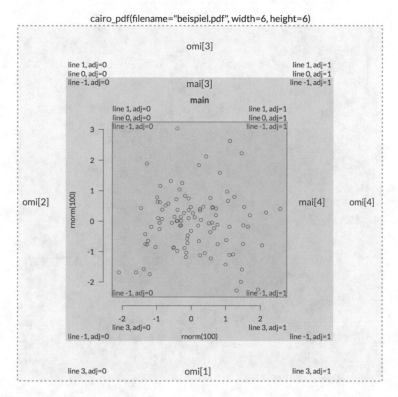

Fig. 3.21 A graphic's margin settings

Figure 3.21 shows the margin settings for a plot() call, and exemplary positions of selected elements. Yellow shows the outer margin, red the inner margin, and green the data area, whose height and width within the inner margin can be specified with fin=c(width, height).

3.3.4 Multiple Charts: Panels with mfrow and mfcol

The differentiation between inner and outer margins is especially useful for multiple charts. If we use

```
par(mfcol=c(2,2))
```

to define a total of four charts (two rows and two columns) in a figure, then we can define inner margins for each of the four charts, but only one outer margin (Fig. 3.22).

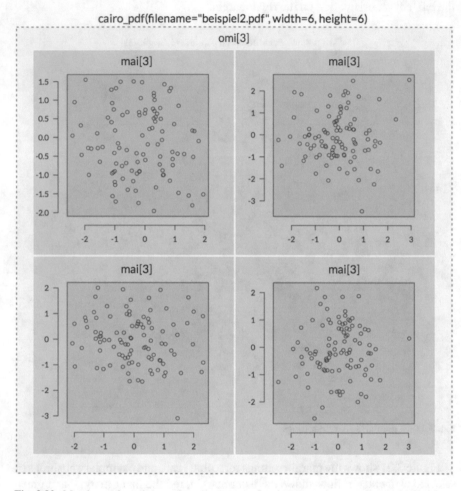

Fig. 3.22 Margin settings in panels—as an example, the respective upper margin has been identified

Labels concerning the entire figure can be placed in the outer margin using the mtext() function and the outer=T parameter.

3.3.5   More Complex Assembly and Layout

Figures with multiple graphics can be created using the parameters mfrow and mfcol, which divide the graphic output into multiple rows and columns. However, this always results in rectangles of identical size. The function layout() offers more flexible options, because the output can be divided into different size rectangles (Fig. 3.23). The R help offers a nice example: this is how scatter plots whose right and upper margin is supplemented by histograms are created:

```
nf <- layout(matrix(c(2,0,1,3),2,2,byrow=TRUE), c(3,1), c(1,3), ⤸
    TRUE)
x <- pmin(3, pmax(-3, stats::rnorm(50)))
y <- pmin(3, pmax(-3, stats::rnorm(50)))
xhist <- hist(x, breaks=seq(-3,3,0.5), plot=FALSE)
yhist <- hist(y, breaks=seq(-3,3,0.5), plot=FALSE)
top <- max(c(xhist$counts, yhist$counts))
par(mai=c(1,1,0.2,0.2))
plot(x, y, xlim=c(-3,3), ylim=c(-3,3), xlab="", ylab="")
par(mai=c(0,1,0.2,0.2))
barplot(xhist$counts, axes=FALSE, ylim=c(0, top), space=0)
par(mai=c(1,0,0.2,0.2))
barplot(yhist$counts, axes=FALSE, xlim=c(0, top), space=0, horiz⤸
    =TRUE)
```

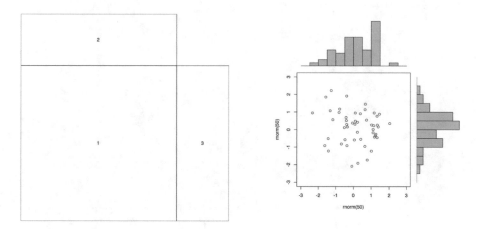

Fig. 3.23 Example for a more complex layout

The left figure shows the created layout, the right one the charts created therein: The layout() function is especially useful to display multiple charts in one panel. To show, e.g., the connection between two variables differentiated by the five attributes of a third variable within a scatter plot, the layout can be designed so all five scatter plots are immediately adjacent to each other. Only the first graphic requires more space, since this is where the label for the y-axis will be placed (Fig. 3.24). Let us look at the following example:

```
layout(matrix(data=c(1,2,3,4,5),nrow=1,ncol=5),
       widths=c(2,1,1,1,1),heights=c(1,1))
par(mai=c(0.5,1,0.5,0),omi=c(0.25,0.25,0.25,0.25))
x<-rnorm(50)
y<-rnorm(50)
plot(x,y,axes=F,col=1,xlim=c(-3,3),ylim=c(-3,3),
     xlab="",ylab="y-axis-\nlabel")
axis(1)
axis(2)
box(lty='solid',col='darkgrey')
par(mai=c(0.5,0,0.5,0))
for (i in 2:5)
{
x<-rnorm(50)
y<-rnorm(50)
plot(x,y,axes=F,col=i,xlim=c(-3,3),ylim=c(-3,3),xlab="")
if (i %% 2 == 0) {axis(3)} else {axis(1)}
box(lty='solid',col='darkgrey')
}
```

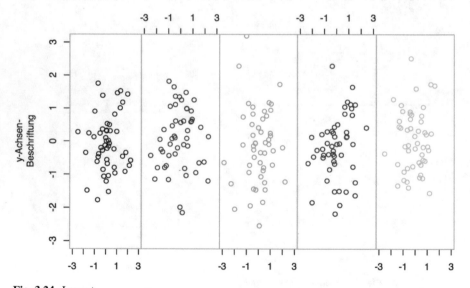

Fig. 3.24 Layout

In this case, we are defining a layout with five columns, but with the first column having twice the width of the other columns. After defining all margins and creating the first plot() and the axes at the left and bottom, the left and right inner margins are set to zero. This remains valid from the time of the call until it is changed. The second to fifth chart are drawn in a loop, with the x-axis alternating between top and bottom.

3.3.6   Font Embedding

When embedding fonts into figures in R, there is an important matter depending on
the used operating system that we will explain in a few examples. Download the free
font "Indie Flower" and install it by double-clicking on the file name. In R, enter

```
par(family="Indie Flower")
plot(1:10, main="Hello World", cex.main=3)
```

If you are working with Linux or Mac OS X, then your output will match the
following figure (Fig. 3.25).

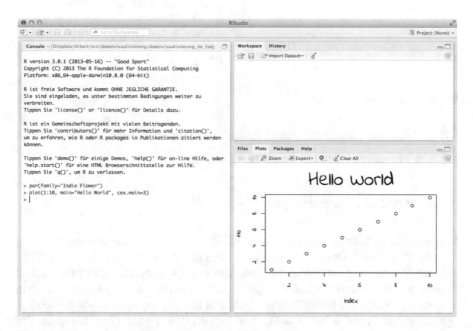

Fig. 3.25 Plot with font embedding in RStudio (Mac OS X)

Doing the same in Windows leads to this result (Fig. 3.26).

If you call warnings as recommended using warnings(), then you will receive
a series of notifications ending with "font family not found in Windows font
database". Fortunately, there is a solution to this problem: the path via the PDF
export of the figures using the cairo_pdf() function.

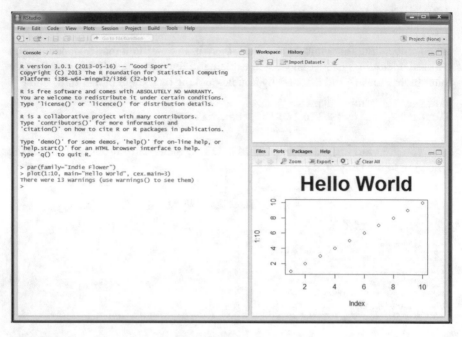

Fig. 3.26 Plot with font embedding in RStudio (Windows)

3.3.7 Output with cairo_pdf

Font embedding without using the detour via PDF export only works if the operating system "plays ball". Once you call a high-level graphic function in R, the result of the call is written in an output window. The exact specifications of this window depend on the respective operating system. If you use R on, e.g., Mac OS X, then a so-called Quartz window will open. Quartz is the graphic layer of the Mac OS X operating system and is based on the PDF format, which means that its capabilities such as the embedding of operating system typefaces or transparent colours can be used in the graphic output in Mac OS X. In a script, a respective output file can be created directly using the quartz() function. However, the disadvantage is that such scripts can only run on computers with Mac OS X, since neither Windows nor Linux can handle the quartz() function. Using Windows, a device called "Windows", which is not based on PDF, will open. The cairo_pdf() function provides a platform-spanning option based on the widely used PDF library. The PDF format is a good choice, since all content including all typefaces and transparency values can be saved in a loss-free file allowing further processing. In contrast to the pdf() function that is also implemented in R, cairo_pdf() has the advantage that all of the operating system's typefaces can be embedded anywhere within a graphic by simply using their name. This cannot be done with pdf(). The respective prerequisite is that the typeface be installed in the operating system. Returning to the last example, calling up the cairo_pdf() function prior to the last two lines will divert output into a PDF file. The dev.off() function closes the output, and R creates the PDF file. If you open this file with, e.g., Reader, then you will see that the typeface was correctly embedded in Windows, too.

3.3.8 Unicode in Figures

When using the cairo_pdf() output, output of the entire scope of Unicode symbols is available in R. To show this, we save the Unicode symbols in an XLSX table (Fig. 3.27).

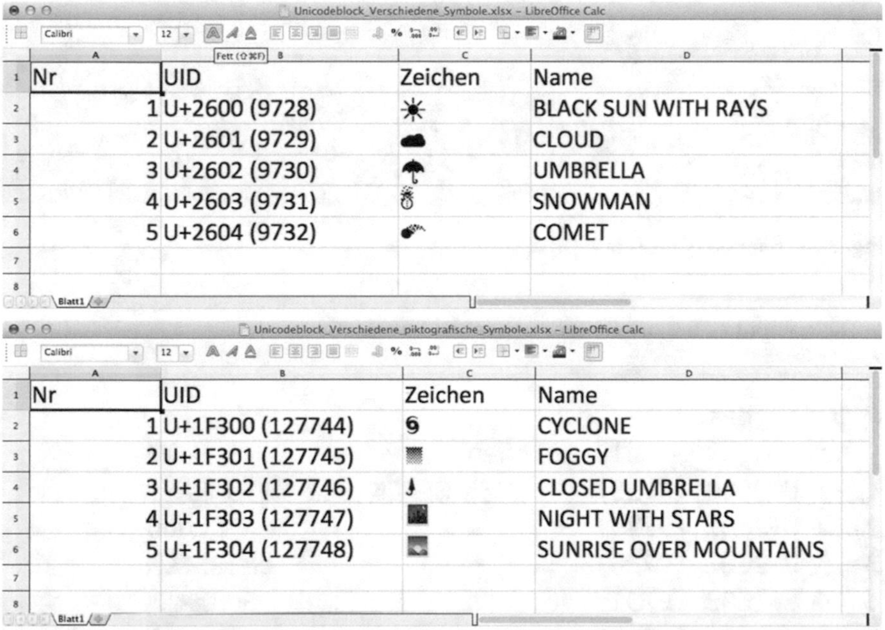

Fig. 3.27 Unicode data in an XLSX file

All Unicode symbols are displayed with their glyphs. If we import these files into R and create a PDF file using cairo_pdf(),

```
cairo_pdf(filename="unicode_symbols_r_xlsx.png", width=9, height↘
    =3)
par(family="Symbola", mfcol=c(1,2), mai=c(0.25, 0, 0.25, 0), omi↘
    =c(0.25,0,0.25,0), bg="aliceblue")
files<-"Unicodeblock_different_symbols.xlsx"
files<-c(files, "Unicodeblock_different_pictographic_symbols.↘
    xlsx")
```

```
for (i in 1:2)
{
data<-read.xls(paste("data/", files[i], sep=""))
print(data)
attach(data)
plot(No, =1:-5, type="n", axes=FALSE, xlab="", ylab="")
text(1, -No, i-1, cex=1.5, xpd=T)
text(1.5, -No, Characters, cex=1.5, xpd=T)
text(2, -No, Name, adj=0, xpd=T)
}
dev.off()
```

then this figure is created without error messages (Fig. 3.28).

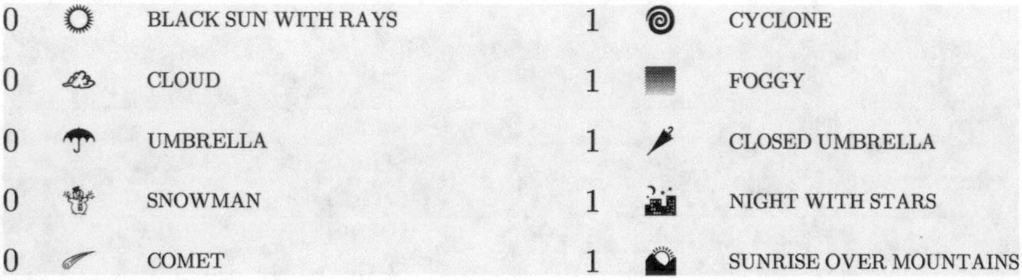

0	☼	BLACK SUN WITH RAYS	1	◉	CYCLONE
0	☁	CLOUD	1	▪	FOGGY
0	☂	UMBRELLA	1	⚡	CLOSED UMBRELLA
0	☃	SNOWMAN	1	⛺	NIGHT WITH STARS
0	☄	COMET	1	⛰	SUNRISE OVER MOUNTAINS

Fig. 3.28 Unicode symbols in R (XLXS)

The output of the print() function looks like this:

```
> print(data)
  Nr           UID Zeichen                    Name
1  1 U+2600 (9728)        *  BLACK SUN WITH RAYS
2  2 U+2601 (9729)        ▲               CLOUD
3  3 U+2602 (9730)        ☂            UMBRELLA
4  4 U+2603 (9731)        ☃             SNOWMAN
5  5 U+2604 (9732)        ☄               COMET
  Nr            UID  Zeichen                       Name
1  1 U+1F300 (127744) \U0001f300               CYCLONE
2  2 U+1F301 (127745) \U0001f301                 FOGGY
3  3 U+1F302 (127746) \U0001f302       CLOSED UMBRELLA
4  4 U+1F303 (127747) \U0001f303      NIGHT WITH STARS
5  5 U+1F304 (127748) \U0001f304 SUNRISE OVER MOUNTAINS
```

As can be seen, the output includes Unicode characters in both cases. The characters of the Basic Multilingual Plane (plane 0) are displayed as glyphs, but those of the Supplementary Multilingual Plane (plane 1) on the other hand, as codes. It is also possible to adopt Unicode symbols from a MySQL database in R. From version 5.5.3, characters from the Supplementary Multilingual Plane are also supported, if the character set utf8mb4 is selected (Fig. 3.29).

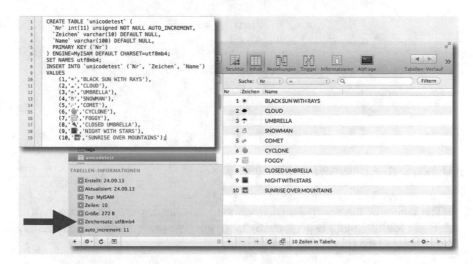

Fig. 3.29 Unicode data from a utf8mb4-encoded table derived from a MySQL database

Before data can be imported into R with this encoding though, this character set has to be selected using SET NAMES utf8mb4 in the current connection.

```
library (RMySQL)
con<- dbConnect(MySQL(),
  user="xxx",
  password="xxx",
  dbname="xxx",
  host="xxx")
sqlset<-"SET NAMES utf8mb4"
do<-dbGetQuery(con,sqlset)

sqldata<-"select * from unicodetest"
data<-dbGetQuery(con,sqldata)
cairo_pdf(filename="unicode_symbols_r_mysql.png", width=9,
    height=3)
par(family="Symbola", mai=c(0.25,0,0.25,0), omi=c(0.25,0,0.25,0)
    , bg="aliceblue")
print(data)
attach(data)
plot(1:1, type="n", axes=F, xlim=c(0,4), ylim=c(-8,0), xlab=" ",
    ylab="")
text(1.5, -No, character, cex=1.5, xpd=T)
text(2, -No, name, adj=0, xpd=T)
dev.off()
```

The result is shown in Fig. 3.30.

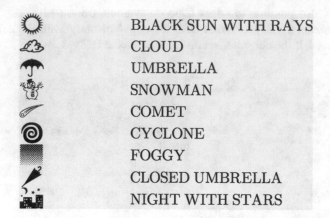

Fig. 3.30 Unicode symbols in R, imported from MySQL

The output of the print() function, on the other hand, shows the characters of the Basic Multilingual Plane (plane 0) as glyphs, but those of the Supplementary Multilingual Plane (plane 1) as code.

```
> print(data)
   Nr    Zeichen                    Name
1   1          *   BLACK SUN WITH RAYS
2   2          ☁               CLOUD
3   3          ☂            UMBRELLA
4   4          ☃             SNOWMAN
5   5          ☄               COMET
6   6 \U0001f300            CYCLONE
7   7 \U0001f301              FOGGY
8   8 \U0001f302    CLOSED UMBRELLA
9   9 \U0001f303   NIGHT WITH STARS
10 10 \U0001f304 SUNRISE OVER MOUNTAINS
```

3.3.9 Colour Settings

There are several ways to set colours in R. We use two of them in the examples in this book: on the one hand, R offers an extensive list of a total of 657 colour names; on the other hand, colours can be defined as RGB values using the rgb() function. With the RGB function, a fourth value can be used to create semi-transparent colours that are supported by the cairo_pdf() function. Additionally, there are a series of ready-made colour palettes in R. Without specific selection of a palette, R uses a standard set of colours that can easily be specified within the plot command using col=1, col=2, etc. You can request the names of the colours like this:

```
> palette()
[1] "black" "red" "green3" "blue" "cyan" "magenta" "yellow" "↘
    gray"
```

To use the RGB values of these colours in a different application, you can use the col2rgb() function:

```
> col2rgb(palette())
      [,1] [,2] [,3] [,4] [,5] [,6] [,7] [,8]
red      0  255    0    0    0  255  255  190
green    0    0  205    0  255    0  255  190
blue     0    0    0  255  255  255    0  190
```

Brewer colour palettes have proven to be very useful for our examples (see also Sect. 2.6); in R, these are implemented in the RColorBrewer Package by Erich Neuwirth. It provides 18 sequential, 8 qualitative, and 9 divergent palettes. Depending on the selected palette, up to 12 matched colours can be selected (Fig. 3.31).

If more colours or other palettes are required, then the colorspace package by Achim Zeileis, Kurt Hornik, and Paul Murrel offers excellent help. This package even contains its own graphic user interface (GUI) to interactively create and try colour palettes. This is a function we have benefited from greatly (Fig. 3.32).

3.4   R Packages and Functions Used in This Book

Below, we briefly describe the packages and functions used in the examples. Packages not embedded in R by default, such as ineq, can be installed as follows:

```
install.packages("ineq", dependencies=TRUE)
```

The call is

```
library(ineq)
```

In RStudio, packages can also be installed using the Tools menu. You can also specify which repository should be used as source. The default is the official CRAN repository. One of the commonest causes of errors when running a script is a missing package. In such cases, install the missing package listed in the error message as described above, and re-run the script.

3.4.1   Packages

Basic Packages

Basic packages are installed by default during R installation.

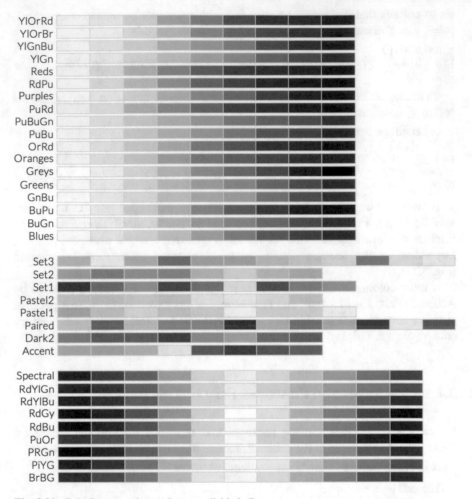

Fig. 3.31 ColorBrewer colour palettes available in R

base
As its name suggests, the base package forms the basis of R. It provides more than a thousand functions for almost every imaginable problem. In this book, we will only use 50 or so.

utils
Another indispensable package for all kinds of problems is utils, which provides more than 250 functions, among them data() or read.cvs() to import data.

stats
The central statistic package contains more than 600 functions with statistical calculations. We will use just under 10 of these in this book.

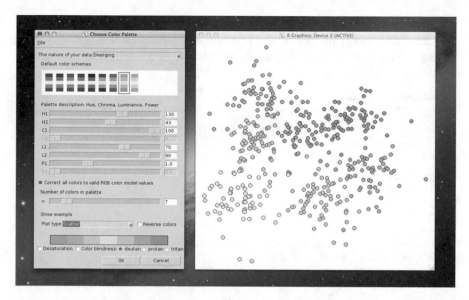

Fig. 3.32 Call of the choose_palette() function from the colorspace package

graphics
Almost 100 functions for the traditional graphic system of R. The central library for this book.

grid
Approximately 200 functions that make up the alternative graphic system of R, which was developed by Paul Murrell. Grid is the basis for the packages lattice and ggplot.

Data Administration and Data Provision

RMySQL
Import of data from MySQL databases. R can also directly address a series of other databases or query them via an ODBC interface; however, we confine ourselves to this package at this point. With RMySQL, data from MySQL databases that support the entire Unicode scope can also be imported (from MySQL version 5.5).

HistData
This package compiled by Michael Friendly contains a series of small data frames that are of great importance for the history of data visualisation, such as the streets of London in 1854, including Charles Snow's notes on the places of cholera outbreaks, Francis Galton's measurements for regression calculations, or the data that formed the basis of Charles Joseph Minard's map for Napoleon's invasion of Russia. The latter are used in Sect. 11.7.

memisc
The memisc package by Martin Elff is a useful compilation of functions for the management of survey data. Especially the use of SPSS data frames is greatly facilitated by this package.

reshape, reshape2
The small but powerful reshape package by Hadley Wickham facilitates restructuring of data. This is done using the melt() and cast() functions, which transform data into a "long" or "wide" format:

```
>x
  ID V1 V2
1  1  3  4
2  2  4  3
3  3  1  9
>y<-melt(x, id="ID")
>y
  ID  variable value
1 1        V1     3
2 2        V1     3
3 3        V1     1
4 1        V2     4
5 2        V2     3
6 3        V2     9
> cast(y)
  ID V1 V2
1  1  3  4
2  2  4  3
3  3  1  9
```

Additionally, statistical formulae can be specified in cast(), which means even complex restructuring of data can be realised in a compact manner. In 2006, this package along with another of Hadley Wickham's packages, ggplot, won the John Chambers Award for Statistical Computing. reshape2 is a reprogrammed version of reshape that is faster, but identical to reshape when it comes to functionality.

sqldf
Normally, a database content request in R is executed using the SQL language. To do this, an SQL command is sent to a database and the result of the query is imported into R. The package sqldf by Gabor Grothendieck makes it possible to phrase SQL commands directly in R and with respect to R data frames. Data are written into a temporary SQL database, where the SQL command is then executed; the result is returned to R, and the database is deleted. This offers very powerful possibilities for data manipulation, which were used here for the creation of the fortune.RData data frame (Sect. 6.3). Graphic

gplots
Provides approximately 25 functions for the creation of graphics or graphic elements. We are using it in our example for creation of a balloon plot in Sect. 6.3.6 and for the barplot2() function in Sect. 8.1.3, providing more functionality than the barplot() function contained in the graphics package.

gridBase
The gridBase package by Paul Murrell only contains a single function, baseView-ports(), which bridges the gap between R's traditional graphic system and its grid system. We make no direct use of it in this book.

plotrix
Jim Lemon's plotrix package contains a colourful mix of almost 150 functions for the creation of diverse graphics or graphic elements. We use it in various examples.

RColorBrewer
The package only provides a function of the same name. It is one of the top ten most downloaded packages for R. The function provides ColorBrewer colour palettes (see Sect. 3.3.9) in R. We make use of this option in several examples.

Hmisc
Hmisc by Frank E. Harrell Jr is also one of the frequently downloaded packages. With more than 130 functions and more than 360 pages of documentation, it is one of the heavy weights within the 4000 R packages. In this book, we use it for the import of SPSS data (which can also be imported using the memisc package), and for the dotchart2() function, which provides more options than dotchart() from the base graphic package (Sect. 6.3).

Other Packages

ellipse
As its name suggests, this package is used to draw ellipses, which we will make some use of in Sect. 9.1.4.

fBasics
The fBasics package belongs to the type of educational packages. It provides functions for market analyses in the finance sector. From this package, we only require the seqPalette() function to create colour gradations of single segments in bar charts.

ineq
The ineq package by Achim Zeileis is suitable for the calculation of data necessary for Lorenz curves (Sect. 7.3.1).

sfsmisc
sfsmisc is a package that was developed at the seminar for statistics, ETH Zurich and spans several topics. In this book, Fig. 2.7 was created using the p.ts() function from this package.

zoo
The exemplary documented zoo ("Zeisel's Ordered Observations") package by Achim Zeisel and Gabor Grothendieck provides multiple options for processing time series data. It is one of the top ten most downloaded packages. We use it for an alternative axis design in Sect. 8.3.6.

For Maps

geoR
The geoR package provides a series of functions for geo-statistical analysis. We use the jitterDupCoords() function from this package in Sect. 10.2.1, which enhances visualisation of overlapping points in a scatter plot or a map.

mapdata
mapdata is a package that does not provide functions, but solely data. As an add-on to the maps package, it provides map data for China, Japan, New Zealand, several rivers, and a world map at higher resolution than in maps.

maps
The maps package provides the map() function for the drawing of maps and corresponding map data in the form of "data bases" for the world, a series of cities, and a small selection of countries.

maptools
The maptools package by Roger Bivand provides approximately 50 functions to support spatial analyses and drawing of maps. For our maps, we use the readShapeSpatial() function from this package to import map data in ESRI Shape format, as well as legend.bubble() to draw legends.

sp
Another package designed by Roger Bivand, and also Edzer Pebesma, is the sp package, which adds classes and methods for spatial data to R. sp expands the plot() function by a map creation method. With spplot, this package also provides its own function for the plotting of maps: however, this function is based on the grid graphic system and is suitable for use in lattice graphics.

rgdal
rgdal is a package that establishes the link between R and Geospatial Data Abstraction Library. We use its spTransform() function for our map examples, which lets us transform maps into a Mercator projection. The package requires installation of GDAL and PROJ.4.

3.4.2   Functions

This list is supposed to serve as a reference section as well. In total, approximately 120 functions are used in this book. A detailed description and concrete examples can be found in the R help by calling up help("function name"). Every use of almost every function is listed in the annex at the end of the book. Usually, the use of a function is explained in more detail at its first utilisation, which is why particularly the first references should be noted. The name of the respective package that contains the function is listed in braces.

`abline {'graphics'}`
Drawing of lines within a graphic. We only use the arguments v and h here to draw vertical and horizontal lines in bar charts, box plots, dot charts, and time series.

`abs {'base'}`
Calculation of absolute values.

`aggregate {'stats'}`
Aggregation of data. Use for box plots.

`agrep {'base'}`
Literally "approximate global regular expression print". Returns the position of elements containing the search term.

```
> x<-c("Peter", "Paul", "Maria", "Paula")
> agrep("Paul", x)
[1] 2 4
```

`arctext {'plotrix'}`
Writes text around an (imaginary) arc. Only used for radial column charts in this book.

`arrows {'graphics'}`
Arrows or rather lines—if the ends are omitted. Used for the multi-coloured population pyramid, two Lorenz curve examples, and two time series examples, among others.

`as.factor{'base'}/as.matrix{'base'}/as.numeric{'base'}`
Functions for conversion of data types. Used in almost half of the examples.

`attach {'base'}`
Adds a data frame to the search path of R. This has solely practical reasons: it means that the name of the data frame no longer has to be prepended to the variable, if a variable of this data frame is used. If a variable is to be used more frequently within a script, then this saves a lot of typing.

`axis {'graphics'}`
Drawing of axes. It is often more flexible to suppress the axes with axes=FALSE during a plot() call, and draw them separately afterwards.

`barp {'plotrix'}`
Drawing of a bar plot. Provides different settings than the conventional barplot() function, especially the staxx argument, which allows two-line axis labels.

`barplot {'graphics'}`
The standard function for drawing bar plots.

`box {'graphics'}`
Drawing of a box around the data. The line type is specified via the lty parameter. With plot(), a box is drawn by default.

boxplot {'graphics'}
Standard function for drawing box plots. The box plot is a rather explorative graphic that addresses specialists. However, we believe that it is quite understandable in its condensed form, as used in our two examples (Sects. 7.1.4 and 7.1.5).

bumpchart {'plotrix'}
Function for drawing bump charts (see Sect. 6.3.3).

c {'base'}
A generic function that combines its arguments. Example: x<-c("A", "B", "C") creates a vector x with the elements A, B, and C. This is not only the function with the shortest name, but also the one most frequently used by us (more than 150 times).

cairo_pdf {'grDevices'}
Opens a PDF output for the following illustration processes. The function allows embedding of all OTF or TTF typefaces installed in the operating system. Output has to be closed with dev.off() to make the file that was named as the argument within the function available. In contrast to the pdf() function, cairo_pdf() does not offer a colormodel parameter, which can be used for the immediate creation of a CMYK file. However, this is no real disadvantage, because the created files can be converted elsewhere if needed (see Sect. 3.3.9).

cbind {'base'}
Combines two objects by columns to create one new object. For combination by row see rbind().

close {'base'}
Closes a connection that was opened with, e.g., url().

colnames {'base'}
Retrieval or setting of column names for a matrix. Accordingly, row names are retrieved or set with rownames().

cumsum {'base'}
Cumulative sums of an argument:

```
> x<-c(3, 7, 2)
> cumsum(x)
[1] 3 10 12
```

curve {'graphics'}
Drawing the plot of a function. Used for a scatter plot example in this book (Sect. 9.1.3).

cut {'base'}
Function for the classification of a continuous variable into n classes. To do this, all class margins are defined, including the uppermost and lowermost and not just the margins between the individual classes. If a value lies below the lowermost or above

the uppermost margin, then a missing value is created:

```
> Z <- stats::rnorm(5)
> K <- cut(Z, breaks = -1:1)
> data<-data.frame(Z)
> data$K<-K
> data
          Z       K
1   0.6427941   (0,1]
2  -0.9235626  (-1,0]
3  -0.4714838  (-1,0]
4   0.6211350   (0,1]
5  -1.3534794    <NA>
```

This is a function we use for choropleth maps, for instance.

data.frame {'base'}
Conversion of an object (usually a matrix) into a data frame.

data {'utils'}
Loading of a data frame available in the system. Files, on the other hand, are loaded with load().

dbGetQuery {'RMySQL'}
Transfer of an SQL statement to a MySQL database.

dev.off {'grDevices'}
Shut down of the graphic output.

dotchart {'graphics'}
Standard function for creation of a dot chart.

dotchart2 {'Hmisc'}
An "improved" version of dotchart() with several additional options. For our purposes, especially the add=TRUE option is useful, which is used in the dot chart example in Sect. 6.1.11.

ellipse {'ellipse'}
Drawing of a confidence region as an ellipse. Used for a scatterplot example in this book (Sect. 9.1.4).

exists {'base'}
Checks if an object exists.

factor {'base'}
Conversion of a variable into a factor (see above).

fitted {'stats'}
Returns the estimated values of a model for the existing data, a regression line, for example.

`floating.pie {'plotrix'}`
Pie chart, whose display position within a coordinate system can be specified. In combination with the possibility to truncate parts protruding from a data region, we can use this to create illustrations for the allocation of seats (semicircles).

`format {'base'}`
Formatting of an object, for example, display of decimal numbers with two decimal digits.

`hist {'graphics'}`
Drawing of a histogram.

`is.na {'base'}`
Check if an element is a missing value.

`lapply {'base'}`
The functions apply(), lapply(), and sapply() each create a kind of loop: a function is applied to every element of an object (vectors, arrays, lists).

`layout {'graphics'}`
Division of the illustration into several partitions, within each of which individual graphics can be drawn. In contrast to par(mfrow=c(n,m)) or par(mfcol=c(n,m)), partitions do not have to be of equal size.

`Lc {'ineq'}`
Calculation of values required for drawing a Lorenz curve.

`legend {'graphics'}`
Addition of a legend.

`length {'base'}`
Determination of the length of an object. Remember: For R, the length of a data frame is the number of its variables or columns.

```
>data1
   V1 V2 V3
1   1  2  3
2   2  3  2
3   3  2  2
4   1  1  3
> length(data1)
[1] 3
```

The number of rows is determined via nrow().

```
>nrow(data1)
[1] 4
```

`levels {'base'}`
Query or definition of a variable's category names.

`library {'base'}`
Loading of a package.

`lines` {`'graphics'`}
Drawing of data as a line.

`list` {`'base'`}
Creation of an object of "list" type.

`lm` {`'stats'`}
Function to fit a linear model to data. Used for calculation of regression lines in a
scatterplot example.

`load` {`'base'`}
Loading of an R data file.

`matrix` {`'base'`}
Creates a matrix from given values.

`merge` {`'base'`} Corresponds to the functionality of the join parameter in an
SQL application: linkage of two data frames by different conditions.

`monthplot` {`'stats'`}
Creation of a so-called seasonal subseries plot.

`mtext` {`'graphics'`}
Margin text for graphics and illustrations.

`na.omit` {`'stats'`}
Omission of missing values.

`nrow` {`'base'`}
Determination of the number of rows in an array or data frame (see also length()).

`order` {`'base'`} Sorting of data. The function returns the position of elements
in a sorted vector:

```
>y<-c("a", "c", "b")
>order(y)
[1] 1 3 2
```

`par` {`'graphics'`}
Setting of graphic parameters.

`paste` {`'base'`}
Concatenation of strings.

`pie` {`'graphics'`}
Drawing of a pie chart.

`plot` {`'graphics'`}
The central function for creation of plots/charts.

`points` {`'graphics'`}
Drawing of data points in a plot.

`polygon` {`'graphics'`}
Drawing of polygons in a plot.

`print` {`'base'`}
Printing or display of an object.

`profile.plot` ...
Drawing of a profile plot (user defined function).

`radial.pie` {`'plotrix'`}
Drawing of a radial pie chart.

`radial.plot` {`'plotrix'`}
Drawing of a radial polygon chart (radar chart).

`rbind` {`'base'`}
Combination of two objects by rows to create one new object. For combination by columns see cbind().

`read.csv` {`'utils'`}
Import of a file in CSV format.

`read.xls` {`'gdata'`}
Import of a file in Excel format.

`readShapeSpatial` {`'maptools'`}
Import of a map shapefile.

`rect` {`'graphics'`}
Drawing of a rectangle.

`rep` {`'base'`}
Replication of an expression.

`residuals` {`'stats'`}
Return of residuals from a previously data-adjusted model.

`rev` {`'base'`}
Reversal of elements of a vector or other object, whose elements can be reversed in their sequence. In a matrix or data frame, the file is reversed by row.

`rgb` {`'grDevices'`}
Creation of colours according to the RGB colour model.

`round` {`'base'`}
Rounding of an object.

`rownames` {`'base'`}
Retrieval or setting of row names for a matrix. Column names are retrieved or set accordingly, using colnames().

`rowSums` {`'base'`}
Computation of row sums for arrays or data frames.

`sapply` {`'base'`}
See lapply().

`scatterplot3d` {`'scatterplot3d'`}
Creation of a 3D scatterplot. Used for creation of a map example in this book
(Sect. 10.2.4).

`seqPalette` {`'fBasics'`}
Function for creation of a sequential colour palette.

`sort` {`'base'`}
Sorting of an object.

`source` {`'base'`}
Integration of a (partial) script from a file.

`spss.system.file` {`'memisc'`}
Import of an SPSS data frame from a file.

`spTransform` {`'sp'`}
Transformation of one map projection into another.

`sqldf` {`'sqldf'`}
Application of an SQL statement on data frames.

`strsplit` {`'base'`}

```
> x<-("Do not agree", "Do agree")
> strsplit(x, " ")
[[1]]
[1] "Do" "not" "agree"
[[2]]
[1] "Do" "Agree"
```

Split a string into substrings.

`strwrap` {`'base'`}

```
> x<-"This sentence contains more than just a few words."
> strwrap(x, width=20)
[1] "This sentence contains" "more than just a few" "words"
```

Wrapping of a string into several substrings.

`subset` {`'base'`}
Selection of a subset of a data frame, matrix or vector.

`substr` {`'base'`}
Return of a substring.

`t` {`'base'`}
Transposition of a data frame or matrix.

`text` {`'graphics'`}
Addition of text to a graphic.

`title` {`'graphics'`}
Addition of a title to a graphic.

`ts` {`'stats'`}
Creation of time series objects.

`unique` {`'base'`}
Return of the different elements of a vector, array or data frame.

`url` {`'base'`}
Opening of a connection with specification of an Internet address (URL).

`which` {`'base'`}
Returns the positions of an object's elements that satisfy the condition handed over as argument

```
> x<-c(4,3,1)
> which(x>1)
[1] 1 2
```

`xline` {`'fields'`}
Plotting of vertical lines into a graphic.

`yline` {`'fields'`}
Plotting of horizontal lines into a graphic.

A Common Look

R provides the possibility to write commands that are to be executed at the start and towards the end of a working session into two profile files. When the program is started, it first checks the contents of a file called Rprofile.site, located in the: R program folder. If a file .Rpofile is located in either the current working folder or in the home folder of the currently active account, then this file is subsequently processed by the program. There are two functions that we can integrate into this file: The .First function and its content is executed at the start of the R session, the commands in the .Last function at its end. We can use this to execute recurring default settings.

```
.First<-function(){
library(grDevices)
library(graphics)
library(gdata)

dev.off(2)
}
.Last<-function(){
# add functions to be executed last.
}
```

3.4.3 Schematic Approach

In conclusion, we want to use the next figure to once more outline the basic steps for the creation of an illustration (Fig. 3.33).

Step 1:
```
cairo_pdf("myfile.pdf", width=9, height=6)
```

Step 2:
```
cairo_pdf("myfile.pdf", width=9, height=6)
par(mfcol=c(1,2))
```

Step 3:
```
cairo_pdf("myfile.pdf", width=9, height=6)
par(mfcol=c(1,2))
barplot(x)
barplot(y)
```
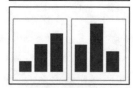

Step 4:
```
cairo_pdf("myfile.pdf", width=9, height=6)
par(mfcol=c(1,2))
barplot(x)
barplot(y)
mtext(side=3,"Hello World", outer=T, adj=0.5)
```
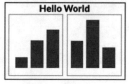

dev.off() ⟶ myfile.pdf

Fig. 3.33 Flow chart of figure creation via scripts

Chapter 4
Beyond R

R provides comprehensive ways for the design of statistical illustrations. However, it is sometimes still sensible to use two additional programs to complement processing of the figures: LaTeX and Inkscape. Both programs are freely available under open-source licences.

4.1 Additions with LaTeX

LaTeX is used to "program" text. This sounds more complicated than it is. True, it takes a bit more getting used to than word processing programs such as Word or OpenOffice, which even beginners can use right away; what applies to our topic has been unequivocally articulated by Nathan Yau: programming knowledge is extraordinarily useful. With LaTeX, typographically pleasing explanations can be added to illustrations with only a few lines of code, as shown in this example (Fig. 4.1).

The advantage of the solution described here is that the explanations are part of the figure and do not have to be treated separately. The LaTeX software is based on the typesetting program TeX. Both TeX and LaTeX have undergone continuous development over the past decades and offer very advanced possibilities for the design of documents. The rules for line or page breaks are superior to most other programs. Current versions of TeX, such as pdfTeX or XeTeX, directly create PDF files as output, also use microtypographic rules such as Adobe InDesign, and offer native Unicode support. These current versions support the use of all TrueType and OpenType fonts installed on your operating system. Free, comfortable editors/work environments are available for Windows, Mac OS X, and Linux; TeXworks , for example, shares identical looks and functions on all three platforms (Fig. 4.2).

© Springer Nature Switzerland AG 2019
T. Rahlf, *Data Visualisation with R*, https://doi.org/10.1007/978-3-030-28444-2_4

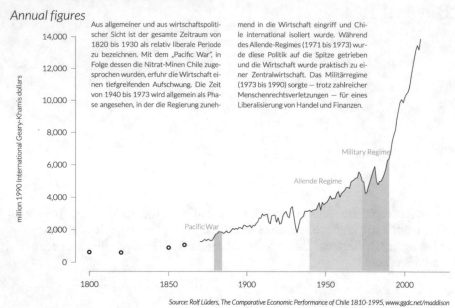

Gross national product of Chile

Annual figures

million 1990 International Geary-Khamis dollars

14,000

12,000

10,000

8,000

6,000

4,000

2,000

0

Aus allgemeiner und aus wirtschaftspolitischer Sicht ist der gesamte Zeitraum von 1820 bis 1930 als relativ liberale Periode zu bezeichnen. Mit dem „Pacific War", in Folge dessen die Nitrat-Minen Chile zugesprochen wurden, erfuhr die Wirtschaft einen tiefgreifenden Aufschwung. Die Zeit von 1940 bis 1973 wird allgemein als Phase angesehen, in der die Regierung zuneh-

mend in die Wirtschaft eingriff und Chile international isoliert wurde. Während des Allende-Regimes (1971 bis 1973) wurde diese Politik auf die Spitze getrieben und die Wirtschaft wurde praktisch zu einer Zentralwirtschaft. Das Militärregime (1973 bis 1990) sorgte — trotz zahlreicher Menschenrechtsverletzungen — für eines Liberalisierung von Handel und Finanzen.

Military Regime

Allende Regime

Pacific War

1800   1850   1900   1950   2000

Source: Rolf Lüders, The Comparative Economic Performance of Chile 1810-1995, www.ggdc.net/maddison

Fig. 4.1 Exemplary illustration with text addition using LaTeX

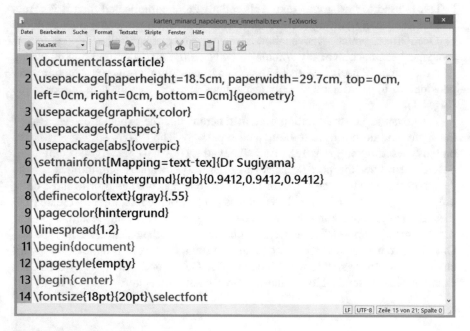

Fig. 4.2 TeXworks in Windows 8.1

R offers a series of options for export into the LaTeX format. However, we will use the software to import PDF files created by R into XeLaTeX files, to add text elements with a few commands, and then again save them as PDF files. We exclusively use LaTeX to create single-page PDF files. The use of LaTeX for this always follows the same pattern (Fig. 4.3).

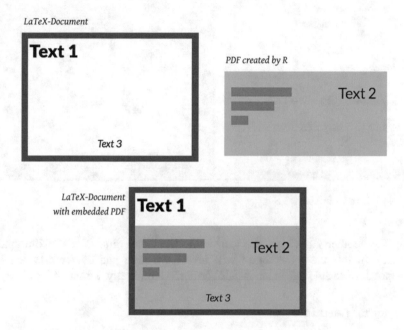

Fig. 4.3 Combination of R and LaTeX

LaTeX commands always begin with a backslash. Parameter contents are put into braces, options in square brackets. In practice, it looks like this (Fig. 4.4).

Let us examine the above example row by row:

```
\documentclass{article}
```

Definition of the document type. This specification is obligatory and causes a series of pre-sets in the document. We do not need a specific document type for our purpose, but because we have to specify one, we choose article.

```
\usepackage[paperheight=21cm, paperwidth=29.7cm, top=0cm, left=0↘
    cm, right=0cm, bottom=0cm]{geometry}
```

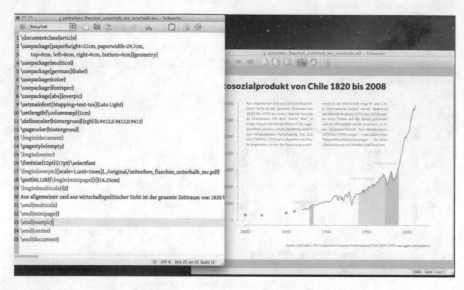

Fig. 4.4 TeXworks in Mac OS X

Using the geometry package, we can define the exact dimensions of the "page" (=figure). In this case, the figure will be 21 cm high and 29.7 cm wide. This corresponds to an A4 page in landscape format. Additionally, we set all four margins to 0.

```
\usepackage{multicol}
```

Loading of the multicol package for multiple-column typesetting.

```
usepackage{color}
```

Loading of the colour package, which allows us to set the document/page background to any desired colour.

```
\usepackage{fontspec}
```

Loading of the fontspec package, with which all system fonts in TTF or OTF format can be embedded into the LaTeX document.

```
\usepackage[abs]{overpic}
```

Loading of the overpic package, with which a figure can be positioned anywhere on a page by specifying coordinates.

```
\setmainfont[Mapping=text-tex]{Lato Light}
```

Specification of the font to be used.

```
\setlength{\columnsep}{1cm}
```

Specification of the distance between columns in the two-column text.

```
\definecolor{mybackground}{rgb}{0.9412,0.9412,0.9412}
```

Definition of a colour called mybackground.

```
\pagecolor{mybackground}
```

Setting of the background colour.

```
\begin{document}
```

Statement indicating the start of the document.

```
\pagestyle{empty}
```

No headers or footers are to be displayed on this page. This is the default for the document class article.

```
\begin{center}
```

The content of the output is centred. A \begin{center} always has to be ended with \end{center}.

```
\fontsize{12pt}{17pt}\selectfont
```

Specification of font size and line spacing.

```
\begin{overpic}[scale=1, unit=1mm]{timeseries_area_underneath_↘
    inc.png}
```

Integration of the figure generated by R.

```
\put(60,120){\begin{minipage}[t]{16.25cm}
```

Absolute positioning of the start of the two-column text with \put(). The width of the text area is defined using the command minipage.

```
\begin{multicols}{2}
```

Definition of two columns.

```
\end{multicols}
\end{minipage}}
\end{overpic}
```

Final markups.

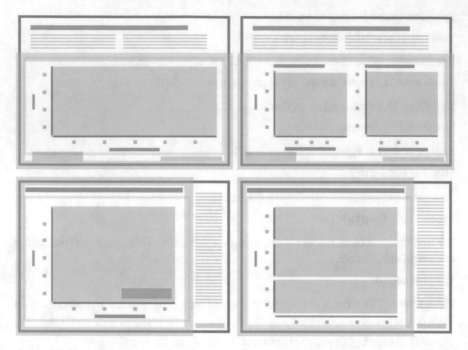

Fig. 4.5 Dimensions of the R figures (green frames) and LaTeX additions

Using this simple method, any combinations of texts and figures can be realised. The only prerequisite is a figure created with pdf_cairo(), with suitable settings for height and width (Sect. 3.4.3); then, only a LaTeX frame has to be "built around it" to position the text anywhere next to or within the figure (Fig. 4.5).

4.2 Manual Post-processing and Creation of Icon Fonts Using Inkscape

4.2.1 Post-processing

The program Inkscape is another useful addition for two reasons: the program is freely available under an open-source licence and, just like Adobe Illustrator, provides the means to process vector graphics. The program can open and edit PDF files, and again save them in PDF format. This means that individual elements of the PDF files created by R, or by R and LaTeX, can be manually corrected if required.

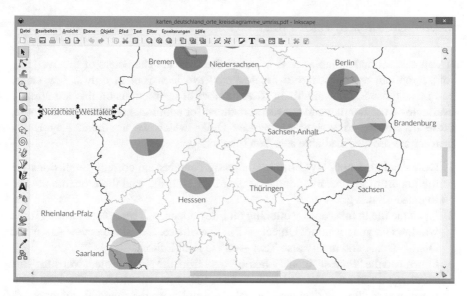

Fig. 4.6 Manual post-processing using Inkscape

You can also regard Inkscape as your "knight in shining armour": everything can still be modified here; the thickness and position of lines and numbers, and colours can be changed, the colour mode can be converted from RGB to CMYK, fonts can be changed, and much more. You can even add entire columns of text (although I much prefer LaTeX for this). The important difference from R is that this is a program that does not require you to program. One could mention though that things can sometimes be done much faster with a WYSIWYG program that provides such practical things as auto alignment.

Especially when supplying an editorial office or a printer, a program such as this provides options to literally adapt everything to the needs of the next entity. However, the decision whether you do your final fine-tuning by clicking in Inkscape or using suitable scripts in R remains with you and depends on the respective situation. For the scope of this book, Inkscape is especially useful for moving labels within scatter plots or maps (Fig. 4.6). Beyond that, Inkscape provides another excellent addition: the possibility to create fonts from icons saved in files in SVG format. Step-by-step instructions on how to do this are presented below.

4.2.2   Creation of Icon Fonts

If you can not find suitable icons in either the Unicode characters or the available fonts, you can use Inkscape to create your own icon font in only a few steps, and make those icons available in R. A number of instructions on this are already available on the Internet and are recommended for additional reading, for example, those by Heydon Pickering[1] or by Kay Hall,[2] which we use here. Kay Hall's instructions are also available as a YouTube video.[3]

1. Download an empty SVG file that already contains the correct dimensions for the import of glyphs, like fontstarter.svg by Kay Hall,[4] and then rename the file to datadesign.svg.
2. Open the file in Inkscape by clicking on File > Open. . ..Then, place another three windows on your desktop: Object → Fill and Stroke. . .. and Object → Align and Distribute. . .., and finally also Text → SVG Font Editor.
3. Download the "Protest" icon by Jakob Vogel from the Noun Project website. This requires you to be registered and confirm that you will attribute the icon with the sentence "Protest designed by Jakob Vogel from the thenounproject.com". The icon will then be downloaded as a zip file. The unzipped file is called noun_project_2376.svg. Import (not: open) the file via File → Import. . .. and enlarge the imported icon by dragging on one of the border arrows while pressing the Ctrl key until its size approaches the size of the canvas. In the Align and Distribute panel, choose Relative to → Page in the pull-down menu, highlight the icon, and click on the symbols to centre vertically and horizontally in the panel. Finally, you have to click on Object → Ungroup and Path → Combine.
4. In the SVG Font Editor above, click on font 1 and replace the default "SVGFont 1" in Family name with "Datadesign". Then click on the Glyphs tab and the first line "glyph1 a" and Get curves from selection (Fig. 4.7). To view the result, you have to replace the entry in Preview Text. Type, e.g., "abc" in there. The Protest icon should now be visible as the first glyph, assigned to the letter "a".
    You can then delete the icon on the canvas.
5. Download an icon created during an Iconathon workshop as a second Protest icon. Its ID is 758. The extracted file is called noun_project_758.svg. Proceed as for the first icon. Once you have selected Ungroup, you have to click on the four sub-elements of the icon while pressing the shift key, and then click on Path → Combine. Click on "glyph2 b" and then again on Get curves from selection. The Preview Text now shows the raised fist as second glyph. Save the file (Fig. 4.8).

---

[1] www.webdesignerdepot.com/2012/01/how-to-make-your-own-icon-webfont/.

[2] cleversomeday.com/inkscape-dings/.

[3] www.youtube.com/watch?v=_KX-e6sijGE.

[4] app.box.com/shared/ohvifhn2ox.

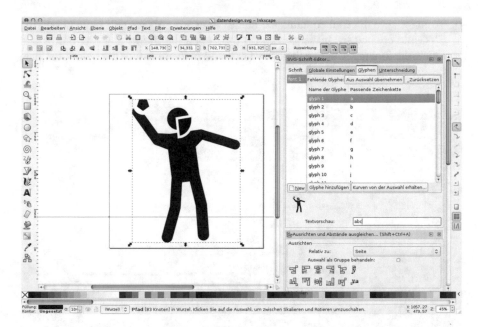

Fig. 4.7 Assignment of the first Protest glyph

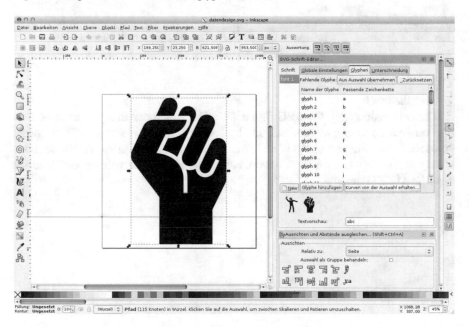

Fig. 4.8 Assignment of the second Protest glyph

6. To make sure the font contains the name "data design", we still have to edit the SVG file created by Inkscape. To do this, open the file in any text editor and locate the font tag. Change the entry here, and in the font-face tag below, to "data design", as shown in Fig. 4.9.

Fig. 4.9 SVG font file in editor

Save the result.

The conversion from the SVG into a TTF file, which can then be installed as a font on your operating system, is done by online conversion services. With those, you select your file and choose "ttf (True Type)" as target format. A few seconds after you have clicked on the Convert button, your font file will be available for download in TTF format.

You can now install your font via a double click in Windows, Mac OS X or Linux (Fig. 4.10).

Fig. 4.10   Installing your font in Mac OS X

Your font is now available to all applications, and the icons can be retrieved by selecting the letters "a" and "b".

Chapter 5
Regarding the Examples

It is in all likelihood impossible to create a consistent and comprehensive systematics. In his semiology, Jaques Bertin distinguished between diagrams with two or three "components" and differentiated non-quantitative and quantitative "problems" for each of these, in addition to "problems that contain more than three components". William S. Cleveland does not base his semiology on the presentation, but on the data, and differentiates between univariate, bivariate, trivariate, hypervariate, and "multiway data". Leland Wilkinson limits his grammar to abstract systematics, and Edward Tufte forgoes all classifications and only discusses individual aspects. In this book, we choose a pragmatic systematisation.

5.1 An Attempt at a Systematics

When attempting a systematisation of statistical visualisation, it is natural to either start with the number of shown variables and their scale of measurement, or with the geometry. Let us look at a rough outline of a few common forms of representation in Fig. 5.1. The first three charts show columns or bars. With those, one or more variables can be represented, including categorical variables such as numbers and metric variables such as averages. The same is true for lines that represent points of a category in the form of a profile chart (4) or for time series (5), which plot the numbers or statistical key parameters over time. Scatter plots (6) correlate two metric variables, but they can also contain a third categorical variable that is identified by colour or different icons. Variation of the point size means a third metric variable can be taken into account. In diagrams on radial axes (7, 8, and 9), lines or areas can be used; so one attribute of several variables as well as several attributes of one variable are shown.

© Springer Nature Switzerland AG 2019

T. Rahlf, *Data Visualisation with R*, https://doi.org/10.1007/978-3-030-28444-2_5

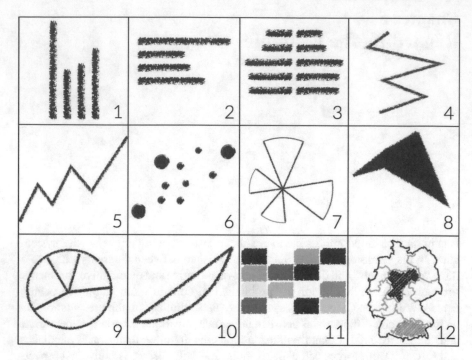

Fig. 5.1 Examples for chart types (schematic)

The Lorenz curve (10) is a line chart that does not fit into either of the previously mentioned categories, and the same goes for a series of visualisations of contingency tables such as so-called heat maps (11). And finally, there are still geodata (12), whose visualisation is also possible in several ways: as areas, with symbols or lines, uni- or multivariate. On top of that, many representation forms can be combined with each other.

We begin with the presentation of categorical data (Chap. 6). This includes bar and column charts (Sect. 6.1), followed by pie charts and radial diagrams (Sect. 6.2). Next is a section on "chart tables" that covers figures based on contingency tables or those that have a tabulated presentation form (Sect. 6.3). Strictly speaking, neither applies to the last two examples of this section, the treemaps. When considering the used data, they are most closely related to bar charts, but when considering their appearance, I felt they fit best into this section. Section 6.4 explains the representation of relationships between units of observation. A popular option, albeit one that is anything but simple to interpret, is visualisation by means of network diagrams. Further options shown here are examples for chord diagrams and riverplots, as well as variants of heat maps and multiple bar charts. After the chart tables comes a chapter on distributions (Chap. 7). In addition to purely statistical presentation forms, such as histograms and box plots, this also includes traditional presentation forms such as population pyramids (Sect. 7.2) or Lorenz curves (Sect. 7.3). Obviously, not only populations can be shown in pyramids, and obviously an inequality, which is what a Lorenz curve commonly depicts, can be shown in a different form, too. We will therefore show alternative examples for both. Chapter 8 is dedicated to time series. As before, there are typical application forms to be differentiated. First, we will look at how to present "short" time series. For this, we will also use columns. There are frequently areas below, between or above time series. Those examples are covered in Sect. 8.2. In our experience, the presentation of daily, weekly or monthly values always proves a bit tricky; Sect. 8.3 deals with those. The chapter concludes with special cases that cannot be allocated to the above groups. Chapter 9 first introduces five variants of scatter plots, each of which illustrates a different aspect and its implementation. This is again followed by "special cases". The last chapter leads into the topic of maps (Chap. 10). After introductory examples, we first distinguish between those that visualise points, icons or entire diagrams in maps, and finally so-called choropleth maps, in which the areas within the maps illustrate the information (Sect. 10.3). As before, we also consider a couple of relevant examples that cannot be assigned to the above-mentioned categories and that therefore have to be listed separately (Sect. 10.4). The next chapter includes a few, rather "illustrative" figures (Chap. 11). With these, the effort put into creating them is potentially higher, and the epistemological value potentially lower than in the preceding chapters.

Finally, Chap. 12 offers an outlook on interactive visualisations with JavaScript. The effort involved in such representations is disproportionately higher, which means that the topic can only be touched on briefly here.

5.2 Getting the Scripts Running

(Note: All scripts are freely available online as Electronic Supplementary Material (https://link.springer.com/chapter/10.1007/978-3-030-28444-2_1)).

The scripts of the following examples require that you have stored the specified data under your working directory in a directory myData. If you are using RStudio and are starting R with it, you can set up the default working directory in the preferences menu (Fig. 5.2).

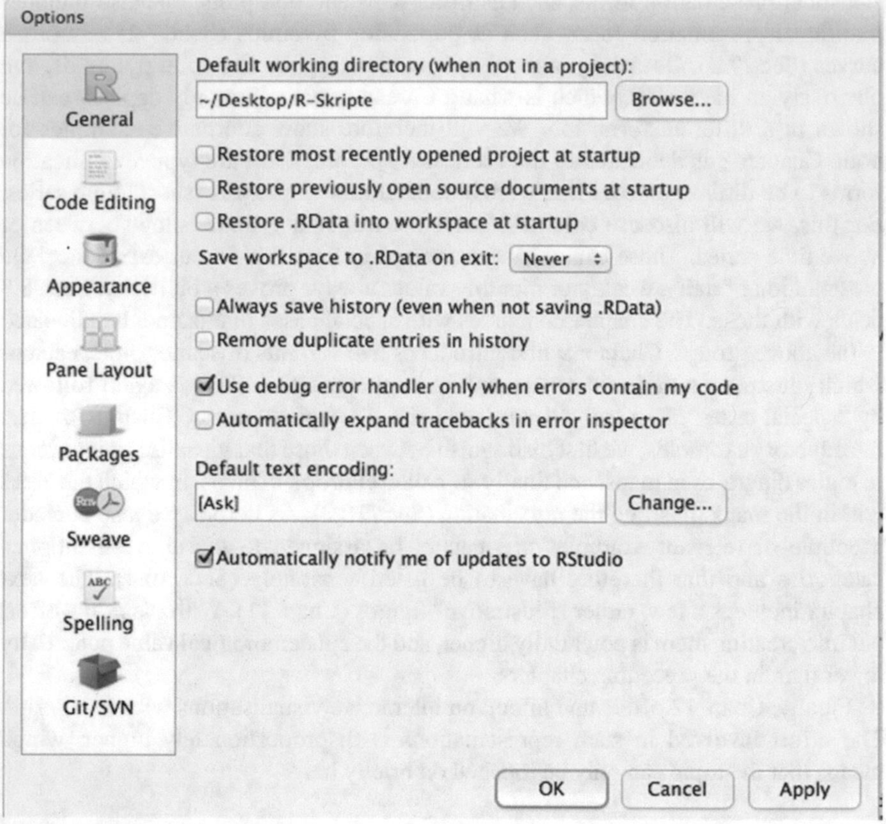

Fig. 5.2 Default settings in RStudio

Below the specified default working directory (which you can change by selecting Browse) is where your data directory should be. Any PDF files created by the scripts are saved in the specified working directory. You can also get your PDFs to be saved into a subdirectory, if you ensure the file name is preceded by the name of the subdirectory in the subsequent examples; in the first example, this means replacing

```
pdf_file<-"bar_chart_simple.pdf"
```

with a statement like

```
pdf_file<-"subdirectory/bar_chart_simple.pdf"
```

In our examples we will use a subdirectory "pdf", so firstly you have to create this subdirectory in your working directory.

Part II
Examples

Chapter 6
Categorical Data

The presentation of simple frequencies or of parameters such as percentages or averages by bar and column charts is certainly one of the most widely used visualisations. Therefore, this is where we want to start. Sometimes though, one might want to also plot parameters that are located on a scale between two statements. In such cases, the use of a profile plot makes sense. If there are a lot of attributes in a bar chart and/or you want to plot several variables simultaneously, then a different illustration form is needed. In such cases, a dot chart comes in handy. The next section explains regular pie charts, but also gives examples of spie charts, radial polygons, and radial column charts. "Chart tables" refers to illustration types in which the arrangement of information has table character. We start with suggestions for two variants of so-called Gantt-charts, then follow with examples for a bump chart, a heat map, a mosaic plot and two examples for tree maps.

6.1 Bar and Column Charts

Attributes of nominal values (frequencies, fractions, etc.) should be presented as bars, while columns should be reserved for ordinal or metrical variables. However, even this should not be considered a rule set in stone. If you think that your concrete case calls for a column chart rather than a bar chart, then you may of course use that. Even though bar charts are omnipresent, their design is anything but easy. The following, seemingly easy examples give an adequate representation of how R can match almost any imaginable design requirements. Starting with the first example, we explain a series of basic principles that we follow throughout this book.

© Springer Nature Switzerland AG 2019
T. Rahlf, *Data Visualisation with R*, https://doi.org/10.1007/978-3-030-28444-2_6

6.1.1   Bar Chart Simple

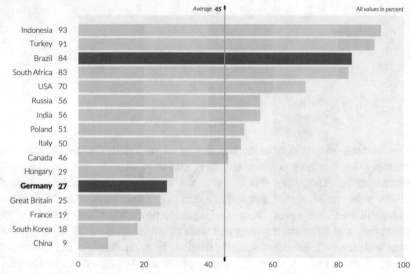

'I Definitely Believe in God or a Supreme Being'

was said in 2010 in:

Average **45**   All values in percent

Indonesia	93
Turkey	91
Brazil	84
South Africa	83
USA	70
Russia	56
India	56
Poland	51
Italy	50
Canada	46
Hungary	29
Germany	**27**
Great Britain	25
France	19
South Korea	18
China	9

0   20   40   60   80   100

Source: www.ipsos-na.com, Design: Stefan Fichtel, ixtract

The figure shows the results of a 2010 survey carried out in different countries: How many percent of the respondents agreed with the statement "I Definitely Believe in God or a Supreme Being"? There is no common rule about the aspect ratio of bar or column charts; the most aesthetically pleasing presentation depends on several factors such as the actual values, the number of attributes, and their range and variance, so that the best value has to be found by trial and error for each individual case. With bar charts, it is usually best to sort bars in descending order. However, this is not obligatory. One could as well sort the bars in alphabetical order. It completely depends on the statement one wants to make. Bars should be styled in one colour. Individual bars/values can be highlighted in another colour. If the bars are left-justified, then the labelling should be right-justified, using visually appealing spacing. Text size should match bar thickness, unless very long labels are used, therefore being wrapped over multiple lines (see Sect. 6.1.3). It is often useful to present the actual values next to the labels. If percent values are plotted instead of numbers, then the repetitive use of "%" can be forgone as long as a single reference reading "all values in percent" is added at a suitable location. A part of the values can be specifically highlighted, here in magenta colour. The colour of the highlight should be clearly distinguishable from the other bars' colour. It is also useful to indicate the highlight in the labelling. To do this, one should choose a typeface that does not only have a regular and a bold cut, but several grades. Here, we use the "light" and "black" cuts of the Lato typeface, which offers five weights. Additionally, it is sometimes useful to also include a trend line as a reference value. This variant is preferred over the use of an artificial category "average" as an extra bar. The line should be explained in a label. As scaling for an x-axis ranging from 0 to 100%, either 20 or 25% steps can be chosen. With either option, a background labelling, whose area matches the markings on the axis, is recommended to facilitate orientation. Alternatively, ledger lines can be used, as shown in the subsequent examples. In the present case, the background area is alternately coloured in light shades of blue. Data are derived from an Ipsos survey that was ordered by the Thompson Reuters News Service and performed between 7 and 23 September 2010 in 24 countries. The participating countries were Argentina, Australia, Belgium, Brazil, Canada, China, France, Germany, Great Britain, Hungary, India, Indonesia, Italy, Japan, Mexico, Poland, Russia, Saudi Arabia, South Africa, South Korea, Spain, Sweden, Turkey, and the USA. The study used an international sample of adults aged between 18 and 64 years from the USA and Canada, and aged between 16 and 64 years from all other countries. The unweighted basis of the respondents numbered 18,531 people. The survey included approximately 1000 people from each country, excluding Argentina, Indonesia, Mexico, Poland, Saudi Arabia, South Africa, South Korea, Sweden, Russia, and Turkey, where the sample size was approximately 500. Data were taken from the Open Mind Journal website and manually entered into an XLSX table.

```
pdf_file<-"pdf/barcharts_simple.pdf"
cairo_pdf(bg="grey98", pdf_file,width=9,height=6.5)

par(omi=c(0.65,0.25,0.75,0.75),mai=c(0.3,2,0.35,0),mgp=c(3,3,0),
   family="Lato Light", las=1)

# Import data and prepare chart

library(gdata)
ipsos<-read.xls("myData/ipsos.xlsx", encoding="latin1")
sort.ipsos<-ipsos[order(ipsos$Percent) ,]
attach(sort.ipsos)

# Create chart

x<-barplot(Percent,names.arg=F,horiz=T,border=NA,xlim=c(0,100),\
      col="grey", cex.names=0.85,axes=F)

# Label chart

for (i in 1:length(Country))
{
if (Country[i] %in% c("Germany","Brasil"))
```

```
   {myFont<-"Lato Black"} else {myFont<-"Lato Light"}
   text(-8,x[i],Country[i],xpd=T,adj=1,cex=0.85,family=myFont)
   text(-3.5,x[i],Percent[i],xpd=T,adj=1,cex=0.85,family=myFont)
   }

# Other elements

rect(0,-0.5,20,28,col=rgb(191,239,255,80,maxColorValue=255),\
      border=NA)
rect(20,-0.5,40,28,col=rgb(191,239,255,120,maxColorValue=255),\
      border=NA)
rect(40,-0.5,60,28,col=rgb(191,239,255,80,maxColorValue=255),\
      border=NA)
rect(60,-0.5,80,28,col=rgb(191,239,255,120,maxColorValue=255),\
      border=NA)
rect(80,-0.5,100,28,col=rgb(191,239,255,80,maxColorValue=255),\
      border=NA)

myValue2<-c(0,0,0,0,27,0,0,0,0,0,0,0,84,0,0)
myColour2<-rgb(255,0,210,maxColorValue=255)
x2<-barplot(myValue2,names.arg=F,horiz=T,border=NA,xlim=c(0,100)\
      ,col=myColour2,cex.names=0.85,axes=F,add=T)

arrows(45,-0.5,45,20.5,lwd=1.5,length=0,xpd=T,col="skyblue3")
arrows(45,-0.5,45,-0.75,lwd=3,length=0,xpd=T)
arrows(45,20.5,45,20.75,lwd=3,length=0,xpd=T)
text(41,20.5,"Average",adj=1,xpd=T,cex=0.65,font=3)
text(44,20.5,"45",adj=1,xpd=T,cex=0.65,family="Lato",font=4)
text(100,20.5,"All values in percent",adj=1,xpd=T,cex=0.65,font\
      =3)
mtext(c(0,20,40,60,80,100),at=c(0,20,40,60,80,100),1,line=0,cex\
      =0.80)

# Titling

mtext("'I Definitely Believe in God or a Supreme Being'",3,line\
      =1.3,adj=0,cex=1.2,family="Lato Black",outer=T)
mtext("was said in 2010 in:",3,line=-0.4,adj=0,cex=0.9,outer=T)
mtext("Source: www.ipsos-na.com, Design: Stefan Fichtel, ixtract\
      ",1,line=1,adj=1.0,cex=0.65,outer=T,font=3)
dev.off()
```

The script begins by setting the window size to a width of 9 inches and a height of 6.5 inches, and setting outer (omi) and inner (mai) margins and the distance of the axis labels (mgp). The left margin is chosen a little wider so we have more space for the labels. Data are read from an XLSX table and assigned to the ipsos variable. Then we sort the values using ipsos[order(ipsos$value),]. This does appear uncommon at first glance: sorting is done by creating a new data area that results from the old one and rearranges the index values. The order() function with the argument ipsos$value returns the order of the sorted rows, here 24, 23, 21, 22.... Since the function is called from within the square brackets of ipsos, followed by a comma and a space, the entire data frame, i.e., all variables, is returned in the sorted order. Note that the barplot() function will plot the bars from bottom to top. This is against our intuition, but does not differ from spreadsheet software. Using attach(), we can append the sorted data frame to R's search path, so that the variables of the data frame can be directly addressed in the subsequent course of the script. This means that the statement Country instead of sort.ipsos$Country suffices. Now, the actual bar chart can be created. R uses the barplot() function for both bar and column charts. The latter is the default setting; to plot bars rather than columns, horiz=T has to be specified. Since we will need the values of the vertical positions of the bars further down the script, we assign the result of barplot() as an object to the variable x. The parameters and values used in barplot are quickly explained: The actual numbers are given first via the value vector, and labels are suppressed by assigning the logical value F to the names.arg parameter. As previously explained, horiz=T creates bars instead of columns, then border=NA means no frames are drawn around the bars; the value axis ranges from 0 to 100, the bar colour is grey, the font size is reduced by the factor 0.85, and no axes are to be drawn. Now we have created a fairly basic bar chart:

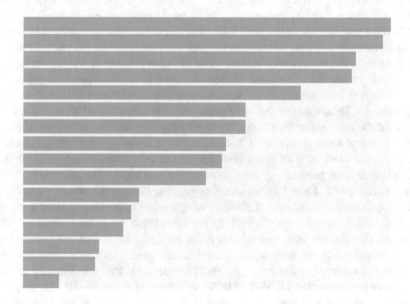

This, however, gives us the opportunity to individualise the missing elements before adding them. First comes the labelling of the bars. To do this, we define a loop that runs through every value of country and changes the font to "Lato Black" if the country is Germany or Brazil. We use the text() function for the labels. The respective country name and the percent value are positioned horizontally at the coordinates -8 and -3.5. The vertical positioning is achieved by referring to the object x, in which we previously saved the result of the barplot() call. The y-position of the ith bar is simply the value x[i]. R automatically extracts the required value from the object, i.e., the vertical position of the bar. Since the x values are located outside of the actual chart (the percent values only start at 0), we have to explicitly enable labelling outside of the chart with xpd=T. Otherwise, R would suppress the output. We also specify right justification (adj=1), a font size, which in turn is decreased by a factor of 0.85, and the font as a vector that contains either the value "Lato Light" or "Lato Black". Background areas are plotted next. This is done through use of the rect() function, which places a shade of blue in different, alternating transparencies over the bar chart. Afterwards, a second co-domain is specified, and a magenta hue as a second colour. To maintain clarity, we use rgb() to assign the colour specification to a variable myColour2. myValue2 only contains the respective percent value at the positions of Germany and Brazil, otherwise its value is zero. x2 causes a second barplot() call that complements the existing bar chart with add=T. Why the detour? Could we not have just plotted two magenta bars in the first bar chart? Technically, this would have been easily feasible; in the first barplot call, the constant grey following the col=... parameter would have to be replaced with a vector containing the colour specifications for each individual bar. However, those bars would then have also been covered by the transparent blue areas. In our case, we first plotted the grey bars, then overlaid them with the transparent blue areas, which were then overlaid by the two non-transparent magenta bars. This is where the basic principle of R's traditional approach towards charts shows, opening up an almost unlimited scope for design. After that, there are only a series of annotations: with three calls of the arrows() function, a blue trend line is plotted at 45%, with black ends for aesthetic reasons. The line is labelled, and "all values in percent" is positioned at the top right. The statement font=3 causes an italic font, font=4 a bold one. Once again, we have to use xpd=T to enable plotting outside the chart area. Labelling of the value axis is done via the mtext() function, which requires specification of the values that are to be plotted and their positions. Finally, the title and subtitle to be plotted above, and the source to be plotted below the chart, have to be defined, once again with the mtext() function. All three are returned in the outer margin. The space for this was initially set with the parameter omi=c(0.65,0.25,0.75,0.75) of the par() function. At the very end, dev.off() completes the graphic output, and then the figure is saved into the pre-defined PDF file.

6.1.2 Bar Chart for Multiple Response Questions: First Two Response Categories

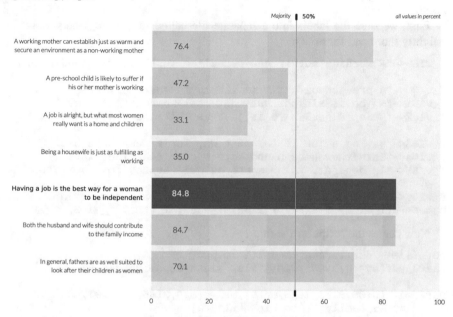

It is often said that attitudes towards gender roles are changing

Agree strongly / agree

Majority ∣ 50%

all values in percent

A working mother can establish just as warm and secure an environment as a non-working mother	76.4
A pre-school child is likely to suffer if his or her mother is working	47.2
A job is alright, but what most women really want is a home and children	33.1
Being a housewife is just as fulfilling as working	35.0
Having a job is the best way for a woman to be independent	84.8
Both the husband and wife should contribute to the family income	84.7
In general, fathers are as well suited to look after their children as women	70.1

0   20   40   60   80   100

Source: European Values Study 2008 Germany, ZA4753. www.gesis.org. Design: Stefan Fichtel, ixtract

The figure shows the results of a survey. The data are based on the European Values Study, a long-term study on values and attitudes of male and female Europeans; shown here is a survey from 2008 to 2010. The study has been conducted since the early 1980s and is repeated every 9 years. Aside from a series of questions concerning value orientation, socio-economic data are also collected. On the topic of "It is often said that attitudes towards gender roles are changing", the respondents were presented with a series of statements. They could respond to each statement with "Agree strongly", "Agree", "Disagree", "Disagree strongly" and "Don't know". The look of the figure almost matches the previous example. However, there are a few differences. The first difference lies in the data: while before the diagram was defined by the individual attributes of a variable, here several variables are combined: each bar shows a variable's value. Such a group of questions on one topic usually means longer labels for the bars, since one wants to show individual statements. The message of the thematic cluster acts perfectly as the title, while the subtitle shows the selections that were chosen from the answers. In the current example, these are the percent values of the first two categories "agree strongly" and "agree". Aside from the repetition of the complete statements that the respondents agreed to, given the extensive labels, it is in this case also useful to write the percent value in the bars. As in the previous examples, the bars are also again complemented by alternating blue areas. For illustration purposes, one question is once more especially highlighted.

Data: see appendix A, ZA4753: European Values Study 2008: Germany (EVS 2008).

First, we save the labels into a separate file called scripts/inc_labels_za4753.r (slightly shortened for print):

```
f_v159<-"A working mother can establish just\nas warm and secure\
    …"
f_v160<-"A pre-school child is likely to suffer if\nhis or her…"
f_v161<-"A job is alright but what most women\nreally want …"
f_v162<-"Being a housewife is just as fulfilling as\nworking…"

f_v163<-"Having a job is the best way for a woman\nto be…"
f_v164<-"Both the husband and wife should contribute\nto …"
f_v165<-"In general, fathers are as well suited to\nlook after …\
    "
f_v166<-"Men should take as much responsibility\nas women …"
names<-c(f_v165, f_v164, f_v163, f_v162, f_v161, f_v160, f_v159)
```

Now comes the script for the figure:

```
pdf_file<-"pdf/barcharts_multiple.pdf"
cairo_pdf(bg="grey98", pdf_file, width=13,height=10.5)

par(omi=c(0.65,0.75,1.25,0.75),mai=c(0.9,3.85,0.55,0),lheight\
    =1.15,family="Lato Light",las=1)
source("scripts/inc_labels_za4753.r")
library(memisc)

# Read data and prepare chart

ZA4753<-spss.system.file("myData/ZA4753_v1-1-0.sav")
myData<-subset(ZA4753,select=c(v159,v160,v161,v162,v163,v164,\
    v165))
attach(myData)
z<-NULL
y<-table(as.matrix(v165))
z<-c(z,100*(y["1"]+y["2"])/sum(y))
y<-table(as.matrix(v164))
z<-c(z,100*(y["1"]+y["2"])/sum(y))
y<-table(as.matrix(v163))
z<-c(z,100*(y["1"]+y["2"])/sum(y))
y<-table(as.matrix(v162))
z<-c(z,100*(y["1"]+y["2"])/sum(y))
y<-table(as.matrix(v161))
z<-c(z,100*(y["1"]+y["2"])/sum(y))
y<-table(as.matrix(v160))
z<-c(z,100*(y["1"]+y["2"])/sum(y))
y<-c(0,table(as.matrix(v159)))
```

```
z<-c(z,100*(y["1"]+y["2"])/sum(y))

# Create chart

bp<-barplot(z,names.arg=F,horiz=T,border=NA,xlim=c(0,100),
  col="grey",axes=F,family="Lato")
myColour<-rgb(255,0,210,maxColorValue=255)
rect(0,-0.1,20,8.6,col=rgb(191,239,255,80,maxColorValue=255),\
      border=NA)
rect(20,-0.1,40,8.6,col=rgb(191,239,255,120,maxColorValue=255),\
      border=NA)
rect(40,-0.1,60,8.6,col=rgb(191,239,255,80,maxColorValue=255),\
      border=NA)
rect(60,-0.1,80,8.6,col=rgb(191,239,255,120,maxColorValue=255),\
      border=NA)
rect(80,-0.1,100,8.6,col=rgb(191,239,255,80,maxColorValue=255),\
      border=NA)
z2<-c(0,0,84.81928,0,0,0,0)
bp<-barplot(z2,names.arg=F,horiz=T,border=NA,xlim=c(0,100),
  col=myColour,axes=F,add=T)

# Other elements

for (i in 1:length(mynames))
{
if (i == 3) {myFont<-"Lato Bold"} else {myFont<-"Lato Light"}
text(-3,bp[i],mynames[i],xpd=T,adj=1,family=myFont,cex=1)
text(10,bp[i],format(round(z[i],1),nsmall=1),family=myFont,cex\
      =1.25,
  col=ifelse(i==3,"white","black"))
}
arrows(50,-0.1,50,8.8,lwd=1.5,length=0,xpd=T,col="skyblue3")
arrows(50,-0.25,50,-0.1,lwd=5,length=0,xpd=T)
arrows(50,8.8,50,8.95,lwd=5,length=0,xpd=T)
text(48,8.9,"Majority",adj=1,xpd=T,cex=0.9,font=3)
text(52,8.9,"50%",adj=0,xpd=T,cex=0.9,family="Lato Bold",font=3)
text(100,8.9,"all values in percent",adj=1,xpd=T,cex=0.9,font=3)
mtext(c(0,20,40,60,80,100),at=c(0,20,40,60,80,100),1,line=0.75)

# Titling

mtext("It is often said that attitudes towards gender roles are \
      changing",3,line=2.2,adj=0,cex=1.8,family="Lato Black",\
      outer=T)
mtext("Agree strongly / agree ",3,line=0,adj=0,cex=1.5,outer=T)
mtext("Source: European Values Study 2008 Germany, ZA4753. www.\
      gesis.org. Design: Stefan Fichtel, ixtract",1,line=0,adj\
      =1,cex=0.95,outer=T,font=3)
dev.off()
```

In the script, a file containing the actual questions that we need for the labelling of the bars has to be read first to create the figures. The questions are not contained in the SPSS file, but only in the original survey available on the GESIS ZACAT portal. We have copy/pasted those questions, assigned them to compact variables (f_v159, f_v160, ...), and fitted a line break where appropriate. The actual data are imported using Martin Elff's memisc package. The advantage of this package lies in the fact that the import does not actually import the entire SPSS data frame, but only the required part. We use the select parameter of the subset() function to import the variables v159, v160, v161, v162, v163, v164, and v165. The attributes of the variables can be viewed using the codebook() function, for example,

```
>codebook(v159)
================================================================

  v159 'working mother warm relationship with children (Q48A)'

  ----------------------------------------------------------------

  Storage mode: double
  Measurement: nominal
  Missing values: -5--1

            Values and labels     N      Percent

  -5 M 'other missing'            0            0.0
  -4 M 'question not asked'       0            0.0
  -3 M 'nap'                      0            0.0
  -2 M 'na'                       0            0.0
  -1 M 'dk'                      67            3.2
   1   'agree strongly'         803     40.0 38.7
   2   'agree'                  782     38.9 37.7
   3   'disagree'               311     15.5 15.0
   4   'disagree strongly'      112      5.6  5.4
```

The values -5, -4, -3, -2, and -1 are defined as missing values and are therefore not taken into account during statistical calculations. With the exception of dk (don't know) and na (no answer), none of these characteristics come to pass in the selected seven variables. If we want to show a bar chart across several variables, we have to initially generate a vector that contains the required values. First, we define a variable z that is pre-assigned the value NIL. Using table(), we can then create a one-row "table" for each variable of our data frame, containing the frequencies of the attributes -2, -1, 1, 2, 3, 4. To do this, we first have to convert the SPSS variable into a matrix using as.matrix(). This also counts -2 and -1 as values. In my opinion, this is useful: the calculation of agreement should also be based on cases that answered "don't know" or are set to na. If you do not share my view, then you would have to exclude those attributes from the calculation. This will not make a visible difference, at least for the latter, since the na attributes are limited to a few cases. A one-row table is now created for every individual variable, the table is assigned the variable y, and the portion of the first two attributes on all attributes is calculated using z<-c(z,100*(y["1"]+y["2"])/sum(y)), and attached to z as an additional value. Note that we start at the back when creating tables, i.e., v165, then v164, etc; this is because the bar chart is plotted from bottom to top (see also Sect. 6.1.1). In v159, the attribute −2 does not occur, which means we have to define this "frequency" as 0 as the first value of the vector y. The call of barplot() is done as usual: we want to plot z, the title is saved in names, horizontal layout, no frames around the bars, x-axis range from 0 to 100, colour "orange", no axes. The object created by the barplot function is saved in the bp variable, so we can use the coordinates of the columns for the text() function in the next line. Each bar's respective value is written as text at position 10, rounded to one digit after the decimal print and complemented with a percent symbol %, in white colour, magnified by a factor of 2. Now all that is left to do, as it were, is to put the titles and captions into the outer margins using mtext().

6.1.3   Bar Chart for Multiple Response Questions: All Response Categories

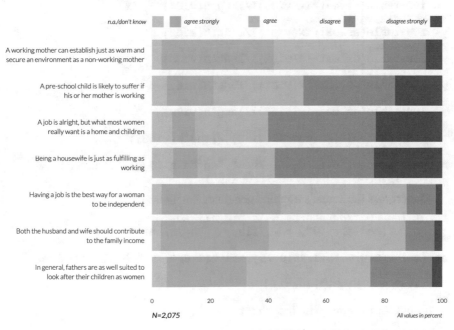

It is often said that attitudes towards gender roles are changing

Source: European Values Study 2008 Germany, ZA4753. www.gesis.org. Design: Stefan Fichtel, ixtract

If you want the figure to show all categories and not only the first two, then you will end up with a bar chart in which all bars have the same length . They only differ in their subdivisions. In this type of chart, colour selection is important. Sensible choices are colour combinations that are easily associated with agreement and disagreement. The category "na/don't know" should be set apart from the selected colour scheme; grey is a good choice here. Since there are now five categories, we also need a legend. In the present chart type, it should be placed above the first bar. The alternative is a separate legend as shown in Sect. 7.1.2 or a "reference bar", similar to that shown in Sects. 7.3.4 and 7.3.5. The width of the legend should correspond to the width of the bars, and it should be easy to associate the elements with the actual data. This means that some category labels are placed to the left and some to the right of the colours.

Data: see annex A, ZA4753: European Values Study 2008: Germany (EVS 2008). We need two files to generate this figure. First, the data are imported. This part was transferred, since we will need it in two other examples.

```
# inc_data_za4753.r
library(memisc)
ZA4753<-spss.system.file("mydata/ZA4753_v1-1-0.sav")
mydata<-subset(ZA4753,select=c(v106,v159,v160,v161,v162,v163,\
      v164,v165))
attach(mydata)
z<-NULL
y<-100*table(as.matrix(v165))/length(v165)
z<-rbind(z,y)
y<-100*table(as.matrix(v164))/length(v164)
z<-rbind(z,y)
y<-100*table(as.matrix(v163))/length(v163)
z<-rbind(z,y)
y<-100*table(as.matrix(v162))/length(v162)
z<-rbind(z,y)
y<-100*table(as.matrix(v161))/length(v161)
z<-rbind(z,y)
y<-100*table(as.matrix(v160))/length(v160)
z<-rbind(z,y)
y<-c(0,100*table(as.matrix(v159))/length(v159))
z<-rbind(z,y)
myresponses<-c("n.a./don't know","agree strongly","agree","\
      disagree","disagree strongly")
```

Then comes the actual script that generates the figure:

```
pdf_file<-"pdf/barcharts_multiple_all.pdf"
cairo_pdf(bg="grey98", pdf_file, width=13,height=10.5)

par(omi=c(0.0,0.75,1.25,0.75),mai=c(1.6,3.75,0.5,0),lheight\
      =1.15,
family="Lato Light",las=1)

# Import data and prepare chart
```

```
source("scripts/inc_labels_za4753.r",encoding="UTF-8")
source("scripts/inc_data_za4753.r",encoding="UTF-8")

myC1<-rgb(0,208,226,maxColorValue=255)
myC2<-rgb(109,221,225,maxColorValue=255)
myC3<-rgb(255,138,238,maxColorValue=255)
myC4<-rgb(255,0,210,maxColorValue=255)
mycolours<-c("grey",myC1,myC2,myC3,myC4)

myData0<-cbind(z[,1]+z[,2],z[,3],z[,4],z[,5],z[,6])
myData1<-t(myData0)

# Create chart

x<-barplot(myData1,names.arg=mynames,cex.names=1.1,horiz=T,  ⤸
      border=NA,xlim=c(0,100),col=mycolours,axes=F)

# Other elements

px<-c(2,8,35,68,98); py<-rep(9,5); tx<-c(-2,23,43,65,95); ty<-⤸
      rep(9,5)
points(px,py,pch=15,cex=4,col=mycolours,xpd=T)
text(tx,ty,myresponses,adj=1,xpd=T,family="Lato Light",font=3)
mtext(c(0,20,40,60,80,100),at=c(0,20,40,60,80,100),1,line=0,cex⤸
      =0.90)

# Titling

mtext("It is often said that attitudes towards gender roles are  ⤸
      changing",3,line=2.2,adj=0,cex=1.8,outer=T,family="Lato  ⤸
      Black")
mtext("All values in percent",1,line=2,adj=1,cex=0.95,font=3)
mtext("Source: European Values Study 2008 Germany, ZA4753. www.⤸
      gesis.org. Design: Stefan Fichtel, ixtract",1,line=4.5,adj⤸
      =1,cex=0.95,font=3)
mtext("N=2,075",1,line=2,adj=0,cex=1.15,family="Lato",font=3)
dev.off()
```

The script starts with the settings for window size and margins, followed by import of first the labels and then the data. Colour definitions match those from the previous example. With the data, the first and second column are combined, then transposed for plotting in the barplot() function. At the end, positions for the legend are specified. Because of the exact placement, we do not use the legend() function; instead, we add the colours and texts using the functions points() and text(). The last step is to add the axes labels, titles, and captions.

6.1.4   Bar Chart for Multiple Response Questions: All Response Categories, Variant

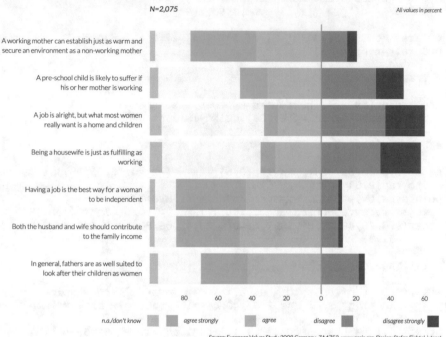

It is often said that attitudes towards gender roles are changing

Source: European Values Study 2008 Germany, ZA4753. www.gesis.org. Design: Stefan Fichtel, ixtract

The figure is a variation of the one from Sect. 6.1.3 and centres the bars around zero. This means that agreements are right-justified as stacked bars, while disagreements are adjacent to those and left-justified. The "no answer/don't know" category should stay separated on the left side. In this case, the legend has been placed below the chart instead of above it. Its layout matches the previous example.

Data: see annex A, ZA4753: European Values Study 2008: Germany (EVS 2008).

```
pdf_file<-"pdf/barcharts_multiple_all_2.pdf"
cairo_pdf(bg="grey98", pdf_file, width=13,height=10.5)

par(omi=c(0.25,0.75,1,0.75),mai=c(1.8,3.75,0.25,0),lheight=1.15,↘
     family="Lato Light",las=1)
library(RColorBrewer)

# Import data and prepare chart

source("scripts/inc_labels_za4753.r",encoding="UTF-8")
source("scripts/inc_data_za4753.r",encoding="UTF-8")

myC1<-rgb(0,208,226,maxColorValue=255)
```

```
myC2<-rgb(109,221,225,maxColorValue=255)
myC3<-rgb(255,138,238,maxColorValue=255)
myC4<-rgb(255,0,210,maxColorValue=255)
colours<-c("grey",myC1,myC2,myC3,myC4)

myData0<-cbind(z[,1]+z[,2],z[,3],z[,4],z[,5],z[,6])
myData1<-t(myData0)

# Create chart

barplot(-rep(100,7),names.arg=mynames,cex.names=1.1,horiz=T, ⬊
        border=par("bg"),xlim=c(-100,70),col=colours[1],axes=F)
barplot(-(100-myData1[1,]),names.arg=mynames,cex.names=1.1,horiz⬊
        =T, border=par("bg"),xlim=c(-100,70),col=par("bg"),axes=F,⬊
        add=T)
barplot(-myData1[3:2,],names.arg=mynames,cex.names=1.1,horiz=T, ⬊
        border=NA,xlim=c(-100,70),col=colours[3:2],axes=F,add=T)
barplot(myData1[4:5,],names.arg=mynames,cex.names=1.1,horiz=T, ⬊
        border=NA,xlim=c(-100,70),col=colours[4:5],axes=F,add=T)

# Other elements

arrows(0,-0.1,0,8.6,lwd=2.5,length=0,xpd=T,col="skyblue3")
px<-c(-98,-87,-41,15,65); tx<-c(-105,-60,-26,8,60); y<-rep(-1,5)
points(px,y,pch=15,cex=4,col=colours,xpd=T)
text(tx,y,myresponses,adj=1,xpd=T,font=3)
mtext(c(80,60,40,20,0,20,40,60),at=c(-80,-60,-40,-20,0,20,40,60)⬊
        ,1,line=0,cex=0.95)

# Titling

mtext("It is often said that attitudes towards gender roles are ⬊
        changing",3,line=2.2,adj=0,cex=1.8,outer=T,family="Lato ⬊
        Black")
mtext("All values in percent",3,line=1,adj=1,cex=0.95,font=3)
mtext("Source: European Values Study 2008 Germany, ZA4753. www.⬊
        gesis.org. Design: Stefan Fichtel, ixtract",1,line=5.2,adj⬊
        =1,cex=0.95,font=3)
mtext("N=2,075",3,line=1,adj=0,cex=1.15,family="Lato",font=3)
dev.off()
```

The first half of the script matches the previous example. The most important difference is that the barplot() function is not called once, but four times. First, seven grey bars ranging from 0 to -100 are plotted. Using add=T, these bars are overlaid with the percent values of the "na/don't know" category as negative difference from

100 in the background colour. After that, only the exact "na/don't know" portion of the grey bars remains visible. The third barplot() call plots the third and second category as stacked bars, again as negatives; the fourth barplot() plots the fourth and fifth category, again as stacked bars, but now in positive direction. Using arrows(), the zero line is highlighted in the colour "skyblue3". The rest matches the previous example; however, the negative labelling of the x-axis has to be replaced with the respective positive values in this example.

6.1.5 Bar Chart for Multiple Response Questions: All Response Categories (Panel)

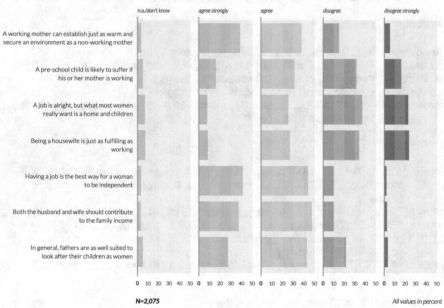

It is often said that attitudes towards gender roles are changing

N=2,075

All values in percent

Source: European Values Study 2008 Germany, ZA4753. www.gesis.org. Design: Stefan Fichtel, ixtract

The figure shows the use of a panel as another alternative, in which the bars for the individual attributes are each drawn separately. In this type of presentation, the differences between the individual attributes stand out clearly. The labels of the individual attributes are now titles, and instead of a single x-axis, a separate one should be provided for each response category.

Data: see annex A, ZA4753: European Values Study 2008: Germany (EVS 2008).

```
pdf_file<-"pdf/barcharts_multiple_all_panel.pdf"
cairo_pdf(bg="grey98", pdf_file, width=13,height=10.5)

par(omi=c(1.25,1.25,1.25,0.25),lheight=1.15,family="Lato Light",\
    las=1)
library(RColorBrewer)

# Import data and prepare chart

source("scripts/inc_labels_za4753.r")
source("scripts/inc_data_za4753.r")
layout(matrix(data=c(1,2,3,4,5),nrow=1,ncol=5), widths=c\
    (2.5,1,1,1,1),heights=c(1,1)))

myData1<-cbind(z[,1]+z[,2],z[,3],z[,4],z[,5],z[,6])
tmyData<-myData1
DD_pos<-c(45,45,45,45,35)
myC1<-rgb(0,208,226,maxColorValue=255)
myC2<-rgb(109,221,225,maxColorValue=255)
myC3<-rgb(255,138,238,maxColorValue=255)
myC4<-rgb(255,0,210,maxColorValue=255)
colours<-c("grey",myC1,myC2,myC3,myC4)

# Create chart

for (i in 1:5) {
if (i == 1)
{
par(mai=c(0.25,2.75,0.25,0.15))
bp1<-barplot(tmyData[ ,i],horiz=T,cex.names=1.6,names.arg=\
    mynames,
    xlim=c(0,50),col=colours[i],border=NA,axes=F)
} else
{
par(mai=c(0.25,0.1,0.25,0.15))
bp2<-barplot(tmyData[ ,i],horiz=T,axisnames=F,xlim=c(0,50),col=\
    colours[i],border=NA,axes=F)
}

# Other elements

rect(0,0,10,8.5,col=rgb(191,239,255,80,maxColorValue=255),border\
    =NA)
rect(10,0,20,8.5,col=rgb(191,239,255,120,maxColorValue=255),\
    border=NA)
```

```
rect(20,0,30,8.5,col=rgb(191,239,255,80,maxColorValue=255),\
     border=NA)
rect(30,0,40,8.5,col=rgb(191,239,255,120,maxColorValue=255),\
     border=NA)
rect(40,0,50,8.5,col=rgb(191,239,255,80,maxColorValue=255),\
     border=NA)

mtext(myresponses[i],3,adj=0,line=0,cex=0.95,font=3)
mtext(c(10,20,30,40,50),at=c(10,20,30,40,50),1,line=1,cex=0.85)
mtext(0,at=0,1,line=1,cex=0.90,family="Lato Bold")
arrows(0,-0.1,0,8.6,lwd=2.5,length=0,xpd=T,col="skyblue3")
}

# Titling

mtext("It is often said that attitudes towards gender roles are \
      changing",3,line=3.5,adj=1,cex=1.8,family="Lato Black",\
      outer=T)
mtext("N=2,075",1,line=3,adj=0.25,cex=1.1,family="Lato",font=4,\
      outer=T)
mtext("All values in percent",1,line=3,adj=1,cex=1.1,font=3,\
      outer=T)
mtext("Source: European Values Study 2008 Germany, ZA4753. www.\
      gesis.org. Design: Stefan Fichtel, ixtract",1,line=5.5,adj\
      =1.0,cex=0.95,outer=T)
dev.off()
```

As in the previous bar chart variants, the script starts with the import of labels and data. Then we define five windows. We use the layout() function for this, since the first window has to be wider than the other four to leave space for the item texts. Otherwise, we could also use the mfcol function. The output window is split so that it is made up of one row and five columns, the first being twice as wide as the others and all being of equal height. The data are compiled as in the previous example. The five bar charts are plotted within the for() loop. Here, we distinguish between the first and the other four bar charts: we require a wider margin for the labels in the first one, while 0.1 suffices for the others. In this case, the first argument of the barplot() function, the data, is column i of the tmyData matrix. The background is defined as in the first example (Sect. 6.1.1). The labels are once more added at the end of the script; the attributes from the response vector are simply left-justified and laid over the existing figure. Using arrows(), a blue line is set at the location of the zero line as visual separator. This finalises the loop; titles and subtitles follow.

6.1.6   Bar Chart for Multiple Response Questions: Symbols for Individuals

It is often said that attitudes towards gender roles are changing

Agree strongly / agree

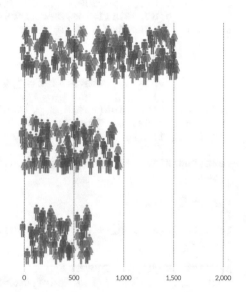

A working mother can establish just as warm and secure an environment as a non-working mother

A pre-school child is likely to suffer if his or her mother is working

A job is alright, but what most women really want is a home and children

0   500   1,000   1,500   2,000

2,075 respondents. Every figure represents 10 people

♀ *Women* ♂ *Men*

Source: EVS 2008 Germany, ZA4753

About the figure: Otto Neurath is probably the most adept in the art of visualising statistics in a way that shows the massiveness of the facts. A characteristic of Otto Neurath's figures is that everything is very orderly. All the persons are always lined up so they are easily countable. Now, while statistics is always an orderly mass in its end result, it does also contain elements that are varied, random, and indistinct. This can be visualised by, e.g., persons that are randomly spread over an area. The number of persons corresponds to the frequency of the measured variable, but one also gets an impression of the statistical character of the presented parameter. The example from Sect. 6.1.1 is reused in the present example, but we limit ourselves to the first three from the group of questions. For explanatory reasons, we also distinguish between men and women. The persons are placed randomly, but so that their number is visible. The legend has to state that every figure represents 10 people and, obviously, also which symbol was chosen for which representation—even if it is apparent from the form of the symbol.

Data: see annex A, ZA4753: European Values Study 2008: Germany (EVS 2008).

```
pdf_file<-"pdf/barcharts_multiple_symbols.pdf"
cairo_pdf(bg="grey98", pdf_file,width=13,height=10.5)

par(omi=c(0.65,0.65,0.85,0.85),mai=c(1.1,5.85,1.55,0),family="\
     Lato Light",las=1)

# Prepare chart

col_f<-rgb(255,97,0,190,maxColorValue=255)
col_m<-rgb(68,90,111,190,maxColorValue=255)
source("scripts/inc_labels_za4753.r")

# Create chart

plot(1:5,type="n",axes=F,xlab="",ylab="",xlim=c(0,20),ylim=c\
     (1,6))
mySymbols<-function(n_f,n_m,y,labelling,... ){
par(family="Symbol Signs")
for (i in 1:n_f)
{
text(runif(1,0,(n_f+ n_m)/10),runif(1,y,y+1),"F",cex=3.25,
  col=col_f)
}
for (i in 1:n_m)
{
text(runif(1,0,(n_f+ n_m)/10),runif(1,y,y+1),"M",cex=3.25,
  col=col_m)
}
par(family="Lato Light")
text(-3,y+0.5,labelling,xpd=T,cex=1.45,adj=1)
}
mySymbols(round(336/10),round(350/10),1,myC_v161)
mySymbols(round(454/10),round(525/10),3,myC_v160)
mySymbols(round(865/10),round(720/10),5,myC_v159)
axis(1,at=c(0,5,10,15,20),labels=c("0","500","1,000","1,500","\
     2,000"),col=par("bg"),col.ticks="grey81",lwd.ticks=0.5,tck\
     =-0.025)

# Other elements

abline(v=c(0,5,10,15,20),lty="dotted")

# Titling

mtext("It is often said that attitudes towards gender roles are \
     changing",3,line=-0.5,adj=0,cex=1.8,family="Lato Black",\
     outer=T)
mtext("Agree strongly / agree",3,line=-3,adj=0,cex=1.8,outer=T,\
     font=3)
mtext("Source: EVS 2008 Germany, ZA4753",1,line=0,adj=1,cex=1.5,\
     outer=T,font=3)
mtext("2,075 respondents. Every figure represents 10 people ",1,\
     line=-2,adj=0,cex=1.5,outer=T,font=3)
```

```
par(family="Lato Light")
mtext(" Women",1,line=1,adj=0.02,cex=1.5,outer=T,font=3)
mtext(" Men",1,line=1,adj=0.12,cex=1.5,outer=T,font=3)
par(family="Symbol Signs")
mtext("F",1,line=1,adj=0,cex=2.5,outer=T,font=3,col=col_f)
mtext("M",1,line=1,adj=0.1,cex=2.5,outer=T,font=3,col=col_m)
dev.off()
```

Within the script, programming differs from the previous examples, because we cannot use the high-level function barplot() or barp(). Instead, we import the labels and then first create an "empty" chart using plot() and the parameter type"n" (that means we only "define" the chart). We do not want to plot axes or axes labels; however, the range of values is specified as 0–20 in x-direction, and as 1–6 in y-direction. This is followed by the definition of a function symbols to which we transfer the parameters n_f, n_m, y, and labelling. Further parameters can be transferred with the "..." expression. These are then passed through to the functions used within the function. Within the symbols function, the font "Symbol Signs" is first selected. Then we use text() in two loops to create text output at positions generated with random numbers. We use the runif() function for number generation. The function is usually called in the form of runif(Number of random numbers, interval start, interval end). Since we always use the function within a loop, the amount of numbers to be generated per run is always 1. The x-position is then a uniformly distributed random number in the interval from 0 to the sum of men and women, divided by 10; the y-position is a uniformly distributed random number in the interval from y to $y + 1$, with y being a value transferred to the function. Font size is greatly increased using cex=2.75; colour is red for women and blue for men, both transparent. After the two loops, the font is switched back to "Lato Light" and the labelling that was also transferred to the function is inserted at position -3, $y+0.5$. Now we can use the symbol function to call up the frequencies divided by ten of men and women as integer values for each variable of the group of questions that is to be displayed. The only reason for the division by 10 is scaling. The transferred parameters are the number of men and women, the y-position, and the labels. The y-positions for the three variables are 1, 3, and 5. Last but not least, the labels and markers have to be attached. We first draw the x-axis with the labels "0", "500", "1,000", "1,500", and "2,000" at the positions 0, 5, 10, 15, and 20, respectively. Dotted lines are added for guidance. Then, titles and legends are attached.

6.1.7   Bar Chart for Multiple Response Questions: All Response Categories, Grouped

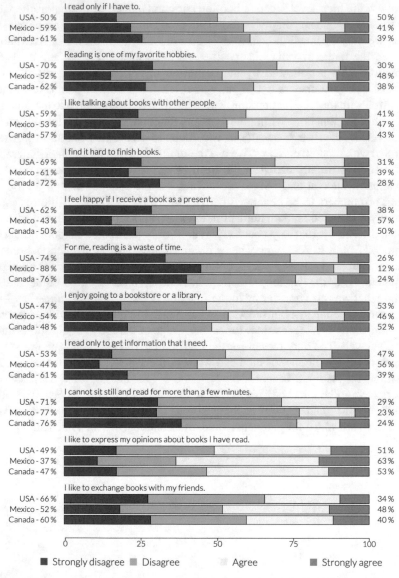

Reading attitude
How much do you disagree or agree with these statements about reading?

Source: PISA 2009 Assessment Framework - Key Competencies in Reading, Mathematics, and Science
© OECD 2009, Data: bryer.org

The figure is a useful extension compared with Sect. 6.1.3, due to the additional consideration of a grouping variable. The present example again deals with presenting a group of questions. Here, it revolves around reading behaviour derived from the PISA study from 2009. There were 11 statements that could once more be answered with "strongly disagree", "disagree", "agree", "strongly agree". Presented are the respective percentages of the response categories, differentiated between responses from US, Mexican, and Canadian students. In addition to Sect. 6.1.3, percentage values are printed to the left and right of bars to show the agreement or disagreement for a certain question and a certain country. In my opinion, the fact that this corresponds to the sums of "strongly disagree" and "disagree" or "agree" and "strongly agree" does not need to be explicitly stated; it is clear from the figure. In contrast to the preceding figures, we have chosen red/light red for the negative and yellow/brown for the positive attributes. The texts of the individual questions of the group are written directly above each bar group. The illustration form is based on a suggestion by Jason Bryer.

The data originate from a survey on the international comparison of school performance study by the Programme for International Student Assessment (PISA) from 2009. This survey, conducted on behest of the Organisation for Economic Co-operation and Development (OECD), recorded the abilities of 15-year-old students in the areas of reading competency, mathematical competency, and scientific competency in the OECD states and in 33 OECD partner countries. Since 2000, the survey has been conducted every 3 years. Data for the USA, Canada and Mexico were provided by Jason Bryer in R file format. The data record comprised 305 variables and 66,690 respondents.

```
pdf_file<-"pdf/barcharts_multiple_all_grouped.pdf"
cairo_pdf(bg="grey98", pdf_file,width=12,height=19)

par(omi=c(1.0,0.5,1.75,0.5),mai=c(0.1,1.45,0.35,0.8),
  family="Lato Light",las=1)

# Import data and prepare chart

load("myData/pisana.rda")
items28=pisana[,substr(names(pisana),1,5) == 'ST24Q']
source("scripts/inc_names_item28.r")

for(i in 1:ncol(items28))
{
items28[,i]=factor(items28[,i],levels=1:4,ordered=T)
}
source("scripts/functions/lickert.r")
library(reshape)
lik=likert(items28,grouping=pisana$CNT)
x<-print(lik); y<-cbind(x[,1],x[,3],x[,4],x[,5],x[,6])
colours<-c("palevioletred4","lightpink","cornsilk1","cornsilk4")
k<-length(y[,1])/length(unique(y[,1]))
par(mfcol=c(k+1,1),las=1)
```

```
for (i in 1:k)
{
z<-y[c(i,i+k,i+2*k),]
prozcan_l<-format(round(z[1,2]+z[1,3],0),nsmall=0)
prozmex_l<-format(round(z[2,2]+z[2,3],0),nsmall=0)
prozusa_l<-format(round(z[3,2]+z[3,3],0),nsmall=0)
prozcan_r<-format(round(z[1,4]+z[1,5],0),nsmall=0)
prozmex_r<-format(round(z[2,4]+z[2,5],0),nsmall=0)
prozusa_r<-format(round(z[3,4]+z[3,5],0),nsmall=0)
b1<-paste("Canada","-",prozcan_l,"%",sep=" ")
b2<-paste("Mexico","-",prozmex_l,"%",sep=" ")
b3<-paste("USA","-",prozusa_l,"%",sep=" ")

# Create chart

barplot(t(z[,2:5]),names.arg=c(b1,b2,b3),cex.names=2,
        col=colours,horiz=T,axes=F)
text(105.5,1.0-0.25,paste(prozcan_r,"%",sep=" "),xpd=T,cex=2)
text(105.5,2.2-0.25,paste(prozmex_r,"%",sep=" "),xpd=T,cex=2)
text(105.5,3.4-0.25,paste(prozusa_r,"%",sep=" "),xpd=T,cex=2)
text(0,4.3,names(items28)[i],cex=2.1,xpd=T,adj=0)
}

# Other elements

par(mai=c(1.1,1.225,0,0.45))
plot(1:2,typ="n",axes=F,xlim=c(0,100),xlab="",ylab="")
axis(1,at=c(0,25,50,75,100),cex.axis=2)
legend(-10,-0.5,pt.cex=4,cex=2.5,pch=15,col=colours,ncol=4,c("\
        Strongly disagree","Disagree","Agree","Strongly agree"),\
        bty="n",xpd=T)

# Titling

mtext("Reading attitude",3,line=5.5,adj=0,cex=3.8,family="Lato \
      Black",outer=T)
mtext("How much do you disagree or agree with these statements \
      about reading?",3,line=2.2,adj=0,cex=2.0,outer=T)
mtext("Source: PISA 2009 Assessment Framework - Key Competencies\
       in Reading, Mathematics, and Science",1,line=1,adj=1.0,\
      cex=1.25,outer=T)
mtext("© OECD 2009, Data: bryer.org",1,line=3.5,adj=1.0,cex\
      =1.25,outer=T)
dev.off()
```

The script first defines a large window (12 by 19 inches) and individual margin settings. The data record is already available in binary R format and is loaded using load(). The following rows are taken from the example script by Jason Bryer for the illustration of his visualisation of a Likert scale: first, the items of question group 28 are extracted by limiting the data record to variables whose name starts with "ST24Q". This is done by addressing the data record with square brackets as a matrix: items28 = pisana[,substr(names(pisana), 1,5) ==, ST24Q']. Within the square brackets, the first character is a comma; this ensures that all rows are carried over into the new selection. Column selection is made via the substr() function containing the stated condition. The variables therby selected are then assigned the statements of the question group 28 as names using names(); then all the variables are converted to "factors" or "categorical variables" in a loop. Now, the likert() function is read, which we downloaded from http://github.com/jbryer/pisa. We still need Hadley Wickham's reshape package for the cast() function that is called by the likert() function. Using print(), the result is saved in the variable x; then y is used to create a matrix that only contains the groups and percent values:

```
>y
       [,1]       [,2]         [,3]         [,4]        [,5]
[1,]    1  25.69810    35.12856    24.883834  14.289507
[2,]    1  26.77758    35.18871    24.636078  13.397637
[3,]    1  25.22917    31.68150    33.470617   9.618706
[4,]    1  31.33106    40.44088    19.698110   8.529946
[5,]    1  23.48092    26.65869    37.827417  12.032974
 .
 .
 .
```

In the next step, every variable is assigned a colour. Here, we use known colours that are included in R: "palevioletred4", "lightpink", "cornsilk1", and "cornsilk4". k is the number of items, here exemplarily calculated from the number of y rows divided by the number of different attributes of the grouping variable. The basic idea of the figure is to draw an individual chart for each item. Additionally, an extra chart window is required to draw the value axes and a legend. Therefore, we use par(mfcol=c(k+1,1), las=1) to split the illustration window into k+1=12 rows of equal size and one column. las=1 causes horizontal labelling. Now follows a loop with k=11 passes. Since the start matrix y is sorted, z is a vector with the three values for each question for Canada, Mexico, and the USA, respectively: in the first run, every 1st, 11th, and 21st value, in the second every 2nd, 12th, and 22nd, etc. b1, b2, and b3 then result in the labels that are made up of the country's name followed by a dash and the sum of the first two item percentages. Numbers are rounded and fitted with a percent symbol. This is followed by the barplot() call. The second to fifth column of z, i.e. the 4% columns, of the transposed matrix are plotted—otherwise the rows would be plotted instead. As per definition, the bars go to 100%, and therefore the sum of the two right percent columns can be written directly after the bar, at position 105.5, and at the height of the centre of the bar using text(). The height of the centre of the bar is: bar width (1), from the second bar onwards additionally the bar distance (0.2), and then back 0.25. The text is positioned outside of the actual chart, so it is important to set the parameter xpd=TRUE, or the text will not appear. Finally, the item is written horizontally at the start of the bar and vertically above the highest bar at position 4.3. Here again, xpd=TRUE has to be set, the font is increased by a factor of 2.1, and text has to be left justified using adj=0 to make sure it really starts at x=0. With the loop, everything is repeated 11 times. In the 12th window, an invisible plot() is drawn—to do this, simply set 1:2 as value, no axes, and type="n"—then an axis with percentages at 0, 25, 50, 75, and 100 is added. Underneath goes a legend for the entire figure. Since this is the first time we have used the legend() function, there should be a short explanation: we are positioning it at coordinates of the invisible plot, −10 and −0.5. These are coordinates outside of the plot area, so once more xpd=TRUE has to be set. The vector c() contains the four attributes of the items, point size is to be increased by a factor of 4 from the preceding value, and the font by a factor of 2.5. We use 15 as a symbol, which is a filled square. Obviously, the colours have to correspond to the items, so the vector colours is used again; ncol=4 means that the legend should be quadrifid. The default is ncol=1, a vertical arrangement. With the box type bty="n", no frame is drawn around the legend; this has to be explicitly stated. In this last window, the margins are set so the plotted x-axis ends flush left- and right-justified with the bars of the other windows. In this case, suitable values were found by trial and error. Finally, as in the previous examples, come the titles—again using mtext() for flexibility's sake, and in the outer margin with the outer=TRUE parameter. The best settings for font size, line position, alignment and, above, for the margins, are once more a matter of trial and error. If we want to first sort according to country and within those according to the item (therefore not 11 times 3, but 3 times 11 bars), then we would have to set k<-unique(y[,1]) and rewrite the loop slightly:

```
for (i in 1:length(k))
{
z<-subset(y,y[,1]==i)
barplot(t(z[,2:5], names.arg=names(items28), horiz=TRUE)
}
```

The preceding examples should cover a large part of the practical applications for bar charts. Now follow examples that each illustrate a single aspect of representation.

6.1.8 Column Chart with Two-Line Labelling

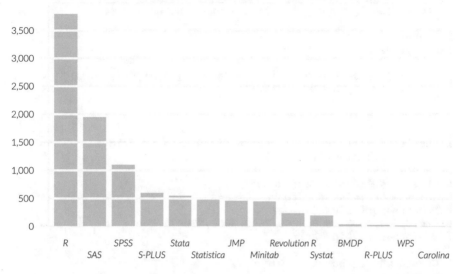

Number of links to homepages of statistical software

Source: r4stats.com/popularity

The figure shows the number of links to homepages of statistical software as a column chart, for illustrative purposes. However, the labels would be too long here and would overlap; alternatively, the bars would have to be spaced far apart, so the figure would not look nice. We have two options: We organise our labels either in two rows or at a 45° angle. Other than that, the chart was purposefully kept simple and follows the "reduced appearance" proposal by Edward Tufte. With bar charts, we should go without grid lines and rather "break" the columns with white lines. Once more, there are no aspect ratios resulting from the presented numbers. Sometimes, the size and ratio has to be chosen dependent on the page layout; that should be the exception though. If the figure is embedded in a text, then an aspect ratio of, e.g., 7:5 is a good choice. With two-line labelling of the x-axis, the individual labels should obviously not be too long. In the present case, this does not pose a problem. Since the largest column is against the y-axis, and the columns are visually subdivided, we can do without tick marks on the axis. Values exceeding 1000 are frequently formatted with a "thousands comma". The common use of a space instead of a comma, especially in official statistical publications, is generally less suited for figures. Scaling of the y-axis does not necessarily have to include the maximal attributes of the values. Data are derived from a figure on http://r4stats. com/popularity.

```
pdf_file<-"pdf/columncharts_labels_tworows.pdf"
cairo_pdf(bg="grey98", pdf_file,width=7,height=5)

library(plotrix)
library(gdata)

par(mai=c(0.95,0.5,0.0,0.5),omi=c(0,0.5,1.0,0),fg=par("bg"),\
      family="Lato Light",las=1)

# Import data and prepare chart

links<-read.xls("myData/listserv_discussion_traffic.xlsx",sheet\
      =2, encoding="latin1")
attach(links)
sort.links<-links[order(-Number) ,]
myNames<-sort.links$Software
myNumber<-sort.links$Number
py<-c(0,500,1000,1500,2000,2500,3000,3500)
fpy<-format(py,big.mark=",")

# Create chart and other elements

barp(myNumber,cex.axis=0.75,names.arg=myNames,border=NA,col="\
      grey",staxx=T,ylim=c(0,4000),height.at=py,height.lab=fpy)
par(col="black")
staxlab(1,1:length(myNames),myNames,nlines=2,top.line=0.55,font\
      =3,cex=0.75)
abline(h=c(500,1000,1500,2000,2500,3000,3500),col=par("bg"),lwd\
      =3)

# Titling

mtext("Number of links to homepages of statistical software",3,\
      line=2,adj=0,cex=1.4,family="Lato Black",outer=T)
mtext("Source: r4stats.com/popularity",1,line=3,adj=1.0,cex\
      =0.65,font=3)
dev.off()
```

In the script, we directly define the aspect ratio of 7:5 as a parameter in inches. The procedure of data import and sorting is the same as described in Sect. 6.1.1. With the minus, sorting is done in descending order. To achieve two-line labelling, we use the plotrix package by Jim Lemon. On the one hand, this package provides the extended barplot function barp(), and on the other hand, the staxlab() function that allows multiple-line axis labels. The barp() arguments largely match those of barplot(). It is important to set staxx=TRUE, which does not output any x-axis labelling, but leaves this to the subsequent staxlab() call. We only have to specify that the parameters refer to axis 1 (x-axis), the positions of the labels (1:length(names)), that the labels are contained within the names vector, that the number of rows equals two, and finally that the distance between the axis and the first row is to be 0.75. The parameter big.mark="," of the format function causes a thousands separator. We used fg=par("bg") at the beginning to hide the line and x-axis labels that the barp() function draws by default. Specification of the background colour using fg also impacts on the staxlab output, so we now have to again set the colour using col="black". Using abline, lines are drawn over the bars at the position of the y-labels (except the zero) in the background colour.

6.1.9   Column Chart with 45° Labelling

The figure is a variant of the reduced illustration from the previous example using 45° labels. Such a form of representation will not convince everybody. In my opinion, it may well be used though—as opposed to vertically aligned labels that completely run against the flow of reading. The advantage over the previous variant is that the labels can be larger. In a confined space for the figure, this type might be an alternative.

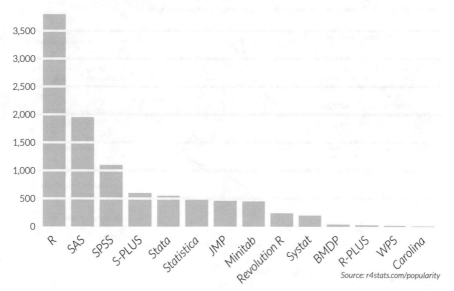

Data are again taken from the respective figure on http://r4stats.com/popularity. The script matches the previous example; the only difference is the staxlab row, which is changed to

```
staxlab(1, 1:length(names), names, srt=45, top.line=1.25)
```

and the file name is adjusted in the first row.

6.1.10   Profile Plot for Multiple Response Questions: Mean Values of the Responses

About the figure: So far, the figures have had in common that they had one label per item. Sometimes though, one might want to also plot parameters that are located on a scale between two statements. In such cases, the use of a profile plot makes sense. The example shows such a profile plot for a series of average estimates that lie between two statements. Here, we are again using a group of questions from the European Values Study 2008. The task was to voice one's opinion on different statements. Two opposing statements were presented to each respondent, and the respondents were asked to rank their answer on a scale between the statements. The connection of the averages plotted as a line chart shows the profile, e.g., for a selected group. Such profiles can be stacked so several groups can be compared. The figure shows profiles for women and men with respect to statements concerning the economy.

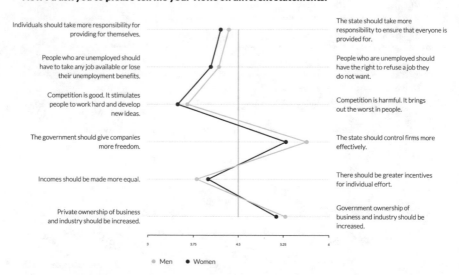

Source: ZA4753: European Values Study 2008: Germany (EVS 2008). N=2,075

Data: see annex A, ZA4753: European Values Study 2008: Germany (EVS 2008).

```
pdf_file<-"pdf/profile_chart.pdf"
cairo_pdf(bg="grey98", pdf_file, width=12, height=8)

source("scripts/functions/profile.plot.r")
par(lheight=1.15,mai=c(0.25,0.25,0.25,0.25), omi=c⟍
    (0.5,0.5,1.1,0.5),
  family="Lato Light",las=1)

# Import labels

text.left<-NULL
text.right<-NULL
text.left<-c(text.left,"Individuals should take more ⟍
    responsibility for\nproviding for themselves.")
text.right<-c(text.right,"The state should take more\⟍
    nresponsibility to ensure that everyone is\nprovided for."⟍
    )
text.left<-c(text.left,"People who are unemployed should\nhave ⟍
    to take any job available or lose\ntheir unemployment ⟍
    benefits.")
text.right<-c(text.right,"People who are unemployed should\nhave⟍
    the right to refuse a job they\ndo not want.")
text.left<-c(text.left,"Competition is good. It stimulates\⟍
    npeople to work hard and develop\nnew ideas.")
text.right<-c(text.right,"Competition is harmful. It brings\nout⟍
    the worst in people.")
text.left<-c(text.left,"The government should give companies\⟍
    nmore freedom.")
text.right<-c(text.right,"The state should control firms more\⟍
    neffectively.")
text.left<-c(text.left,"Incomes should be made more equal.")
text.right<-c(text.right,"There should be greater incentives\⟍
    nfor individual effort.")
text.left<-c(text.left,"Private ownership of business\nand ⟍
    industry should be increased.")
text.right<-c(text.right,"Government ownership of\nbusiness and ⟍
    industry should be\nincreased.")

# Import and prepare data

library(Hmisc)
ZA4753<-spss.get("myData/ZA4753_v1-1-0.sav",use.value.labels=F)
myVariables<-c("v302","v194","v195", "v196","v197","v198","v199"⟍
    )
myResult<-dim(2)
myData<-ZA4753[, myVariables]
for (i in 2:length(myVariables))
{
  mySelection<-subset(myData[, c(1, i)], myData [, i] >= 1 & ⟍
      myData [, i] <= 10)
```

```
  myValues<-t(aggregate(mySelection [, 2], by=list(mySelection ⬐
      [, 1]), FUN=mean, na.rm=T))
  myResult<-rbind(myResult, myValues[2, ])
}

# Create chart

colnames(myResult)=c("Men","Women")
myC1<-"skyblue"
myC2<-"darkred"
profile.plot(myResult,text.left,text.right,colours=c(myC1,myC2),⬐
    legend.n.col=2)

# Titling

mtext("Now I'd ask you to please tell me your views on different⬐
      statements.",3,line=3,adj=0,cex=1.5,family="Lato Black",⬐
      outer=T)
mtext("Source: ZA4753: European Values Study 2008: Germany (EVS ⬐
      2008). N=2,075",1,line=1,adj=1.0,cex=1.1,font=3,outer=T)
dev.off()
```

The script is essentially confined to a few lines, since we use the function profile.plot() created by Patrick R. Schmid. We have only slightly adjusted the margin settings within the function. After importing the source file and the function, a large part of the script is made up of defining two vectors containing the statement texts. Both, text.left and text.right, are first initialised with NULL and then extended statement by statement, by appending the statements as additional elements to the vector. After import of the data, a list of variables is processed in a loop. For every variable, the aggregate() function calculates the mean value for men and women and appends it to result. After specification of a colour for each group, the actual function is then called, data, texts, and colours are handed over as parameters, and a two-column legend is created. At the end come the titles and captions.

6.1.11 Dot Chart for Three Variables

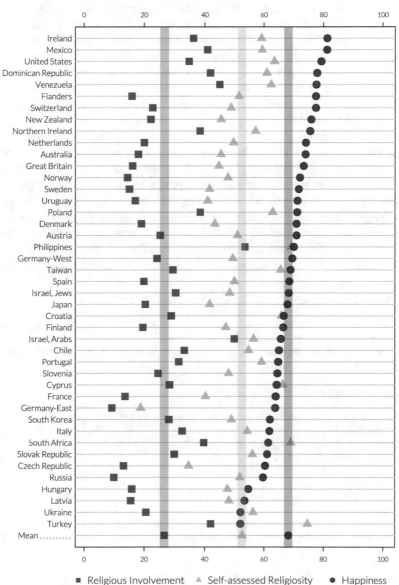

Religious Involvement, Self-assessed Religiosity, and Happiness

(Mean values scaled from 0 to 100, Data from 2008)

■ Religious Involvement ▲ Self-assessed Religiosity ● Happiness

Source: ISSP Data Report Religious Attitudes and Religious Change

About the figure: if there are a lot of attributes in a bar chart and/or you want to plot several variables simultaneously, then a different illustration form is needed. In such cases, the dot chart that was already used as an example in Fig. 2.5 comes in handy. In the variant presented here, three parameters are plotted simultaneously. Data are once more derived from a survey data record, the International Social Survey Programme: Religion III-ISSP 2008. The chart shows three indicators for each country: "Religious Involvement", "Self-assessed Religiosity", and "Happiness". The average value for all countries is shown by a transparent line for each indicator, the individual values as red, blue, and yellow dots. The data record was sorted by the third variable "Happiness". We can see from the figure that there is no linear correlation between religiosity—either personal or ecclesiastic—and personal happiness. The data are derived from the data record "International Social Survey Programme: Social Inequality IV—ISSP 2009". The ISSP is a social-economic survey program that has conducted yearly surveys since 1985. These days, institutions from 48 countries take part in its implementation. Apart from social and demographic questions, different main topics that can be repeated every few years are chosen on a yearly basis. In 2009, the main focus was the topic "social inequities". The data record contains 350 variables and 55,238 units (respondents). The record is available from gesis.org (requires registration) under the title ZA5400 (doi:10.4232/1.11506). It is supplied in different formats; for our example, the record was downloaded in SPSS format. The data record is access class A (data and documents are free for academic research and teaching).

```
pdf_file<-"pdf/dotcharts_overlay.pdf"
cairo_pdf(bg="grey98", pdf_file,width=7,height=10)

library(Hmisc) # because of dotchart2, only there add=T can be ⬎
      used
par(omi=c(0.15,0.75,0.95,0.75),mai=c(0.9,1.75,0.25,0),family="⬎
      Lato Light",las=1)

# Import data and prepare chart

myData<-read.xls("myData/Bechert_Graph.xlsx",sheet=1, encoding="⬎
      latin1")
row.names(myData)<- myData$Countries
myData$Countries<-NULL

total<- myData["Mean . . . . . . . . . . ",]
myData<- myData[rownames(myData)!="Mean . . . . . . . . . . ",]

myDatasort <- myData [order(-myData$Happiness),]
myDatasort <-rbind(myDatasort,total)
attach(myDatasort)

myC1<-rgb(255,165,0,190,maxColorValue=255)
myC2<-rgb(0,0,139,190,maxColorValue=255)
myC3<-rgb(100,0,0,190,maxColorValue=255)

myC1g<-rgb(255,165,0,60,maxColorValue=255)
```

```
myC2g<-rgb(0,0,139,60,maxColorValue=255)
myC3g<-rgb(100,0,0,60,maxColorValue=255)

# Create chart and other elements

dotchart2(Religiosity.,labels=row.names(myDatasort),pch=17,\
      dotsize=4,cex=0.6,cex.labels=0.75,xlab="",col=myC1,xaxis=F\
      ,xlim=c(1,100))
dotchart2(Involvement,labels=row.names(myDatasort),pch=15,\
      dotsize=4,cex=0.6,xlab="",col=myC2,xaxis=F,add=T)
dotchart2(Happiness,labels=row.names(myDatasort),pch=19,dotsize\
      =4,cex=0.6,xlab="",col=myC3,xaxis=F,add=T)

axis(1)
axis(3)

abline(v=total,col=c(myC1g,myC2g,myC3g),lwd=12)
legend(-5,-2.6,c("Religious Involvement"),ncol=1,pch=15,col=myC2\
      ,bty="n",cex=1.5,pt.cex=1.5,xpd=T)
legend(35,-2.6,c("Self-assessed Religiosity"),ncol=1,pch=17,col=\
      myC1,bty="n",cex=1.5,pt.cex=1.5,xpd=T)

legend(80,-2.6,c("Happiness"),ncol=1,pch=19,col=myC3,bty="n",cex\
      =1.5,pt.cex=1.5,xpd=T)


# Titling

mtext("Religious Involvement, Self-assessed Religiosity, and \
      Happiness ",3,line=3.75,adj=0,cex=1.05,family="Lato Black"\
      ,outer=T)
mtext("(Mean values scaled from 0 to 100, Data from 2008)",3,\
      line=1.25,adj=0,cex=0.90,font=3,outer=T)
mtext("Source: ISSP Data Report Religious Attitudes and \
      Religious Change",1,line=-1,adj=1,cex=0.90,font=3,outer=T)

dev.off()
```

The script first defines an illustration window that is twice as high as wide, and specific margin settings. The Hmisc library by Frank E. Harrell Jr is embedded because we require its dotchart2() function. The advantage this has over the original dotchart() function is that it lets us specify dot size, x-axis layout, and above all the stacking of multiple dot charts. We import data from an XLS sheet, define the content of the "Country" column as row name, and the vector total that contains the row of means. Once this is done, this row is removed from the data, the data record is sorted in descending order by the "Religiosity" variable, and the total row is appended. The attach() function saves us from having to spell out the name of the data record before the variables in the future. This is followed by three definitions of transparent colours for the dots and another three even more transparent ones for the mean lines. The dotchart2() function then plots the dot charts. The function is called three times, with add=TRUE for the second and third call so they are stacked on top of the existing one. For better orientation, we draw x-axes at both the top and bottom using axis(1) and axis(3). This is followed by the three vertical lines specifying the means using abline(). "v" means "vertical"; total is the vector with the position of the means. We use the value 12, thereby choosing very wide lines, because their width will then approximately match the dots' diameter. The legend is essentially the same as in Sect. 6.1.7. As before, the script ends with the titles and captions.

6.1.12   Column Chart with Shares

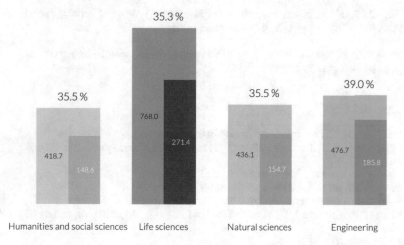

DFG grants in 2010

Individual grants by science sector, values in million Euro. Percent value: approval quota

Source: DFG Information Cards, www.dfg.de

About the figure: With bar charts, there is also the option to show shares. However, there is one thing to be mindful of: in the commonly used illustration form of "stacked" areas, a share cannot be distinguished from an addition without further explanation. In these cases, it makes sense not to use the entire width of the bars or columns for the presentation of the share, but only half of them. However, to maintain proportions, the height of this area has to be doubled. This example shows such an illustration form for the sums of requested and granted applications for research grants submitted to the German Research Foundation [Deutsche Forschungsgemeinschaft (DFG)], split into four scientific sectors. Since the granted amounts are real parts of the resources requested—not one Euro that wasn't requested is granted—the granted sums can be plotted as parts of the sums requested. In the figure, column height reflects the amount requested, and the lower area the fraction of that amount that was granted. For better orientation, the respective sums are written inside the columns, and the grant rates (granted divided by requested) over the columns. The figure's essential messages are that the grant rates are almost identical across all four science sectors and that significantly more is requested and therefore also granted in the life sciences than in the other three fields. The data can be directly extracted from a PDF publication "DFG Information Cards 2011" (card 9, "Research Grants") on the DFG website and was manually entered into the R script.

```
pdf_file<-"pdf/columncharts_shares_1x4.pdf"
cairo_pdf(bg="grey98", pdf_file,width=11,height=7)

par(cex=0.9,omi=c(0.75,0.5,1.25,0.5),mai=c(0.5,1,0.75,1),mgp=c⟍
      (3,2,0),family="Lato Light",las=1)

# Import data

source("scripts/inc_data_dfg.r", encoding="latin1")

# Create charts and other elements

barplot(x,col=c(myC1a,myC1a,myC2a,myC2a,myC3a,myC3a,myC4a,myC4a)⟍
      ,beside=T,border=NA,axes=F,names.arg=c("","","",""))
barplot(2*y,col=c(myC1a,myC1b,myC2a,myC2b,myC3a,myC3b,myC4a,⟍
      myC4b),beside=T,border=NA,axes=F,add=T,names.arg=labelling⟍
      ,cex.names=1.25)
z<-1
for (i in 1:4)
{
text(z+0.25,x[1,i]/2,format(round(x[1,i],1),nsmall=1),adj=0)
text(z+1.25,y[2,i],format(round(y[2,i],1),nsmall=1),adj=0,col="⟍
      white")
text(z+0.65,x[1,i]+50,paste(format(round(100*y[2,i]/x[1,i],1),
  nsmall=1),"%",sep=" "),adj=0,cex=1.5,xpd=T)
z<-z+3
}

# Titling

mtext("DFG grants in 2010",3,line=4,adj=0,family="Lato Black",⟍
      outer=T,cex=2)
mtext("Individual grants by science sector, values in million ⟍
      Euro. Percent value: approval quota",3,line=1,adj=0,cex⟍
      =1.35,font=3,outer=T)
mtext("Source: DFG Information Cards, www.dfg.de",1,line=2,adj⟍
      =1.0,cex=1.1,font=3,outer=T)
dev.off()
```

The script first embeds the data: first four colours, twice, then the actual data, and finally the labels. Note that the application sums are read twice. This makes the bars twice as wide. The grant data on the other hand are only read once; the respective first values are set to zero. This is followed by two barplot() calls. Since we do not set horiz=TRUE this time, not bar, but column charts are plotted. The second call is done with add=TRUE, so that these columns are plotted on top of the first ones. To maintain correct proportions, grant data (y) have to be multiplied by 2. Alternatively, column width could have been adjusted accordingly within the barplot() calls. Finally follow four times three labels: halfway up, the values of the requested and granted sums, respectively (requested sums in the left columns), centrally over the columns the ratio of granted to requested sums in percent. The term format(round(x[1,i],1),nsmall=1),adj=0) ensures output of a decimal place even with integers. Further examples for bar charts will follow in Sect. 7.3.

6.2   Pie Charts and Radial Diagrams

Pie charts are pretty to look at—nothing more. But no less for that, which is why I do not want to discourage their use. The frequently used argument that human perception finds it easier to compare, e.g., points on lines (dot charts) remains unchallenged. However, if the magnitude of individual data is so close together that they are hard to sort by their exact difference in magnitude, then the message of these data is that they are similar and small differences are therefore unimportant. In this section, we will not only explain regular pie charts, but also give examples of spie charts and radial polygons, and in Chap. 11 also radial column charts. The types differ in the variability of their radii and/or angles:

Type	Radius	Angle
Pie chart	Constant	Variable
Spie chart	Variable	Variable
Radial polygon	Variable	Constant
Radial column chart	Variable	Constant

6.2.1 Simple Pie Chart

Global energy mix (including sea and air transport)

Shares of energy sources in the primary energy supply* in percent, 2008

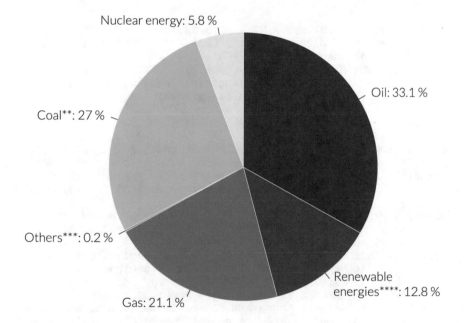

* Primary energy sources = primary energy production + imports - exports +/- stock changes
** Including peat
*** Bio matter, biodegradable waste (excluding industrial waste), water power, geothermal energy, solar, wind, and marine power.
**** Industrial waste and flammable waste that can serve as energy sources and are non-biodegradable

Source: German Federal Agency for Civic Education: keyword 'Enegiemix' [energy mix], www.bpb.de [website in German]

The figure shows data concerning the "Global energy mix". Drawing a pie chart is more difficult than you think. With a pie chart, you can—and should—vary the order of the individual segments, as well as the rotation. Ideally, a segment should start at 0°, but that will not always be possible. Depending on the space you want to assign to the figure, and the attributes of the concrete data and the concrete labels, one should play around a little to find the optimal order of segments and rotation for the given case. In the present case, three or four iterations were necessary, i.e., changes of segment order and rotations, to achieve the presented result. If there are no substantive reasons for the use of multiple colours, one should use a colour palette with one colour for pie charts. The data are derived from a page of the Bundeszentrale für politische Bildung (German Federal Agency for Political Education).

```
pdf_file<-"pdf/piecharts_simple.pdf"
cairo_pdf(bg="grey98", pdf_file,width=11,height=11)

par(omi=c(2,0.5,1,0.25),mai=c(0,1.25,0.5,0.5),family="Lato Light\
     ",las=1)
library(RColorBrewer)

# Create chart

pie.myData<-c(5.8,27.0,0.2,21.1,12.8,33.1)
energytypes<-c("Nuclear energy:","Coal**:","Others***:","Gas:","\
     Renewable\nenergies****:","Oil:")
names(pie.myData)<-paste(energytypes,pie.myData,"%",sep=" ")
pie(pie.myData,col=brewer.pal(length(pie.myData),"Reds"),border\
     =0,cex=1.75,radius=0.9,init.angle=90)

# Titling

mtext("Global energy mix (including sea and air transport)",3,\
     line=2,adj=0,family="Lato Black",outer=T,cex=2.5)
mtext("Shares of energy sources in the primary energy supply* in\
      percent, 2008",3,line=-0.75,adj=0,cex=1.65,font=3,outer=T\
     )
mtext("* Primary energy sources = primary energy production + \
     imports - exports +/- stock changes",1,line=2,adj=0,cex\
     =1.05,outer=T)
mtext("** Including peat",1,line=3.2,adj=0,cex=1.05,outer=T)
mtext("*** Bio matter, biodegradable waste (excluding industrial\
      waste), water power, geothermal energy, solar, wind, and \
     marine power.",1,line=4.4,adj=0,cex=1.05,outer=T)
mtext("**** Industrial waste and flammable waste that can serve \
     as energy sources and are non-biodegradable",1,line=5.6,\
     adj=0,cex=1.05,outer=T)
mtext("Source: German Federal Agency for Civic Education: \
     keyword 'Enegiemix' [energy mix], www.bpb.de [website in \
     German]",1,line=8,adj=1,cex=1.25,font=3,outer=T)
dev.off()
```

In the script, we start with the individual margin settings and then load the RColorBrewer package. Data are defined directly within the script, as are the labels for the segments. In the next line, these are complemented with their percent values and the percent symbol. The pie chart is plotted with pie(); we are using a Brewer colour palette for the segment colours, and the number of colour gradations results from the size of the vector containing the data. Suitable values for radius and initial angle (init.angle) can be found through trial and error. The labels follow at the end, as usual.

6.2.2 Pie Charts, Labels Inside (Panel)

About the figure: illustrations with multiple figures are obviously also an option with pie charts. Here, we once more refer to the data from the example in Sect. 6.1.12. As the granted sums are parts of the requested sums, pie charts are a suitable choice. The size of the circle symbolises the magnitude of the requested sum of the respective science sector. The approval rate, the ratio between granted and requested sums, becomes our heading, and the science sector the caption. The absolute numbers are written within the slices, which is explained in the subheading. The colours match the ones the DFG uses for the respective science sectors. Note that you have to use the square root for the radius, since doubling of the radius will quadruple the area.

DFG grants 2010

Individual grants by science sector, values in million Euro. Percent values: approval ratio

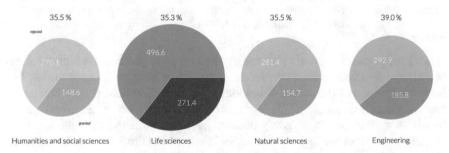

Source: DFG Information Cards 2011, www.dfg.de

The data can be directly extracted from a PDF published on the DFG website "DFG Information Cards 2011" (card 9 "Research Grants"), and were manually entered into the R script.

```
pdf_file<-"pdf/piecharts_1x4.pdf"
cairo_pdf(bg="grey98", pdf_file,width=14,height=6)

library(plotrix)
par(omi=c(0.5,0.5,1,0.5),mai=c(0,0,0,0),xpd=T,mfcol=c(1,4),\
    family="Lato Light",las=1)

# Import data

source("scripts/inc_data_dfg.r")

# Define charts and other elements

for (i in 1:4)
{
plot(1:5,type="n",axes=F,xlab="",ylab="")
values<-c(x[2,i]-y[2,i],y[2,i])
```

```
myCircle<-floating.pie(3,3,values,border="white",radius=2.1*sqrt↘
      (x[1,i]/max(x[1,])),col=c(myColours1[i],myColours2[i]))
pie.labels(3,3,myCircle,values,bg=NA,border=NA,radius=x[1,i]/max↘
      (x[1,]),cex=2,col="white")
if (i==1) pie.labels(3,3,myCircle,c("rejected","granted"),bg=NA,↘
      border=NA,radius=1.95,font=3)
text(3,4.7,cex=2,adj=0.5,paste(format(round(100*y[2,i]/x[1,i],1)↘
      ,nsmall=1),"%",sep=" "))
text(3,1.2,labelling[i],cex=2,adj=0.5)
}

# Titling

mtext("DFG grants 2010",3,line=4,adj=0,family="Lato Black",outer↘
      =T,cex=2)
mtext("Individual grants by science sector, values in million ↘
      Euro. Percent values: approval ratio",3,line=1,adj=0,cex↘
      =1.35,font=2,outer=T)
mtext("Source: DFG Information Cards 2011, www.dfg.de",1,line=2,↘
      adj=1.0,cex=1.1,font=3,outer=T)
dev.off()
```

This script uses the plotrix package by Jim Lemon to position the four pie charts at different locations within one figure using the floating.pie() function. We first import the data and the colour definitions as described in Sect. 6.1.12, and create four windows using mfcol(). In each of these, we first define an empty plot(), in whose middle (i.e. at x- and y-position 3) the pie chart is then plotted using floating.pie(). Since we then want to set the labels within the pie chart, floating.pie() does not only plot the result, but also saves the object into the variable myCircle. The radius of the pie chart is meant to symbolise the magnitude of the requested sums. Therefore, we calculate it using x[1,i]/max(x[1,]) as the ratio between the value of the requested sum of the i-th circle and the largest requested sum. All radii are then multiplied by the factor 2.1 to maximise the size of the circles within the figure. The next line plots the labels that are to go in the charts. We can achieve this by not multiplying the radius by the factor 2.1 here. The ratio, however, is set as before so the labels appear "centred" within the slices. Finally follows one condition: In the first pie chart, the slice contents should contain the additional labels "rejected" and "approved" in italics (font=3), effectively as a legend. Since all the pie charts look very similar, labelling one is enough. Now follow the headings and captions, comprising the approval rates and the science sectors, as in Sect. 6.1.12. Just like before, we use the text() function to do this. The title which was stored in the inc_title_dfg.r file, is plotted last.

6.2.3 Seat Distribution (Panel)

About the figure: this presentation type is often found in publications of election results. It involves semi-circles, i.e. pie charts in which the individual slices do not add up to an entire circle but only a "semi-circle"—or more precisely a "half donut". This chart type is very useful for seat distributions, as it is a stylised parliament: it adequately reflects the impression of the plenary chamber. The figure compares the seat distribution in the German Bundestag during the 16th and 17th election period. It particularly shows the severe loss of seats for the Social Democratic Party of Germany (SPD) from 222 to 146.

Seat distribution in the German Bundestag

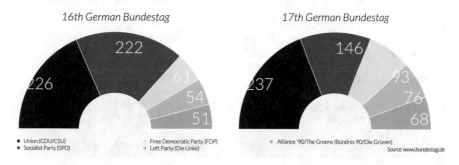

Data can be extracted from the Bundestag homepage (http://www.bundestag.de), and were manually entered into the script.

```
pdf_file<-"pdf/piecharts_allocation_of_seats_1x2.pdf"
cairo_pdf(bg="grey98", pdf_file,width=10,height=3.75)

par(omi=c(0.5,0.5,1,0.5),mai=c(0,0,0,0),xpd=T,mfcol=c(1,2),\
    family="Lato Light")
library(plotrix)

# Define chart

plot(1:5,type="n",axes=F,xlab="",ylab="",xlim=c(1,5),ylim=c\
    (1,10))
mySeats<-c(51,54,61,222,226)
myDes<-c(mySeats,""); mySlices<-50*mySeats /sum(mySeats)
myValues<-c(mySlices,50); myDisc<-100
MyColour<-c("white", "white", "black", "white", "white")

# Create chart

mySemiCircle<-floating.pie(3,1,myValues,border="white",radius\
    =1.9,xpd=F,col=c("green","pink","yellow","red","black",par\
    ("bg")))
pie.labels(3,1,mySemiCircle,myDes,bg=NA,border=NA,radius=1.5,cex\
    =2,col=MyColour)
floating.pie(3,1,myDisc,border="white",col=par("bg"),radius=0.7,\
    xpd=F)
mtext("16th German Bundestag",3,line=0,adj=0.5,font=3,cex=1.3)
```

```
par(xpd=T)
legend(1,0.5,c("Union (CDU/CSU)","Socialist Party (SPD)","Free ⬎
     Democratic Party (FDP)","Left Party (Die Linke)"," ⬎
     Alliance '90/The Greens (Bündnis 90/Die Grünen)"),border=F⬎
     ,pch=15,col=c("black","red","yellow","pink","green"),bty="⬎
     n",cex=0.7,xpd=NA,ncol=3)
par(xpd=F)

# Define chart

plot(1:5,type="n",axes=F,xlab="",ylab="",xlim=c(1,5),ylim=c⬎
     (1,10))
mySeats<-c(68,76,93,146,237)
myDes<-c(mySeats,""); mySlices <-50*mySeats/sum(mySeats)
myValues<-c(mySlices,50); myDisc<-100

# Create chart

semicirlce<-floating.pie(3,1,myValues,border="white",radius=1.9,⬎
     xpd=F,col=c("green","pink","yellow","red","black",par("bg"⬎
     )))
pie.labels(3,1,mySemiCircle,myDes,bg=NA,border=NA,radius=1.5,cex⬎
     =2,col=MyColour)
floating.pie(3,1,myDisc,border="white",col=par("bg"),radius=0.7,⬎
     xpd=F)
mtext("17th German Bundestag",3,line=0,adj=0.5,font=3,cex=1.3)

# Titling

mtext("Seat distribution in the German Bundestag",3,line=3,adj⬎
     =0,family="Lato Black",outer=T,cex=1.8)
mtext("Source: www.bundestag.de",1,line=1,adj=1.0,cex=0.7,font⬎
     =3,outer=T)
dev.off()
```

About the script: to our knowledge, R does not offer a direct option to draw semi-circle pie charts. However, we can manage with a trick. We first embed the plotrix package by Jim Lemon. As seen in the previous example, this package allows us to draw pie charts at a specific position within a plot. If we now use this method to place a pie chart with its centre exactly on the border of such a plot, and simultaneously ensure that the part outside said border is not displayed and that the non-displayed part is just a slice comprising half the actual values, then we are pretty close to our desired result. For the presented figure, we first define a window with one row and two columns using mfcol=c(1,2), i.e. two figures next to each other. Then an "empty" plot spanning 1–5 in x-plan and 1–10 in y-plan is defined using plot(). The number of seats for the individual parties is stored in the mySeats variable; myDes matches the mySeats variable, but we append an empty element at the end. Since we only require a semi-circle and not a full one, we also define a variable mySlices that contains the halved shares (50*mySeats/sum(mySeats)). The values for the pie chart are saved in myValues: these are the segment values from slices that add up to 50 (percent), complemented by the value 50 for the part that will then be invisible. For aesthetic reasons, we then superimpose a "disc", a pie chart consisting of a single slice. The call for the pie chart is done via the floating.pie() function, which draws the pie chart at position 3, 1 into the "empty" plot, so that the centre point is exactly on the y-axis. The radius was chosen as 1.9, so that the width is adequately filled. With xpd=FALSE, the lower half of the pie chart is intentionally cut off. The segment colours are supposed to match the parties' colours. Since we want to label the slices, we store the object created with floating.pie() into the mySemiCircle variable. Labelling follows in the next line using pie.labels(), but we choose a slightly smaller radius here, so the labels fit completely within the slices. Thirdly, the white pie chart disc that only consists of one value is drawn on top with radius 0.7; it has the effect that the first pie chart becomes a donut that resembles the appearance of a parliament. After the title comes the legend, preceded by xpd=TRUE, so it can be drawn into the bottom corner; then we reset the value to FALSE. Then comes a second call for the three already known pie chart elements for the second chart window, this time with the seat distribution during the 17th German Bundestag. As usual, the joint heading and captions in the outer margin follow last.

6.2.4   Spie Chart

The Cost of Getting Sick

The Medical Expenditure Panel Survey. Age: 60, Total Costs: 41.4 Mio. US $

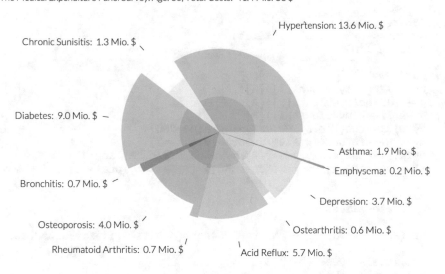

Inside: Personal Costs. Outside: Insurer Costs.   *visualization.geblogs.com/visualization/health_costs/*

About the figure: Contrary to a common pie chart, a spie chart is an extension in which the statistical information is reflected not only by the magnitude of the angles but also by those of the radius. The present example, adapted from a conception by Ben Fry, shows insurance costs caused by sickness across the USA, according to the Medical Expenditure Panel Survey (MEPS) for the age group of 60-year-olds. Pictured are 11 different illnesses, with the radius reflecting the cost per insured patient and the angle the number of people affected by this illness. Therefore, slice area reflects the total cost for the illness. Personal costs are additionally shown as a second level, so that the area of the total cost that is not overlaid reflects the insurer's cost. Slice labelling should be in the form of a circle and therefore based on the longest slice. Data were manually entered into an XLSX spreadsheet.

```
pdf_file<-"pdf/piecharts_spiechart.pdf"
cairo_pdf(bg="grey98", pdf_file,width=15,height=11)

par(omi=c(0.5,0.5,0.75,0.5),mai=c(0.1,0.1,0.1,0.1),family="Lato ⬐
    Light",las=1)
library(RColorBrewer)

# Import data and prepare chart

x<-read.xls("myData/Healthcare_costs.xlsx",1,encoding="latin1")
attach(x)
n<-nrow(x)
myFactor<-max(sqrt(Acosts60))/0.8

# Define chart and other elements

plot.new()
myC0<-rep(NA,n)
myColours<-brewer.pal(n,"Set3")
for (i in 1:n)
{
  par(new=T)
  r<-col2rgb(myColours[i])[1]
  g<-col2rgb(myColours[i])[2]
  b<-col2rgb(myColours[i])[3]
  myC0[i]<-rgb(r,g,b,190,maxColorValue=255)
  myValue<-format(Total60/1000000,digits=1)
  myTotal<-paste(Disease,": ",myValue," Mio. $",sep="")
  if (Acosts60[i] == max(Acosts60)) {myDes<-myTotal} else {myDes⬐
      <-NA}

  # Create slices

  pie(Patients60,border=NA,radius=sqrt(Acosts60[i])/myFactor,col⬐
      =myC0,
      labels=myDes,cex=1.8)
  par(new=T)
  r<-col2rgb(myColours[i])[1]
  g<-col2rgb(myColours[i])[2]
  b<-col2rgb(myColours[i])[3]
```

```
  myC0[i]<-rgb(r,g,b,maxColorValue=255)
  pie(Patients60,border=NA,radius=sqrt(Pcosts60[i])/myFactor,col⟍
      =myC0,labels=NA)
  myC0<-rep(NA,n)
}

# Titling

mtext("The Cost of Getting Sick",3,line=-1,adj=0,cex=3.5,family=⟍
      "Lato Black",outer=T)
mtext("The Medical Expenditure Panel Survey. Age: 60, Total ⟍
      Costs:  41.4 Mio. US $",3,line=-3.6,adj=0,cex=1.75,outer=T⟍
      )
mtext("Inside: Personal Costs.  Outside: Insurer Costs.",1,line⟍
      =0,adj=0,cex=1.75,outer=T,font=3)
mtext("visualization.geblogs.com/visualization/health_costs/",1,⟍
      line=0,adj=1.0,cex=1.75,outer=T,font=3)
dev.off()
```

About the script: R offers functions for the creation of spie charts in different packages, such as spie() in the caroline package. However, we are only using basic functions included in the standard package, which do not limit our layout options.

After importing the data, we first construct a factor from the maximum of the square root of Acosts60. To adjust to the dimensions of the illustration, the value is divided by 0.8. After definition of the illustration using plot.new(), we define the vector myC0, which will later be filled with colour parameters. For this we use a Brewer colour palette as a basis. Then a loop is used to draw a one-slice pie chart for each data record, with individual radii depending on Acosts60. The trick lies in the fact that the colour myC0 is always set to NA at the end of the loop. In each run of the loop, only one slice is defined by colour: MyC0[i]. This means that two times i circles are drawn, of which only one slice each is visible. Every label is taken from the circle with the maximal radius, so that all labels are at the same distance from the centre. Since we are drawing two levels that should be colour-matched, we plot the lower levels as a transparent value of the chosen palette. To do this, we dissolve the selected colour from the Brewer palette into its RGB components. The lower circle slice is then plotted with intensity of 190, and the upper one with intensity of 255.

6.2.5 Radial Polygons (Panel)

World energy mix
Shares of different energy types in total energy use

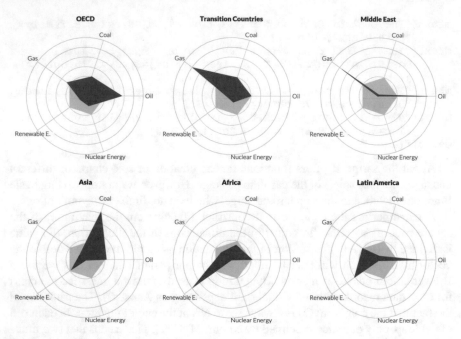

Source: German Federal Agency for Civic Education: keyword 'Enegiemix' [energy mix], www.bpb.de [website in German]

About the figure: If there are multiple variables that have been measured in the same dimension, such as school grades or a person's characteristics on a scale from 1 to 10 or, such as here, shares of different energy types in the total energy mix in individual regions (see Sect. 6.2.1), then radial polygons are a good choice for presentation. Sometimes, the names spider plot, radar chart or net chart are also used. The values of the variables are plotted on radii at identical angles from a circle centre and then connected; the lengths of the radii correspond to the attributes.

Data: see annex A, worldenergymix.xlsx.

```
pdf_file<-"pdf/radial_polygons_2x3.pdf"
cairo_pdf(bg="grey98", pdf_file,width=12,height=12)

par(mfcol=c(2,3),omi=c(1,0.5,1,0.5),mai=c(0,0,0,0),cex.axis=0.9,↘
    cex.lab=1,xpd=T,col.axis="green",col.main="red",family="↘
    Lato Light",las=1)

library(plotrix)
```

```
library(gdata)

# Import data and prepare chart

myRegions<-read.xls("myData/worldenergymix.xlsx", encoding="↘
    latin1")
row.names(myRegions)<-myRegions$Region
myRegions$Region<-NULL
myLabelling<-c("Oil","Coal","Gas","Renewable E.","Nuclear Energy↘
    ")

myRegions<-myRegions[, c(1,2,3,4,5)]
myLabelling<-myLabelling[c(1,2,3,4,5)]

# Create charts

for (i in 2:nrow(myRegions))
{
radial.plot(rep(100/length(myRegions),length(myRegions)),labels=↘
    myLabelling,rp.type="p",main="",line.col="grey",show.grid=↘
    T,show.grid.labels=F,radial.lim=c(0,55),poly.col="grey")
radial.plot(myRegions[i,],labels="",rp.type="p",main="",line.col↘
    ="red",show.grid=F,radial.lim=c(0,55),poly.col="red",add=T↘
    )
mtext(row.names(myRegions)[i],line=2,family="Lato Black")
}

# Titling

mtext("World energy mix",line=2,cex=3,family="Lato Black",outer=↘
    T,adj=0)
mtext(line=-1,"Shares of different energy types in total energy ↘
    use",cex=1.5,font=3,outer=T,adj=0)
mtext(side=1, "Source: German Federal Agency for Civic Education↘
    : keyword 'Enegiemix' [energy mix], www.bpb.de [website in↘
    German]",line=2,cex=1.3,font=3,outer=T,adj=1)
dev.off()
```

In the script, radial polygons are plotted using the radial.plot() function from the plotrix package by Jim Lemon. To do this, we first create row names with the regions from the data imported from an XLSX spreadsheet. We carry out the labelling directly in the script. The first line of data (the "world") is ignored, and we then use a loop to draw two radial polygon charts on top of each other for each region: the first, in grey colour, draws a polygon of equal edge lengths as a background. To do this, we transfer the parameter 100/length(regions) to the radial.plot() function; the number is the number of featured regions. The labels, too, are output here. The rp.type parameter specifies whether lines, polygons (areas) or symbols (points) are to be plotted. With show.grid=T, the drawing of guide circles is enabled. The radius is set with c(0,55). In the second call of radial.plot(), we use add=T to plot the actual data over this "reference polygon". The parameters correspond to the ones from the first call. At the end of the loop, the heading is set above the polygons.

6.2.6   Radial Polygons (Panel): Different Column Arrangement

The degree of dependence between the form of the illustration and the position of the columns can be seen if the second and third columns are swapped.

```
myRegions<-myRegions[, c(1,3,2,4,5)]
myLabelling<-myLabelling[c(1,3,2,4,5)]
```

The result, looking at e.g. the polygon for the Middle East, leaves a completely different impression, since the area of the polygon is now much bigger.

World energy mix
Shares of different energy types in total energy use

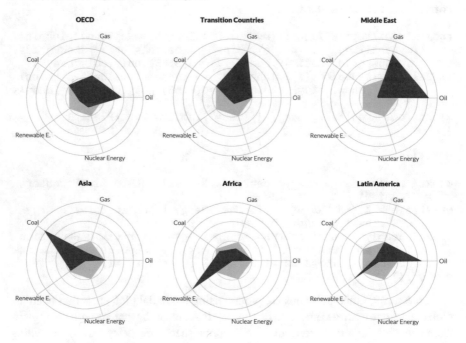

Source: German Federal Agency for Civic Education: keyword 'Enegiemix' [energy mix], www.bpb.de [website in German]

Therefore, this form of representation should be treated with caution. In my opinion, however, it is useful for the comparison of multiple persons, countries etc. with identically arranged categories.

6.2.7   *Radial Polygons Overlay*

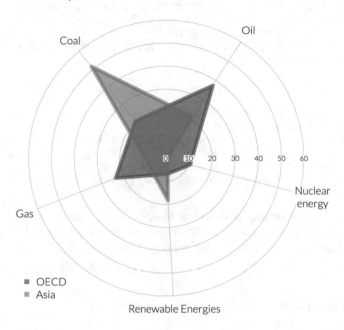

Energy mix: OECD and Asia by comparison
All values in percent

Coal
Oil
0 10 20 30 40 50 60
Nuclear energy
Gas
■ OECD
■ Asia
Renewable Energies

Source: German Federal Agency for Civic Education: keyword 'Enegiemix' [energy mix], www.bpb.de [website in German]

About the figure: aside from a panel type, an illustration form in which graphics are stacked once more comes in handy. In this case, a coordinate system can also be added to the chart. If the areas occupied by the polygons are filled, then transparent colours should be chosen.

Data: see annex A, wordenergymix.xlsx.

```
pdf_file<-"pdf/radial_polygons_overlay.pdf"
cairo_pdf(bg="grey98", pdf_file,width=10,height=10)

par(omi=c(1,0.25,1,1),mai=c(0,2,0,0.5),cex.axis=1.5,cex.lab=1,↘
    xpd=T,family="Lato Light",las=1)
library(plotrix)

# Import data and prepare chart

myRegions<-read.xls("myData/worldenergymix.xlsx", encoding="↘
    latin1")
myC1<-rgb(80,80,80,155,maxColorValue=255)
myC2<-rgb(255,97,0,155,maxColorValue=255)
```

```
myRegions$Region<-NULL
myLabelling<-c("Oil","Coal","Gas","Renewable Energies","Nuclear\
    nenergy")
```

```
# Create chart
```

```
radial.plot(myRegions[2:3,],start=1,grid.left=T,labels=\
    myLabelling,rp.type="p",main="",line.col=c(myC1,myC2),poly\
    .col=c(myC1,myC2),show.grid=T,radial.lim=c(0,55),lwd=8)
legend("bottomleft",c("OECD","Asia"),pch=15,col=c(myC1,myC2),bty\
    ="n",cex=1.5)
```

```
# Titling
```

```
mtext(line=3,"Energy mix: OECD and Asia by comparison",cex=2.5,\
    adj=0,family="Lato Black")
mtext(line=1,"All values in percent",cex=1.5,adj=0,font=3)
mtext(side=1,line=2,"Source: German Federal Agency for Civic \
    Education: keyword 'Enegiemix' [energy mix], www.bpb.de [\
    website in German]",cex=1.05,adj=1,font=3,outer=T)
dev.off()
```

The structure of the script does not essentially differ from the previous one. Instead of one, two columns of the data record are transferred as data. This is also the reason for the specification of two colours. The colours should be transparent.

6.3   Chart Tables

Here, "chart tables" refers to illustration types in which the arrangement of information has table character. Strictly speaking, this pragmatic definition also applies to bar charts, but there are a series of illustration forms that differ significantly from the form of bar charts and therefore justify their own category. We start with suggestions for two variants of so-called Gantt-charts, then follow with examples for a bump chart (Sect. 6.3.3), a heat map (Sect. 6.3.4), a mosaic plot (Sect. 6.3.5) and two examples for tree maps (Sect. 6.3.7 and 6.3.8).

Gantt charts are named after their inventor Henry L. Gantt, who developed this illustration form for visualisation of the individual operational steps within projects. The individual project steps are reflected by row as the span from the planned start to the planned end time of the project section. Additional optional elements include dependencies in the form of connecting lines between different spans as well as horizontal brackets for task groups. A common Gantt chart looks like Fig. 6.1 (built with LaTeX package gantt.sty by Martin Kumm).

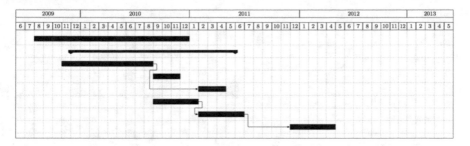

Fig. 6.1 Simplified Gantt chart

For the creation of Gantt charts, R provides e.g. the gantt.chart function within the plotrix package. Here we present a possibility of creating Gantt charts with the functions lines() and points().

6.3.1   Simplified Gantt Chart

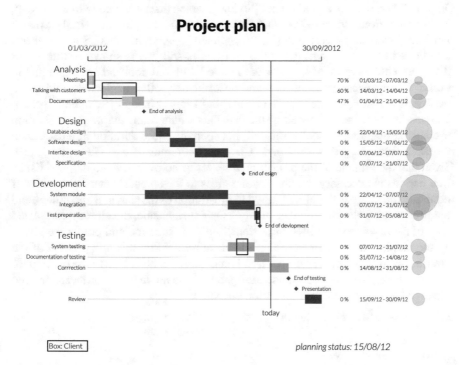

	A	B	C	D	E	F	G	H	I	J	K	L	M	N
1	Milestone	when	Group	what	from	to	Duran	who	done	PAN	PAG	AG_from	AG_to	Persons
2			Analysis											
3				Meetings	01.03.12	07.03.12	6	Schmitz	70	80	20	01.03.12	07.03.12	1
4				Talking with customers	14.03.12	14.04.12	31	Schmitz	60	80	20	14.03.12	14.04.12	1
5				Documentation	01.04.12	21.04.12	20	Schmitz	47	100	0			1
6	End of analysis	21.04.12												1
7			Design											
8				Database design	22.04.12	15.05.12	23	Dent	45	100	0			3
9				Software design	15.05.12	07.06.12	23	Meyer	0	100	0			1
10				Interface design	07.06.12	07.07.12	30	Meyer	0	100	0			2
11				Specification	07.07.12	21.07.12	14	Müller	0	100	0			1
12	End of esign	21.07.12												
13			Development											
14				System module	22.04.12	07.07.12	76	Meyer	0	100	0			2
15				Integration	07.07.12	31.07.12	24	Meyer	0	100	0			2
16				Test preparation	31.07.12	05.08.12	5	Dent	0	80	20	02.08.12	05.08.12	2
17	End of devlopment	05.08.12												
18			Testing											
19				System testing	07.07.12	31.07.12	24	Dent	0	50	50	15.07.12	25.07.12	1
20				Documentation of testing	31.07.12	14.08.12	14	Schmitz	0	100	0			1
21				Corrrection	14.08.12	31.08.12	17	Meyer	0	100	0			1
22	End of testing	31.08.12												1
23	Presentation	07.09.12												
24				Review	15.09.12	30.09.12	15	Schmitz	0	100	0			1
25														

Fig. 6.2 Data for project planning

The figure is based on the "classic" Gantt chart, but with a few variations. The basis is an XLSX spreadsheet in which data were collected (Fig. 6.2).

The figure groups the different tasks of the fictional project into "Analysis", "Design", "Development" and "Test" blocks. Each group is assigned its own colour. On top of that, there is a final point "Review", which does not belong to any group. Within each group, three to four tasks are defined and their duration is shown as a bar. The already completed part (the "completed" column in the XLS spreadsheet) is indicated by grey colour, while the part that is still due is shown in the colour of the respective group. Milestones are marked with red dots and are explained in a text. With projects, an important and unfortunately often neglected topic is the involvement of clients. Their involvement should appear in a project planning chart. One possibility is "superimposing" the task bars with a frame marking the duration of client participation. Usually, this will not span the entire duration of a task. Data are taken from the "CL_from" and "CL_until" columns of the XLS spreadsheet. Given that the graphic of the project plan serves as an overview, no detailed list of time periods is necessary; for these, references to the spreadsheet can be added if required. The figure is therefore limited to illustration of the start and end points, and to a line marking the current date "today". On the right side, the already completed parts of the tasks are identified in percent, and their start and end points are given. On the far side, the weight of the task is visualised. To do this, the product of the duration and number of persons per task is shown as a bubble. Due to the overlap for extensive tasks, these are best made transparent. Data are fictional and were entered into an XLS spreadsheet for illustration purposes.

```
pdf_file<-"pdf/tablecharts_gantt_simplified.pdf"
c0<-"black"; c1<-"green"; c2<-"red"; c3<-"blue"; c4<-"orange"; ⤵
     c5<-"brown"
myColour_done<-"grey"
myColour<-c(c0,c1,c1,c1,c0,c0,c2,c2,c2,c2,c0,c0,c3,c3,c3,c0,c0,⤵
     c4,c4,c4,c0,c0,c5)
source("scripts/inc_gantt_simplified.r")
```

```
dev.off()
```

and

```
# inc_gantt_simplified.r

library(gdata)

cairo_pdf(bg="grey98", pdf_file,width=11.7,height=8.26)

par(lend=1,omi=c(0.25,1,1,0.25),mai=c(1,1.85,0.25,2.75),family="\
    Lato Light",las=1)
mySchedule<-read.xls("mydata/projectplanning.xlsx", encoding="\
    latin1")
n<-nrow(mySchedule)
myScheduleData<-subset(mySchedule,nchar(as.character(mySchedule$\
    from))>0)
myBegin<-min(as.Date(as.matrix(myScheduleData[,c('from','to')]))\
    )
myEnd<-max(as.Date(as.matrix(myScheduleData[,c('from','to')])))
attach(mySchedule)

plot(from,1:n,type="n",axes=F,xlim=c(myBegin,myEnd),ylim=c(n,1))
for (i in 1:n)
{
if (nchar(as.character(Group[i]))>0)
{
text(myBegin-2,i,Group[i],adj=1,xpd=T,cex=1.25)
}
else if (nchar(as.character(what[i]))>0)
{
x1<-as.Date(mySchedule[i,'from'])
x2<-as.Date(mySchedule[i,'to'])
x3<-x1+((x2-x1)*mySchedule[i,'done']/100)
x<-c(x1,x2)
x_done<-c(x1,x3)
y<-c(i,i)
segments(myBegin, i, myEnd, i, col="grey")
lines(x,y,lwd=20,col=myColour[i])
points(myEnd+90,i,cex=(mySchedule[i,'Persons']*mySchedule[i,'\
    Durance'])**0.5,pch=19,col=rgb(110,110,110,50,\
    maxColorValue=255),xpd=T)
if (x3-x1>1) lines(x_done,y,lwd=20,col=myColour_done)
if (mySchedule[i,'PAG'] > 0)
{
x4<-as.Date(mySchedule[i,'AG_from'])
x5<-as.Date(mySchedule[i,'AG_to'])
x_ag<-c(x4,x5)
rect(x4,i-0.75,x5,i+0.75,lwd=2)
}
text(myBegin-2,i,what[i],adj=1,xpd=T,cex=0.75)
text(myEnd+25,i,paste(done[i],"%",sep=" "),adj=1,xpd=T,cex=0.75)
text(myEnd+35,i,paste(format(x1,format="%d/%m/%y"),"-",format(x2\
    ,format="%d/%m/%y"),sep=" "),adj=0,xpd=T,cex=0.75)
}
```

```
else # Milestone
{
x3<-as.Date(mySchedule[i,'when'])
myHalf<-(myEnd-myBegin)/2
if (x3-x1<myHalf)
{
points(as.Date(mySchedule[i,'when']),i,pch=18,cex=1.25,col="red"↘
    )
text(as.Date(mySchedule[i,'when'])+5,i,Milestone[i],adj=0,xpd=T,↘
    cex=0.75)
} else
{
points(as.Date(mySchedule[i,'when']),i,pch=18,cex=1.25,col="red"↘
    )
text(as.Date(mySchedule[i,'when'])-5,i,Milestone[i],adj=1,xpd=T,↘
    cex=0.75)
}
}
}
axis(3,at=c(myBegin,myEnd),labels=c(format(myBegin,format="%d/%m↘
    /%Y"),format(myEnd,format="%d/%m/%Y")))
myToday<-as.Date("15.08.2012", "%d.%m.%Y")
abline(v=myToday)

mtext("today",1,line=0,at=myToday)

# Titling

mtext("Project plan",3,line=2,adj=0,cex=2.25,family="Lato Black"↘
    ,outer=T)
mtext(paste("planning status: ",format(myToday,format="%d/%m/%y"↘
    ),sep=""),1,line=4,at=myEnd+20,cex=1.25,font=3)
rect(myBegin-36, n+5, myBegin, n+4, xpd=T,lwd=2)
text(myBegin-35, n+4.5, "Box: Client",xpd=T, adj=0)
```

The script first defines the colours for the individual groups (c0 to c5). Subsequently, the colouring of the individual bars are defined as elements of the colour vector. For ease of use, this was done manually. However, data could also be imported from a data table. First, rows containing date information are extracted from the XLSX spreadsheet, and the first and last occurring dates are identified. In the next step, the chart is scaled (not drawn) using plot() and the group name is written to the left of the start using text(). If the column "what" contains an entry, then the time periods and the completed part are determined, and time periods are plotted. On the right border, a point (a "bubble") is plotted to indicate the size of the project part, i.e. the duration multiplied by the number of people involved. Since we are using the radius as a variable, we have to use the square root to ensure that area enlargement is proportionally correct. In the next step, the completed parts are plotted over the time periods. If the project client (PCL) has to be involved, a rectangle is drawn around the bars, with length and circumference taken from the data. Then follow the labelling of the rows at the beginning (with the "what") and at the end with the percentage of the completed task part and the time period of the task. If there is no entry in the from and to columns, but in the when column, then it is a "milestone". In those cases, they are plotted as points and labelled. Finally, an axis is plotted at the upper margin.

6.3.2   Simplified Gantt Chart: Colours by People

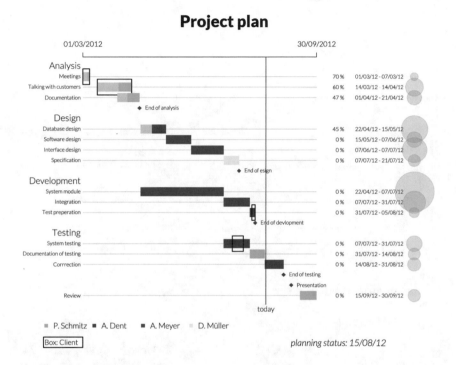

The figure essentially matches the previous one. Only the colouring is different here: it is based on the people involved in the project. Data are fictional and were entered into an XLS spreadsheet for illustration purposes.

```
pdf_file<-"pdf/tablecharts_gantt_simplified_who.pdf"
f0<-"black"; f1<-"green"; f2<-"red"; f3<-"blue"; f4<-"yellow"
farbe_erl<-"grey"
farbe<-c(f0,f1,f1,f1,f0,f0,f2,f3,f3,f4,f0,f0,f3,f3,f2,f0,f0,f2,↘
    f1,f3,f0,f0,f1)
source("scripts/inc_gantt_simplified.r")
legend(anfang-40,n+2,c("P. Schmitz","A. Dent","A. Meyer","D. Mü↘
    ller"),pch=15,col=c(f1,f2,f3,f4),bty="n",cex=1.1,horiz=T,↘
    xpd=T)
dev.off()
```

In the script, we only have to slightly redefine the colours for this case, and add a legend with the names of the people involved.

6.3.3   Bump Chart

Revenue development of Fortune 500 enterprises

in billion Euro

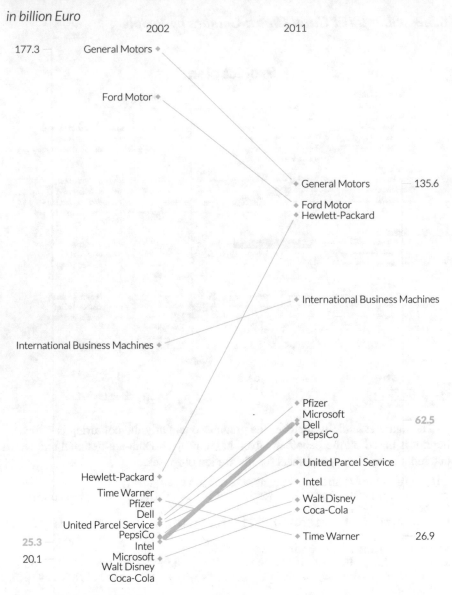

Source: money.cnn.com/magazines/fortune/fortune500/

The figure uses the bumpchart() function from the plotrix package by Jim Lemon. It would also be possible to directly use the matplot() function and add labels right-justified on the left side and left-justified on the right side using text(); the y-coordinates for the function would then simply be the revenue numbers. Depending on the value combination, however, labels could then overlap. The advantage of Jim Lemon's solution is that, in such cases, labels are automatically moved vertically by the appropriate amount.

About the figure: a bump chart usually compares two or more points in time for multiple numerical parameters; characteristic labels for these parameters are written at the ends of the connecting lines. There are two variants: one uses only the ranks and plots these on an ordinal scale; the other plots the actual values on an interval scale. In this example, revenue development of the Fortune 500 enterprises in the USA between 2002 and 2011 is compared. Here, the use of the actual values is a lot more informative than their ranks. It should be noted though, that the labels of individual points might overlap, depending on the data. In such cases, the labels should be attached with minimal, but identical line spacing, but in such a way that at least rank is maintained. On the left side, this is the case for all enterprises whose revenue is less than Hewlett-Packard's. As an example, Microsoft's revenue has been highlighted here. The data were first taken from a CNN website by year and copied into an XLS spreadsheet; a new sheet was used for each year. The data were then reorganised using the sqldf package and then saved as a binary R file fortune_revenue.RData.

```
library(sqldf)
f2011<-read.xls("data/fortune100.xlsx", sheet = 1)
f2010<-read.xls("data/fortune100.xlsx", sheet = 2)
f2009<-read.xls("data/fortune100.xlsx", sheet = 3)
f2008<-read.xls("data/fortune100.xlsx", sheet = 4)
f2007<-read.xls("data/fortune100.xlsx", sheet = 5)
f2006<-read.xls("data/fortune100.xlsx", sheet = 6)
f2005<-read.xls("data/fortune100.xlsx", sheet = 7)
f2004<-read.xls("data/fortune100.xlsx", sheet = 8)
f2003<-read.xls("data/fortune100.xlsx", sheet = 9)
f2002<-read.xls("data/fortune100.xlsx", sheet = 10)


total<-sqldf("select enterprise from f2011
union select enterprise from f2010
union select enterprise from f2009
union select enterprise from f2008
union select enterprise from f2007
union select enterprise from f2006
union select enterprise from f2005
union select enterprise from f2004
union select enterprise from f2003
union select enterprise from f2002")

x<-sqldf("select
total.enterprise,
f2002.revenue r2002
f2002.revenue r2003
```

```
f2002.revenue r2004
f2002.revenue r2005
f2002.revenue r2006
f2002.revenue r2007
f2002.revenue r2008
f2002.revenue r2009
f2002.revenue r2010
f2002.revenue r2011
from total
left join f2002 on total.enterprise=f2002.enterprise
left join f2003 on total.enterprise=f2003.enterprise
left join f2004 on total.enterprise=f2004.enterprise
left join f2005 on total.enterprise=f2005.enterprise
left join f2006 on total.enterprise=f2006.enterprise
left join f2007 on total.enterprise=f2007.enterprise
left join f2008 on total.enterprise=f2008.enterprise
left join f2009 on total.enterprise=f2009.enterprise
left join f2010 on total.enterprise=f2010.enterprise
left join f2011 on total.enterprise=f2011.enterprise
")

row.names(x)<-x$enterprise
x$enterprise<-NULL
y<-t(x)
save(y, file="fortune_revenue.RData.")
```

The saved data can now be used in the script:

```
pdf_file<-"pdf/tablecharts_bumpchart.pdf"
cairo_pdf(bg="grey98", pdf_file,width=9,height=12)

par(omi=c(0.5,0.5,0.9,0.5),mai=c(0,0.75,0.25,0.75),xpd=T,family=\
      "Lato Light",las=1)
library(plotrix)
library(gdata)


# Import data and prepare chart

z1<-read.xls("myData/bumpdata.xlsx", encoding="latin1")
rownames(z1)<-z1$name
z1$name<-NULL
myColours<-rep("grey",nrow(z1)); myLineWidth<-rep(1,nrow(z1))
myColours[5]<-"skyblue"; myLineWidth[5]<-8
par(cex=1.1)


# Create chart

bumpchart(z1,rank=F,pch=18,top.labels=c("2002","2011"),col=\
      myColours,lwd=myLineWidth,mar=c(2,12,1,12),cex=1.1)

# Titling
```

```
mtext("Revenue development of Fortune 500 enterprises",3,line↘
    =1.5,adj=0,family="Lato Black",outer=T,cex=2.1)
mtext("Source: money.cnn.com/magazines/fortune/fortune500/",1,↘
    line=0,adj=1,cex=0.95,font=3,outer=T)

# Other elements

axis(2,col=par("bg"),col.ticks="grey81",lwd.ticks=0.5,tck=-↘
    0.025, at=c(min(z1$r2002), max(z1$r2002)),c(round(min(z1$↘
    r2002)/1000,digits=1), round(max(z1$r2002)/1000, digits=1)↘
    ))
axis(4,col=par("bg"),col.ticks="grey81",lwd.ticks=0.5,tck=-↘
    0.025, at=c(min(z1$r2011), max(z1$r2011)),c(round(min(z1$↘
    r2011)/1000,digits=1), round(max(z1$r2011)/1000, digits=1)↘
    ))

mtext("in billion Euro",3,font=3,adj=0,cex=1.5,line=-0.5,outer=T↘
    )

par(family="Lato Black")
axis(2,col=par("bg"),col.ticks="grey81",col.axis="skyblue",lwd.↘
    ticks=0.5,tck=-0.025,at=z1[5,1],round(z1[5,1]/1000, digits↘
    =1))
axis(4,col=par("bg"),col.ticks="grey81",col.axis="skyblue",lwd.↘
    ticks=0.5,tck=-0.025,at=z1[5,2],round(z1[5,2]/1000, digits↘
    =1))

dev.off()
```

In the script, we need the plotrix package for the bump chart. Data are read from a XLSX spreadsheet, and row names are created from the name column. Then, the name column is deleted so that the data frame only comprises the data to be plotted. A vector with identical values is defined for each colour and line width, and individually modified for the 5th data frame. Margins are set with the mar parameter within the bumpchart() function. At the end, we add two axis labels on both the left and right, stating the range of values. The first call sets the minima and maxima, the second the value for Microsoft.

6.3.4 Heat Map

About the figure: A heat map is a two-dimensional matrix, in which the cells are coloured depending on their value. It may be a table with individual data or aggregated values. There is no rule stating how the rows or columns should be arranged for illustration. If the order of both rows and columns is random or does not contain required information, then a cluster method can be used to group "similar" rows and/or columns. Additionally, dendrograms on the sides can show the grouping at different levels. Another variant is sorting the data. If comparable statistics can be done for the columns, then sorting by these is possible as well. An example for this is school grades. The figure shows a heat map of (fictional) school grades of a (fictional) class. Here, the best pupils are arranged at the top and the subjects with the best grades on the left. This gives a good impression of the grade distribution within a class. Of course, a comparison with one or several other classes would be an obvious option.

Heat map of school grades within a fictional class

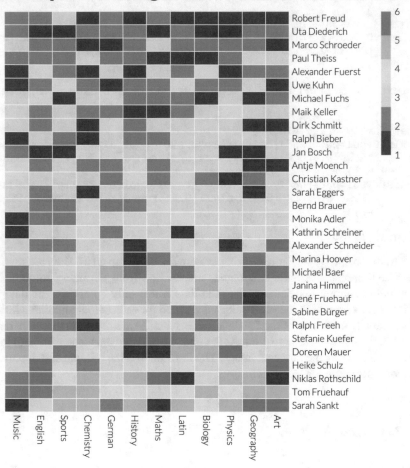

Fictional data, names generated with de.fakenamegenerator.com

Data were generated with the help of the http://de.fakenamegenerator.com site, and entered into an XLS spreadsheet.

```
pdf_file<-"pdf/tablecharts_heatmap.pdf"
cairo_pdf(bg="grey98", pdf_file,width=7,height=8)

library(RColorBrewer)
library(pheatmap)
par(mai=c(0.25,0.25,0.25,1.75),omi=c(0.25,0.25,0.75,0.85),family↘
    ="Lato Light",las=1)

# Import data and prepare chart

myGrades<-read.xls("myData/grades.xlsx", encoding="latin1")
x<-as.matrix(myGrades[,2:13])
rownames(x)<-myGrades$names
x<-x[order(rowSums(x)), ]
x<-x[,order(colSums(x))]

# Create chart

plot.new()
pheatmap(x,col=brewer.pal(6,"Spectral"),cluster_rows=F,cluster_↘
    cols=F,cellwidth=25,cellheight=14,border_color="white",↘
    fontfamily="Lato Light")

# Titling

mtext("Heat map of school grades within a fictional class",3,↘
    line=1,adj=0.2,cex=1.75,family="Lato Black",outer=T)
mtext("Fictional data, names generated with de.fakenamegenerator↘
    .com",1,line=-1,adj=1,cex=0.85,font=3,outer=T)
dev.off()
```

About the script: The standard package stats provides a heatmap() function that can be used to create the corresponding charts. However, there is only limited potential to modify their appearance. The gplots package provides an extended heat map function called heatmap.2(); however, it uses the layout() function and can therefore not be used in a panel illustration defined with mfcol or mfrow. For this reason, we use the pheatmap() function from the package of the same name by Raivo Kolde for our purposes. Here, cell height and width can be defined individually. We forgo the use of dendrograms on the sides (cluster_rows=F, cluster_cols=F). Data are read from the grades.xlsx file, then row names are created from the "names" column and the column is deleted. Data are then sorted first by row and then by column, so that the best pupils are shown at the top and the best grades on the left. Then follows the call of the pheatmap() function. Before that, plot.new() has to be called though, since pheatmap() is not a high-level function from R's traditional base graphic environment, but is based on grid.

6.3.5   Mosaic Plot (Panel)

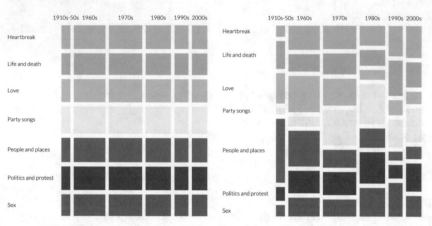

1000 songs to hear before you die

Guardian 1000 Songs Distribution

Source: www.stubbornmule.net

About the figure: in a mosaic plot, cells of a contingency table are shown in the form of rectangles, with the size of the rectangle corresponding to the frequency of the cell. This is generally also possible for multi-dimensional data and offers statistics specialists great help for gaining insights. Without prior knowledge, this illustration form requires some getting used to, but is useful in individual cases. It should be noted that the area varies in two dimensions and not independently, as is the case with a "bubble plot", a scatter plot with different size dots. Here, increasing one length will cause a shift of the subsequent elements. We are using an example from Sean Carmody that is also used in Wikipedia. The principle of constructing a two-dimensional mosaic plot is this: first, the rectangle of the entire tables is split in vertical slices, so that column width corresponds to the relative frequencies of the margin distribution of the column variable. In our case, this is the distribution of the number of songs within the individual epochs. In a second step, the area for each epoch is cut horizontally in such a way that the heights correspond to the relative frequencies of the row variable (in our case: the topics) in the respective epochs. In a two-dimensional case, the result is therefore a stacked 100% bar chart, in which column width corresponds to the relative frequencies of a second categorical variable. In my opinion, such a representation is especially useful if an independency table is set next to it for comparison, i.e. one that assumes identical frequencies for the row variable in those categories. The data are taken from a list that was compiled by The Guardian. A CSV file with the data is available on http://www.stubbornmule.net.

```
pdf_file<-"pdf/tablecharts_mosaicplot_1x2.pdf"
cairo_pdf(bg="grey98", pdf_file,width=10,height=6)

par(mai=c(0.25,0.0,0.0,0.25),omi=c(0.5,0.5,1.25,0.5),las=1,mfcol↴
    =c(1,2),family="Lato Light",las=1)
library(RColorBrewer)

# Import data and prepare chart

data<-read.csv("myData/1000.csv",as.is=c(F,T,F,T,T),sep=";")
data$DEKADE<-floor(data$YEAR/10) * 10
data$KDEKADE<-paste(data$DEKADE,"s",sep="")
data$KDEKADE[data$DEKADE < 1960]<-"1910s-50s"
tab<-table(data$KDEKADE,data$THEME)
utab<-chisq.test(tab)

# Create chart

mosaicplot(utab$expected,col=brewer.pal(7,"Accent"),main="",↴
    border=par("bg"))
mosaicplot(tab,col=brewer.pal(7,"Accent"),main="",border=par("bg↴
    "))

# Titling

mtext("1000 songs to hear before you die",3,line=3,adj=0,cex↴
    =1.5,family="Lato Black",outer=T)
mtext("Guardian 1000 Songs Distribution",3,line=1.5,adj=0,cex↴
    =0.9,font=3,outer=T)
mtext("Source: www.stubbornmule.net",1,line=1,adj=1.0,cex=0.85,↴
    font=3,outer=T)
dev.off()
```

In the script, we add an extra column DECADE to the period name in the data for illustration purposes, and an "s" is appended; then the dates before 1960 are summarised. For the mosaic plot, a table is generated from the data. We use the chis.test function from the stats package for the illustration of the "independency table". The mosaicplot() function is part of the graphics package, which means that we do not have to load an additional package. The left mosaic plot shows the data under the assumption of independence, the right one the distribution of the actual data. In both cases, we use a qualitative Brewer palette and a margin-free background. No other settings have to be made. Extended features are provided by the vcd package, with which numerous variants of mosaic plots are possible. These graphics are based on grid though, not on the traditional graphic.

6.3.6 Balloon Plot

Titanic - Passenger and Crew Statistics

Balloon Plot for Age, Sex by Class, Survived

Class	Survived	Child Male	Child Female	Adult Male	Adult Female	
1st	Yes	5	1	57	140	203
	No	0	0	118	4	122
2nd	Yes	11	13	14	80	118
	No	0	0	154	13	167
3rd	Yes	13	14	75	76	178
	No	35	17	387	89	528
Crew	Yes	0	0	192	20	212
	No	0	0	670	8	673
		64	45	1667	425	2201

Area is proportional to Number of Passengers   *Source: R library gplots*

About the figure: The name "balloon plot" is slightly misleading, as it is frequently used to describe a scatter plot with variable dot size (a "bubble chart"). Here, it refers to a specific, graphically supplemented variant of contingency tables that has to date occasionally been used in biostatistics or mineralogy. The illustration form was implemented into R with the plots package by Gregory R. Warnes. The data are an example data frame frequently employed in R, the Titanic passengers, classified by sex, age (children and adults), their on-board status (1st, 2nd, 3rd class or crew) and their survival of the sinking (yes, no). The figure shows the data in the form of a bivariate frequency table, where the rows contain the status on the first level and the survival on the second level, the columns the age on the first level and the sex on the second level. In addition to the numbers showing cell and margin frequencies, cell frequencies are highlighted with a dot whose size is proportional to the number. The colour of the dots differentiates the survivors from the drowned.

Margin frequencies are reflected in the header rows and columns as bar or column portions of 100% in the respective row or column. This type of illustration gives a better impression of the distribution than a "naked" contingency table would. Data are taken from the "Titanic" data frame supplied with R.

```
pdf_file<-"pdf/tablecharts_ballonplot.pdf"
cairo_pdf(bg="grey98",pdf_file,width=9,height=9)

par(omi=c(0.75,0.25,0.5,0.25),mai=c(0.25,0.55,0.25,0),family="\
      Lato Light",cex=1.15)
library(gplots)

# Import data and prepare chart

data(Titanic)
myData<-as.data.frame(Titanic) # convert to 1 entry per row \
      format
attach(myData)
myColours<-Titanic
myColours[,,,"Yes"]<-"LightSkyBlue"
myColours[,,,"No"]<-"plum1"
myColours<-as.character(as.data.frame(myColours)$Freq)

# Create chart

balloonplot(x=list(Age,Sex),main="",
            y=list(Class=Class,
                Survived=gdata::reorder.factor(Survived,new.order=c\
                    (2,1))),
            z=Freq,dotsize=18,
            zlab="Number of Passengers",
            sort=T,
            dotcol=myColours,
            show.zeros=T,
            show.margins=T)

# Titling

mtext("Titanic - Passenger and Crew Statistics",3,line=0,adj=0,\
      cex=2,family="Lato Black",outer=T)
mtext("Balloon Plot for Age, Sex by Class, Survived",3,line=-2,\
      adj=0,cex=1.25,font=3,outer=T)
mtext("Source: R library gplots",1,line=1,adj=1.0,cex=1.25,font\
      =3,outer=T)
mtext("Area is proportional to Number of Passengers",1,line=1,\
      adj=0,cex=1.25,font=3,outer=T)
dev.off()
```

The script is the example from the documentation of the ballonplot() function, in which only the colour selection and dot size have been adjusted. Data were loaded from the data frame "Titanic" that is supplied with R and converted into a data frame (the original data are an object of table type). The original table colours are used for the creation of the colours for the balloon plot. Data from the data frame are transferred to the function in the form of a list. We choose 18 as point size. Headings are created as before.

6.3.7   Tree Map

Tree maps are useful for the presentation of proportions. The New York Times used tree maps for the presentation of Obama's 2012 budget. A neat example with German data can be found offenerhaushalt.de/haushalt/bund/. Here, a breakdown of the German federal budget can be found in tabular form and as a tree map, for both the entire budget and the individual categories. Data can be exported in JSON or RDF format.

Federal Budget 2011
Shares of Expenditure

Source: bund.offenerhaushalt.de

About the figure: A tree map shows the attribute of a cardinally scaled variable as nested rectangles. The size and order of the rectangles are calculated so that, with preset outer dimensions, the large rectangle is completely filled and the areas of the individual rectangles correspond to the size of the variables. There are different algorithms for calculation of the rectangles that each optimise different aspects of the subdivision. Mostly, a procedure is used that produces the maximum number of rectangles with aspect ratio approximating 1. Since the outer margins are always set, even hierarchies can be shown with tree maps: a created rectangle can again be considered the outer margin for a new subdivision of the attribute. The first example shows the shares of individual expenditures of the federal budget in 2011 as a tree map, which clearly depicts the unequal distribution of expenditures. The different

continent and within those nested by population size. As an additional variable, the gross national income (GNI) was colour-coded in three classes. Data are provided online by the World Bank. Data were filtered for the tree map examples used in this book, connected to the continent data and saved as a binary R file hnp.RData. The file is available for download as Electronic Supplementary Material available at https://link.springer.com/chapter/10.1007/978-3-030-28444-2_1.

```
pdf_file<-"pdf/tablecharts_treemap_2a_inc.pdf"
cairo_pdf(bg="grey98", pdf_file,width=11.69,height=7.5)

par(omi=c(0.65,0.25,1.25,0.75),mai=c(0.3,2,0.35,0),family="Lato ⤸
    Light",las=1)
library(treemap)
library(RColorBrewer)

# Read data and prepare chart

load("myData/hnp.RData")
myData<-subset(daten,daten$gni>0)
attach(myData)
popgni<-pop*gni
myData$popgni<-popgni
myContinents<-aggregate(cbind(pop,popgni) ~ kontinent,data=⤸
    myData,sum)
kgni<-myContinents$popgni/myContinents$pop
myContinents$kgni<-kgni
kkgni<-cut(kgni,c(0,5000,10000,100000))
levels(kkgni)<-c("low","middle","high")
myContinents$kkgni<-kkgni
myContinents$nkkgni<-as.numeric(kkgni)

# Create chart and other elements

plot(1:1,type="n",axes=F)
treemap(myContinents,title="",index="kontinent", vSize="pop",⤸
    vColor="kgni",type="value",palette="YlOrBr",aspRatio=2.5,⤸
    position.legend="none",inflate.labels=T)
legend(0.35,0.6,levels(kkgni)[1:3],cex=1.65,ncol=6,border=F,bty=⤸
    "n",fill= brewer.pal(5,"YlOrBr")[3:5],text.col="black",xpd⤸
    =NA)

# Titling

mtext("Population and Gross National Income",3,line=2,adj=0,cex⤸
    =2.4,outer=T,family="Lato Black")
mtext("Size: population - Colour: GNI per capita. Atlas method (⤸
    current US $), 2010",3,line=0,adj=0,cex=1.75,outer=T,font⤸
    =3)
mtext("",1,line=1,adj=1.0,cex=1.25,outer=T,font=3)
dev.off()
```

In the script, contrary to the previous example, two separate figures are created that are then connected in LaTeX. For the first figure, data from the hnp data frame are filtered so that only the countries with an entry for gni are kept. Then the product of pop and gni is calculated and added to the data frame. Then a second data frame continents is created by aggregation, in which the population and the product of population and the GNI are added up. This gives us the correct continent-related GNIs called cgni. From this, we create three classes "low", "middle" and "high" called ccgni. Lastly, for the tree map, we require this as a numerical variable nccgni. As in the previous example, we have to call plot() to define the figure prior to plotting the tree map, since treemap() is based on grid. The tree map graphic is then plotted with the function of the same name. In contrast to the previous example, we set type=value, since the colour represents the attribute of an additional variable in this case. This variable is defined using vColor=cgni. We also specify that no legend be created, since we want to draw that separately at the end. The size of the rectangle is defined with vSize=pop. We then draw a legend using the legend() function. When doing this, we have to be careful to use the correct colours from the selected Brewer palette, as the treemap() function apparently assumes a fiver palette. The second tree map for the countries within the continents is created essentially like the first one. The only differences are that we define two variables using index=c("continent", "iso3"), because we want to keep the first level (continents) as order criteria. We also have to use fontsize.labels=c(0.1,20) to set specific font sizes for the two levels: 0.1 makes the labels of the first level invisible. The result is a tree map of all countries, but sorted within their continents.

```
pdf_file<-"pdf/tablecharts_treemap_2b_inc.pdf"
cairo_pdf(bg="grey98", pdf_file,width=11.69,height=7.5)

par(omi=c(0.65,0.25,1.25,0.75),mai=c(0.3,2,0.35,0),family="Lato ↘
      Light",las=1)
library(treemap)
library(RColorBrewer)

# Daten einlesen und Grafik vorbereiten

load("myData/hnp.RData")
myData<-subset(daten,daten$gni>0)
attach(myData)
kgni<-cut(gni,c(0,40000,80000))
levels(kgni)<-c("low","middle","high")
myData$kgni<-kgni
myData$nkgni<-as.numeric(kgni)

# Grafik definieren und weitere Elemente

plot(1:1,type="n",axes=F)
```

```
treemap(myData,title="",index=c("kontinent","iso3"), vSize="pop"↘
    ,vColor="nkgni",type="value",palette="Blues",aspRatio=2.5,↘
    fontsize.labels=c(0.1,20),position.legend="none")
legend(0.35,0.6,levels(kgni)[1:3],cex=1.65,ncol=3,border=F,bty="↘
    n",fill= brewer.pal(9,"Blues")[7:9],text.col="black",xpd=↘
    NA)

# Betitelung

mtext("Within Continent: Country Level",3,line=2,adj=0,cex=2.4,↘
    outer=T,family="Lato Black")
mtext("Size: population - Colour: GNI per capita. Atlas method (↘
    current US $), 2010",3,line=0,adj=0,cex=1.75,outer=T,font↘
    =3)
mtext("Source: data.wordlbank.org",1,line=1,adj=1.0,cex=1.25,↘
    outer=T,font=3)
dev.off()
```

The last step is to combine the two tree maps into one figure. To do that, both are sequentially embedded into LaTeX. Finally, the LaTeX embedding:

```
\documentclass{article}
\usepackage[paperheight=26.7cm, paperwidth=21cm, top=0cm, left=0↘
    cm, right=0cm, bottom=0cm]{geometry}
\usepackage{color}
\usepackage[abs]{overpic}
\definecolor{myBackground}{rgb}{0.9412,0.9412,0.9412}
\pagecolor{myBackground}
\begin{document}
\pagestyle{empty}
\begin{center}
\begin{overpic}[scale=0.70,unit=1mm]{../pdf/tablecharts_treemap_↘
    2a_inc.pdf}
\put(60,128){}
\end{overpic}
\begin{overpic}[scale=0.70,unit=1mm]{../pdf/tablecharts_treemap_↘
    2b_inc.pdf}
\put(60,28){}
\end{overpic}
\end{center}
\end{document}
```

The examples so far have focused on visualisations of "variables". Section 6.4 explains the representation of relationships between units of observation. A popular option, albeit one that is anything but simple to interpret, is visualisation by means of network diagrams.[1] Further options shown here are examples for chord diagrams and riverplots, as well as variants of heat maps and multiple bar charts.

6.4   Network Relationships

Networks can be visualised for both exploration and presentation purposes. The objective is to reflect relationships between individual sampling units: for example, the friendship relationships between people who use Facebook, trading relationships between individual states, migration flows between regions of the world, etc. Examples of questions that arise in network analysis are:

- Which of the objects are particularly important in the sense that they have a particularly high number of relationships or the relationships to them are particularly strong?
- How are the relationships structured? How is their overall strength reflected?
- Are there individuals or subgroups that are visibly distinguishable from other groups or individuals?

The relationships can be symmetrical or asymmetrical. Depending on the content, they can also be measured based on the individual objects themselves—for example, migration within a country. This means that for n objects, there are a maximum n^2 possible relationships; in the case of symmetrical relationships without "internal" measurements, the maximum possible is $((n^2 - n)/2$. When representing networks, we initially differentiate between nodes (the units of observation) and edges (the relationships between the units of observation themselves). The following variations are possible:

- The size of the nodes can reflect the frequency of the relationships to this node or an "internal" size.
- The colour of the nodes can represent a classifying variable—continents for countries, gender for persons, etc.
- The colour of the edges can reflect the direction of the relationships. In this case, the outgoing relationship is indicated by the colour of the node. Alternatively, or in addition, the directions can be indicated by arrows rather than lines.

---

[1] A very helpful technical introduction to this topic can be found on the website Network visualization with R, by Katya Ognyanova: http://kateto.net/network-visualization. For general aspects, the following is recommended: Richard Brath/David Jonker (2015): Graph Analysis and Visualization: Discovering Business Opportunity in Linked Data, Indianapolis, IN: Wiley.

- In the same way as with colour, groups of nodes and edges can be differentiated using different shapes and patterns.
- The thickness of the lines can reflect the strength of the relationship.
- Labels can be applied to individual or all nodes or, even better, labels can be used instead of the nodes.

The most important property of network visualisations, which is also the most difficult property to understand, is the position of the nodes. There are a whole series of different algorithms for the positioning. What they all have in common is that they try to reduce a high-dimensional space to a two-dimensional space. Therefore, we often find visualisations that apply a "strength-based" design. The first step involves arranging the nodes (i.e. the units of observation or the cases) randomly in a two-dimensional space. From this starting point, an optimisation algorithm is used to consider the "physical strengths" between all points in an attempt to achieve an arrangement in which the strength of all relationships acts as a force of attraction or force of repulsion. However, there is no one unique solution for this; the end result is instead heavily dependent on the starting point.[2]

6.4.1   Undirected Network

We begin with an example of an undirected network in which the relationships between the objects are symmetrical. The data used here are fixtures in the football Champions League from 2013 to 2016. The data can be downloaded at www. weltfussball.de. In total, there are 375 fixtures. For anyone who is not interested in football, the epistemic value of these data is not particularly high, but in my opinion, the data are very suitable as an introductory example of network visualisations. Let us first look at the data in Fig. 6.3.

---

[2]See https://lists.nongnu.org/archive/html/igraph-help/2010-04/msg00076.html.

	links ×	
	⬚ ▽ Filter	
	X1	**X2**
1	Manchester United	Bayer Leverkusen
2	Real Sociedad	Shakhtar Donetsk
3	Shakhtar Donetsk	Manchester United
4	Bayer Leverkusen	Real Sociedad
5	Bayer Leverkusen	Shakhtar Donetsk
6	Manchester United	Real Sociedad
7	Shakhtar Donetsk	Bayer Leverkusen
8	Real Sociedad	Manchester United
9	Bayer Leverkusen	Manchester United
10	Shakhtar Donetsk	Real Sociedad
11	Manchester United	Shakhtar Donetsk
12	Real Sociedad	Bayer Leverkusen
13	Galatasaray	Real Madrid
14	FC København	Juventus
15	Juventus	Galatasaray

Showing 1 to 16 of 375 entries

Fig. 6.3 Table with matches

The first option for getting an overview of the relationships between the units of observation is a matrix in which all units of observation are arranged in a relationship table, both as rows and columns. The cells then contain the frequencies of the occurrence of individual combinations/fixtures. To create the relationship table, we use the R package igraph. Using the function graph_from_data_frame() from this package we create the network structure and then, using get.adjacency(), from the network structure, we create the relationship matrix.

```
library(gdata)
library(igraph)

X2013_2014 <- read.csv("myData/2013_2014.txt",sep="\t", head=↩
     FALSE)
X2014_2015 <- read.csv("myData/2014_2015.txt",sep="\t", head=↩
     FALSE)
X2015_2016 <- read.csv("myData/2015_2016.txt",sep="\t", head=↩
     FALSE)
links<-rbind(X2013_2014, X2014_2015, X2015_2016)
beztab<-as.matrix(get.adjacency(graph_from_data_frame(links)))
```

The result is the relationship table shown in Fig. 6.4.

	Manchester United	Real Sociedad	Shakhtar Donetsk	Bayer Leverkusen	Galatasaray	FC København	Juventus	Real Madrid	Olympiak Piräus
Manchester United	0	1	1	1	0	0	0	0	
Real Sociedad	1	0	1	1	0	0	0	0	
Shakhtar Donetsk	1	1	0	1	0	0	0	1	
Bayer Leverkusen	1	1	1	0	0	0	0	0	
Galatasaray	0	0	0	0	0	1	1	1	
FC København	0	0	0	0	1	0	1	1	
Juventus	0	0	0	0	1	1	0	2	
Real Madrid	0	0	1	0	1	1	2	0	
Olympiakos Piräus	1	0	0	0	0	0	1	0	
SL Benfica	0	0	0	1	1	0	0	0	
Paris Saint-Germain	0	0	1	1	0	0	0	1	
RSC Anderlecht	0	0	0	0	1	0	0	0	
Bayern München	1	0	1	0	0	0	1	1	
Viktoria Plzeň	0	0	0	0	0	0	0	0	

Showing 1 to 15 of 53 entries

Fig. 6.4 Relationship table (extract)

This relationship table forms the basis for our network illustration:

Champions League - Matches
Base: all matches 2013-2016

Source: http://www.weltfussball.de/alle_spiele/champions-league-2015-2016/

The example shows the teams with their names instead of points. The German teams are highlighted in colour. We can see that the teams that are shown large and centrally in the middle are those we would expect: Bayern München (Bayern Munich), Atlético Madrid, Real Madrid. Other, more exotic teams are located in the outer areas. On the other hand, Malmö FF is also arranged centrally in the middle. This contradicts the fact that Malmö is not particularly important in the Champions League.

```
pdf_file<-"pdf/networks_undirected_network.pdf"
cairo_pdf(bg="grey98", pdf_file,width=8,height=7)

par(mai=c(0.25,0.25,0.25,0.5),omi=c(0.25,0.25,0.25,0.25), family⤸
    ="Lato Light",las=1)

library(igraph)
library(sqldf)
library(gdata)

# Import data and prepare chart
```

```
X2013_2014 <- read.csv("data/2013_2014.txt",sep="\t", head=FALSE↘
    )
X2014_2015 <- read.csv("data/2014_2015.txt",sep="\t", head=FALSE↘
    )
X2015_2016 <- read.csv("data/2015_2016.txt",sep="\t", head=FALSE↘
    )
links<-rbind(X2013_2014, X2014_2015, X2015_2016)

teams<-as.data.frame(unique(c(links$V1, links$V2)))
teams<-sqldf("select team, count(*) games from (select V1 team ↘
    from links union all select V2 team from links) a group by↘
    team")
teams$col<-"grey55"
teams$col[c(11, 12, 13, 14, 24, 51)]<-"#f768a1"

mySeed = as.POSIXlt(Sys.time())
mySeed = 1000*(mySeed$hour*3600 + mySeed$min*60 + mySeed$sec)
mySeed

set.seed(56313585)
net2 <- graph_from_data_frame(d=links, directed=F, vertices=↘
    teams)
net2simp<-simplify(net2, edge.attr.comb=list(weight="sum","↘
    ignore"))

# Crete chart

plot(net2simp, vertex.shape="none", vertex.label=V(net2simp)$↘
    media, vertex.label.font=2, vertex.label.color=teams$col, ↘
    vertex.label.cex=0.7*sqrt(teams$games/23), edge.color="↘
    grey80", vertex.label.family=ifelse(teams$col=="grey95", "↘
    Avenir Next Condensed Ultra Light", "Avenir Next Condensed↘
    Demi Bold"))

# Titling

mtext("Champions League - Matches", line=-1.5, adj=0, cex=2, ↘
    family="Lato Black", col="grey40", outer=T)
mtext("Base: all matches 2013-2016", line=-2.75, adj=0, cex=0.9,↘
    family="Lato Bold", col="grey40", outer=T)
mtext("Source: http://www.weltfussball.de/alle_spiele/champions-↘
    league-2015-2016/", side=1, line=-1, adj=1, cex=0.9, font↘
    =3, outer=T)

dev.off()
```

In the script, the first step is the integration of the libraries required. The data are then imported and appended to one another by season. These data form the basis for the representation of the edges. We also need a second file that provides the nodes. For this purpose, we define a single-column data record teams which contains each team once only. To do this, we use the package sqldf so that we can compose the data in SQL notation. The result is a list to which the number of games has been added for each team. In a further column, we add the colour of the nodes: we generally select grey, and for the German teams, a shade of red. We use Avenir Next Condensed as the font.

To make it possible to reproduce the illustration, we have to define a fixed value for the random generator otherwise, every time we run the script, we would receive different visualisations despite the data being identical.[3]

Here we select the following procedure: we set the current second as the starting value for the random generator. We then output this immediately so that we can note it from the R console. We can now run the script multiple times until a visualization that we like is created. We can then reproduce this visualisation by defining the value output by R as the starting value for the random generator. In our specific case, this is the value 56313585. We create the data structure we need for the visualisation using the functions graph_from_data_frame() and simplify(). For the numerous setting options, see the extensive igraph manual. The data can be copied from the website http://www.weltfussball.de/alle_spiele/champions-league-2015-2016 (or 2013–2014 and 2014–1015). The data have been saved here in TXT files.

Network diagrams are a very popular although not unproblematic form for visu-alising relationships between units of observation. They are suitable for exploring relationships and for recognising structures and analysing them further. In our view, there are more suitable forms of representation for presenting and illustrating data, and we present these below.

6.4.2 Chord Diagram

A visualisation that has recently gained a lot of attention is the chord diagram. In this type of diagram, the units of observation or categories are arranged on a circle and the relationships between them are visualised using connecting lines within the circle. The following illustration, which was presented in an article in the journal *Science* in 2014, shows an example of a chord diagram depicting migration movements between regions of the world from 2005 to 2010.[4] The 15 world regions form segments of a circle and the migration flows are indicated by chords. These regions are connected not linearly but in the form of a curve, which gives a clearer overview. Scales are shown at the segments—the width of the connections at the starting points corresponds to the quantitative size of the migration. Internal migration is also depicted: these are the connections that point back to the respective segment. In contrast to the original illustration in *Science*,

---

[3]http://stackoverflow.com/questions/40927099/r-igraph-save-layout.

[4]Abel, Guy J. / Sander, N., Quantifying Global International Migration Flows, in: Science 343 (6178), S. 1520–1522.

here, G. Abel has developed the information depicted further[5]: for example, here, the direction of the migration is shown via arrows at the ends of the lines.

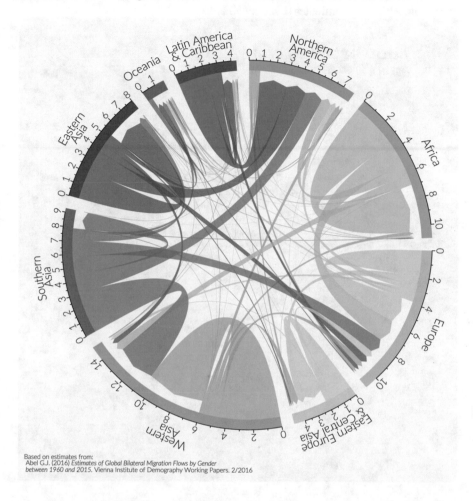

Based on estimates from:
Abel G.J. (2016) *Estimates of Global Bilateral Migration Flows by Gender between 1960 and 2015.* Vienna Institute of Demography Working Papers. 2/2016

This type of illustration is very suitable for representing relationships between objects. However, note an important aspect: in the same way as for the radial polygons (Sects. 6.2.5 and 6.2.6), the perception of the information is heavily dependent on the arrangement of the segments. If two regions are in neighbouring segments of the circle, this creates a very different impression compared to that of a different arrangement in which the regions are opposite each other, for example: in that case, with the same content, more or less area is visible. In the following, we want to use an example on the migration between federal states in Germany in

[5]https://github.com/gjabel/migest/blob/master/demo/cfplot_reg2.R.

2010 to explain the construction of a chord diagram. The diagram does not show all the movements; instead, it is limited to a specific threshold and excludes internal migration, which predominates based on size.

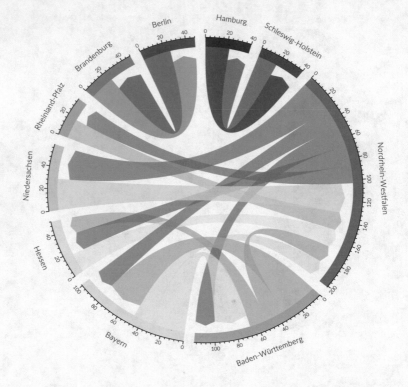

Migration between states 2010

all migrations over 15,000 persons, without internal migration, data in thousands

From the illustration we can see—not unsurprisingly—that migration takes place in particular into neighbouring federal states. Here too, we have to consider the arrangement. The impression of migration from Hamburg to Schleswig-Holstein is different to that which we get for the migration from Nordrhein-Westfalen (North Rhine-Westphalia) to Niedersachsen (Lower Saxony) because these two federal states are arranged opposite each other rather than next to each other. We can also clearly see that the influx and outflux balance out overall.

```
pdf_file<-"pdf/networks_chord_diagram.pdf"
cairo_pdf(bg="grey98", pdf_file, width=8,height=8)
par(omi=c(0.25,0.25,0.5,0.25), mai=c(0,0,0,0), family="Lato ⤸
    Light")
```

```r
library(circlize)
library(RColorBrewer)
library(readr)
library(sqldf)

# Import data and prepare chart


df0 <- read_csv("data/2010 GIM data.csv")
df1 <- read_csv("data/2010 GIM data lookup.csv")

bulas <- read_csv("data/bulas.csv")
kreise<-sqldf("select min(kreis_destatis/1000) bula, hk_id from ↘
     df1 group by hk_id")

df_bula<-sqldf("select a.bula bula_destination, b.* from kreise ↘
     a, df0 b where a.hk_id = b.destination_hk")
df_bula<-sqldf("select a.bula bula_origin, b.* from kreise a, df↘
     _bula b where a.hk_id = b.origin_hk")
df_bula<-sqldf("select bula_origin, bula_destination, sum(t_↘
     total) t_total from df_bula group by bula_origin, bula_↘
     destination")
df_bula<-sqldf("select a.bula bula_origin_name, b.bula_↘
     destination, b.t_total from bulas a, df_bula b where a.nr ↘
     = b.bula_origin")
df_bula<-sqldf("select b.bula_origin_name, a.bula bula_↘
     destination_name, b.t_total/1000 t_total from bulas a, df_↘
     bula b where a.nr = b.bula_destination")
df_bula<-df_bula[df_bula$t_total>15, ]
df_bula<-df_bula[df_bula$bula_origin!=df_bula$bula_destination, ↘
     ]


circos.clear()
circos.par(start.degree = 90, gap.degree = 4, track.margin = c(-↘
     0.1, 0.1), points.overflow.warning = FALSE)
par(mar = rcp(0, 4))

# Create chart

chordDiagram(x = df_bula, grid.col = brewer.pal(length(unique(df↘
     _bula$bula_origin)), "Spectral"), transparency = 0.25,
             directional = 1,
             direction.type = c("arrows", "diffHeight"), ↘
                 diffHeight  = -0.04,
             annotationTrack = c("grid", "name", "axis"), ↘
                 annotationTrackHeight = c(0.05, 0.1),
             link.arr.type = "big.arrow", link.sort = TRUE, link↘
                 .largest.ontop = TRUE)

# Titling

mtext("Migration between states 2010", adj=0, cex=2, family="↘
     Lato Black", col="grey40", outer=T)
```

```
mtext("all migrations over 15,000 persons, without internal ⟍
      migration, data in thousands", line=-1.25, adj=0, cex=0.9,⟍
      family="Lato Bold", col="grey40", outer=T)
mtext("Source: www.nikolasander.net/news/diy", side=1, line=-1, ⟍
      adj=1, cex=0.9, font=3, outer=T)

circos.clear()
dev.off()
```

In the script, we first load the library circlize from Zuguang Gu, which was also used to create the afore-mentioned representation in *Science*. We also need RColorBrewer for the colours, readr to import the data, and sqldf because we have to first prepare the data. Just like in the previous example, the data consist of two data records. The data record df0 contains the actual migration data at county level: origin_hk is the ID of the source county, destination_hk is that of the target county, and t_total is the number of migrating persons. We do not need the other variables. The data record df1 contains the source data for the circle segments. In a network representation, this would be the nodes. The first step is to aggregate the data at federal state level. To do this, we need a third file containing the IDs and names of the federal states. We have created this file manually. Again, we use SQL notation here. Firstly, we create a data frame that assigns the federal state ID to each county. Then, in multiple steps, we create the data required for the illustration. The result is a data frame, df_bula, which aggregates the migrations to the federal states. To reduce the representation to the main information and thus make it clearer, in the next two steps we restrict the data to migrations of more than 15,000 persons and exclude internal migration. We can now start the actual drawing. First, we reset the layout parameters using circos.clear(). We then define some basic parameters for the representation and set the borders of the illustration to 0. We draw the chord diagram using the function chordDiagram() The basis for this is the data frame df_bula,We then define a number of further parameters here. There are no specifications as to what should be defined in advance via the function circos.par() and what should be set directly in chordDiagram() With link.arr.type we can use arrow tips to indicate the direction—a real improvement compared to the original type of representation in *Science*.

At the end, we define headers and footers and reset the parameters to the original values.

The **data** were collated by Nikola Sander and are available on her website.

6.4.3 Directed Network

With the chord diagram, we have learned about a type of illustration that represents relationships between units of observation or categories clearly and more understandably than an "unsorted" network visualisation. One particular problem was in the representation of internal movements. Therefore, as an alternative, we want to return to a network visualisation but this time, to a variant in which the position of the nodes is defined. To do this, we can use a circular layout. As an example, we select the migration flows between world regions which are the basis for the illustration in *Science*.

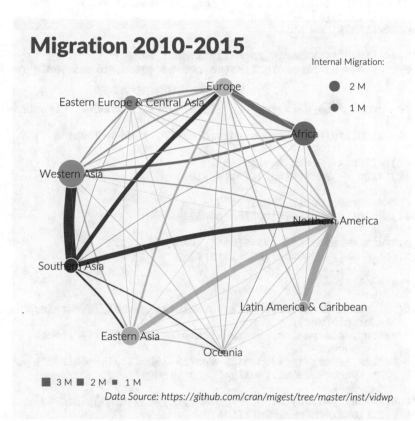

In this example, we use a circular (i.e. fixed) network layout that reflects the individual world regions as nodes. Each world region is assigned a separate colour. The same colour is used respectively for those connecting lines that reflect the emigration. The thickness of the lines reflects the scope of the migration, and the size of the node the internal migration.

```
pdf_file<-"pdf/networks_directed_network.pdf"
cairo_pdf(bg="grey98", pdf_file,width=6,height=6)

par(mai=c(0.25,0.25,0.25,0.5),omi=c(0.25,0.25,0.25,0.25),
  family="Lato Light",las=1)
library(igraph)
library(RColorBrewer)
```

```
# Import data and prepare chart

nodes <- read.csv("data/reg_plot.csv", header=T, as.is=T)
links <- read.csv("data/reg_flow.csv", header=T, as.is=T)

links <- links[order(links$orig_reg, links$dest_reg),]
colnames(links)[3] <- "weight"
rownames(links) <- NULL

binnen<-links[links$orig_reg==links$dest_reg, ]
nodes$inside<-binnen$weight[match(nodes$region, binnen$orig_reg)↘
      ]

net <- graph_from_data_frame(d=links, vertices=nodes, directed=T↘
      )
net <- simplify(net, remove.multiple = F, remove.loops = T)

E(net)$width <- E(net)$weight*5
V(net)$size <- sqrt(V(net)$inside*100)

colrs <- brewer.pal(9, "Paired")
V(net)$color <- colrs[V(net)$order1]

edge.start <- ends(net, es=E(net), names=F)[,1]
edge.col <- V(net)$color[edge.start]

# Create chart

plot(net, edge.arrow.size=0, edge.color=edge.col,layout=layout_↘
      in_circle(net),
     vertex.color=colrs, vertex.frame.color="#ffffff", edge.↘
          curved=.1,
     vertex.label=V(net)$media, vertex.label.color="black", ↘
          vertex.label.family="Lato Light")

legend(x=0.8, y=1.25, c("", "   2 M","", "   1 M"), pch=19,xpd=T↘
      ,title="Internal Migration:",
       col="#777777", pt.cex=c(0, sqrt(4),0,sqrt(2)), cex=.8, ↘
            bty="n", ncol=1)

legend(x=-1.25, y=-1.15, c(" 3 M"," 2 M", " 1 M"), pch=15,xpd=T,↘
      horiz=T,
       col="#777777", pt.cex=c(sqrt(3),sqrt(2),sqrt(1)), cex=.8,↘
            bty="n", ncol=1)

# Titling

mtext("Migration 2010-2015", line=-1.5, adj=0, cex=2, family="↘
      Lato Black", col="grey40", outer=T)
mtext("Data Source: https://github.com/cran/migest/tree/master/↘
      inst/vidwp", side=1, line=-1, adj=1, cex=0.9, font=3, ↘
      outer=T)
dev.off()
```

In the **script**, the first step is the import of the package igraph—the package which executes the actual network visualisation. Just like in the previous example, the data are imported in two files. The first file, reg_plot.csv contains the nodes and, in a variable, defines the colours. The second file, reg_flow.csv, contains the migrations between the individual world regions (Fig. 6.5).

	region	order1	col1	reg1	reg2
1	Northern America	1	#40A4D8	Northern	America
2	Africa	2	#33BEB7	Africa	
3	Europe	3	#B2C224	Europe	
4	Eastern Europe & Central Asia	4	#FECC2F	Eastern Europe	& Central Asia
5	Western Asia	5	#FBA127	Western	Asia
6	Southern Asia	6	#F66320	Southern	Asia
7	Eastern Asia	7	#DB3937	Eastern	Asia
8	Oceania	8	#A463D7	Oceania	
9	Latin America & Caribbean	9	#0C5BCE	Latin America	& Caribbean

	orig_reg	dest_reg	flow
1	Africa	Africa	3.498892
2	Africa	Eastern Asia	0.007214
3	Africa	Eastern Europe & Central Asia	0.036820
4	Africa	Europe	1.744666
5	Africa	Latin America & Caribbean	0.022516
6	Africa	Northern America	0.825409
7	Africa	Oceania	0.107072
8	Africa	Southern Asia	0.000869
9	Africa	Western Asia	0.675984
10	Eastern Asia	Africa	0.021043
11	Eastern Asia	Eastern Asia	2.178887
12	Eastern Asia	Eastern Europe & Central Asia	0.085849
13	Eastern Asia	Europe	0.565937
14	Eastern Asia	Latin America & Caribbean	0.040990
15	Eastern Asia	Northern America	1.593311

Fig. 6.5 View of the data for the world regions

Initially, the data are sorted by region of origin and then destination region. We then create a data frame binnen, which filters out the internal migration. We then link this internal migration with the data frame nodes and thus define a new variable inside there. Using the function graph_from_data_frame() from the package igraph, we now define the network structure. The parameter remove.loops = T of the function simplify() removes self-references. The package igraph contains two very useful functions, E() and V(), which can be used to edit the edges and nodes of a network or apply properties to them. Here we define the width of the edges (= connecting lines), which is stored in the predefined variable width , with the variable width. multiplied by 5. The size of the nodes, defined by the variable size, is determined by inside. Due to the scaling, the size of the nodes is multiplied by 100. As we are dealing with an area, we have to calculate the root. The colour is defined via a ColorBrewer palette; the order results from the order1 variable from the data frame nodes. The colours of the lines are defined such that the emigration corresponds to the nodes. Once these preparations are complete, we can call up the function plot() which igraph has extended with the method of the visualisation of networks, if the object is of the class "igraph". As the layout, we define layout_in_circle, in which the position of the nodes is defined on a circle.

At the end, we add two legends (one for the nodes, one for the edges), as well as a header and footer.

The **data** correspond to the introductory example in Sect. 6.4.2 and can be downloaded at https://github.com/cran/migest/tree/master/inst/vidwp.

6.4.4 Riverplot

In my opinion, the visualisation of the migration data by means of a circular network layout is a practical alternative to a chord diagram. In the former, the internal migration is clearer than in the latter, with the difference between internal and external migration visually larger. Ultimately, the choice of the type of illustration is a matter of personal preference. Another variant, which is more closely oriented on the chord diagram but which, unlike the chord diagram and the circular network, is not based on a circle arrangement, is the representation referred to as a riverplot. A "riverplot" package was created for R by January Weiner.[6] Just like in a chord diagram, the information is represented with curved lines. However, the categories are arranged not in a circle, but as opposite pairs. In the example[7] we are looking at here, the internal migration is visualised by means of horizontal lines, and the migration into other regions with curved lines. The source regions are shown on the left and the destination regions on the right. Here, too, we have the problem that the visual impression is heavily dependent on the arrangement of the countries (which is again random). Compared to the circular arrangement, the internal migration may well be clearer, but each person may see that differently.

---

[6]My thanks go to Stefan Fichtel for suggesting this type of representation. Riverplots offer a wide range of other options—for details, see the documentation of the package. For example, riverplots are suitable for representing a Sankey diagram for the development of biological processes. See also the illustration created with the riverplot package in Kenthirapalan, Sanketha et al. (2016): Functional profiles of orphan membrane transporters in the life cycle of the malaria parasite, in: Nature Communications 7, http://dx.doi.org/10.1038/ncomms10519.

[7]Based on http://stackoverflow.com/questions/41088751/riverplot-package-in-r-output-plot-covered-in-gridlines-or-outlines.

Migration 2010-2015

All figures in millions. Data Source: https://github.com/cran/migest/tree/master/inst/vidwp

In the script,[8] the first step is the integration of the riverplot and RColorBrewer packages, and then the data are imported. Numerically, the data are identical with those of the previous examples. However, the source and destination regions already have the required labels. We use these data to define the data structure required by riverplot. We do so by naming the columns "N1", "N2" and "Value" and by adding an ID. We create the nodes (regions) directly in the script. The next step is to define a classification variable x and a column for the colour. edges2 reduces the data frame to the actual regions available. The ID must then be defined as a character. notes2 is then sorted into the order required for the representation and a column for the colour is added. In the next step, we select the lines with the internal migration and transfer these to the node_styles list. Using the function makeriver(), we create the data table required for the visualisation (Fig. 6.6).

---

[8]Based on a template http://stackoverflow.com/questions/41088751/riverplot-package-in-r-output-plot-covered-in-gridlines-or-outlines. My heartfelt thanks at this point to the author of the riverplot package, January Weiner, for his valuable help in adapting the script.

	from Western Asia	from Southern Asia	from Oceania	from Northern America	from Latin America and Caribbean	from Europe	from Eastern Asia	from Africa	to Western Asia
x	−0.02327260	−0.0232726	−0.0232726	−0.0232726	−0.0232726	−0.0232726	−0.0232726	−0.0232726	1.0232726
top	0.06842138	0.2836770	0.3537152	0.4298161	0.5553264	0.6815283	0.8440249	1.0400000	0.1323167
center	0.01421069	0.2078139	0.3504608	0.4235304	0.5243360	0.6501921	0.7945413	0.9737771	0.0461583
bottom	−0.04000000	0.1319508	0.3472064	0.4172446	0.4933455	0.6188558	0.7450577	0.9075543	−0.0400000
lpos	0.01421069	0.2078139	0.3504608	0.4235304	0.5243360	0.6501921	0.7945413	0.9737771	−0.0400000
rpos	−0.04000000	0.1319508	0.3472064	0.4172446	0.4933455	0.6188558	0.7450577	0.9075543	0.0461583
yscale	0.01924350	0.0192435	0.0192435	0.0192435	0.0192435	0.0192435	0.0192435	0.0192435	0.0192435

Fig. 6.6 Structure of the data matrix created by riverplot

We then use riverplot() to create the actual illustration and save it in a variable myplot. To avoid white lines in the representation, we set fix.pdf=T.[9] The labels are now created and output with text(), and then the header and footer are created. No scaling is specified in this example.

```
pdf_file<-"pdf/networks_riverplot.pdf"

cairo_pdf(bg="white", pdf_file, width=12,height=12)
par(omi=c(1,1.95,1,1.95), mai=c(0,0,0,0), family="Lato Light")

library(riverplot)
library(RColorBrewer)

# Import data and prepare chart

xreg_flow<-read.csv("data/xreg_flow.csv", stringsAsFactors=F)

edges = rep(xreg_flow, col.names = c("N1","N2","Value"))
edges     <- data.frame(edges, stringsAsFactors=F)
edges$ID <- 1:81

regionen<-c("from Latin America and Caribbean","from Northern ⬎
      America","from Africa","from Europe","from Eastern Europe"⬎
      ,"from Western Asia","from Southern Asia","from Eastern ⬎
      Asia","from Oceania","to Latin America and Caribbean","to ⬎
      Northern America","to Africa","to Europe","to Eastern ⬎
      Europe","to Western Asia","to Southern Asia","to Eastern ⬎
      Asia","to Oceania")

nodes <- data.frame(ID = regionen, stringsAsFactors=F)

nodes$x = c(1,1,1,1,1,1,1,1,1,2,2,2,2,2,2,2,2,2)

edges2 <- edges[ edges$N1 %in% nodes$ID & edges$N2 %in% nodes$ID⬎
      , ]
```

---

[9]http://stackoverflow.com/questions/41088751/riverplot-package-in-r-output-plot-covered-in-gridlines-or-outlines/42294324#42294324.

```
edges2$ID <- as.character(edges2$ID)
edges2 <- edges2[nrow(edges2):1,]

nodes2 <- nodes[ match(unique(c(edges2$N1, edges2$N2)), nodes$ID⬎
      ), ]
nodes2$col <- rep(brewer.pal(9, "Blues")[2:9], 2)
sel <- gsub("from ", "", edges2$N1) == gsub("to ", "", edges2$N2⬎
      )
node_styles <- sapply(edges2$ID[ sel ], function(x) list(⬎
      horizontal=TRUE), simplify=F)
r <- makeRiver( nodes2, edges2, node_labels = "", node_styles=⬎
      node_styles)

# Create chart

par(lty=0)
myplot <- riverplot(r, col = nodes2$col, srt=0 , plot_area = 1, ⬎
      fix.pdf=T)

colnames(myplot) <- gsub("and", "\\\nand", colnames(myplot))
oldpar <- par(xpd=NA)
sel <- grep("^from", colnames(myplot))
text(myplot["x",sel] - strwidth("X"), myplot["center", sel], ⬎
      colnames(myplot)[sel], pos=2)
sel <- grep("^to", colnames(myplot))
text(myplot["x",sel] + strwidth("X"), myplot["center", sel], ⬎
      colnames(myplot)[sel], pos=4)

# Titling

mtext("Migration 2010-2015", side=3, outer=T, cex=3, line=1, col⬎
      ="grey30", family="Lato Black")
mtext("All figures in millions. Data Source: https://github.com/⬎
      cran/migest/tree/master/inst/vidwp",1,line=2, font=3, ⬎
      outer=T)

dev.off()
```

The data correspond to the introductory example in Sect. 6.2.4.

6.4.5   Heat Map for Relationships

The previous visualisations in this section are based on representation as a network.
However, it also makes sense to reflect the relationship matrix directly. There are
two options that we can use here: firstly, a heat map, which we became familiar
with in Sect. 6.3.4; secondly, a multiple bar chart. Let us start with the heat map.

Migration 2010-2015

0.11	0.01	0.21	0.17	0.03	0.00	0.04	0.04	0.21	Northern America
0.83	3.50	1.74	0.04	0.68	0.00	0.01	0.11	0.02	Africa
0.45	0.18	2.01	0.60	0.22	0.00	0.01	0.22	0.17	Europe
0.04	0.00	0.57	1.35	0.09	0.00	0.03	0.00	0.00	Eastern Europe & Central Asia
0.32	0.24	0.43	0.40	4.52	0.04	0.01	0.07	0.01	Western Asia
1.42	0.05	1.17	0.04	3.11	1.31	0.51	0.32	0.01	Southern Asia
1.59	0.02	0.57	0.09	0.38	0.00	2.18	0.36	0.04	Eastern Asia
0.05	0.01	0.08	0.02	0.01	0.00	0.01	0.18	0.00	Oceania
2.20	0.00	0.29	0.01	0.01	0.00	0.01	0.02	0.68	Latin America & Caribbean

Columns: Northern America, Africa, Europe, Eastern Europe & Central Asia, Western Asia, Southern Asia, Eastern Asia, Oceania, Latin America & Caribbean

All figures in millions. Data Source: https://github.com/cran/migest/tree/master/inst/vidwp

Here, we make a couple of changes compared to the representation in Sect. 6.3.4: firstly, we write the actual sizes in the cells. We select the colour gradations such that only "relevant" cells are highlighted. The direction of the migration is highlighted by an arrow.

```
pdf_file<-"pdf/networks_heatmap.pdf"
cairo_pdf(bg="grey98", pdf_file,width=7,height=6)

par(mai=c(0.25,0.25,0.25,1.75),omi=c(0.25,0.25,0.75,0.85),
  family="Lato Light",las=1)

library(pheatmap)
library(RColorBrewer)
library(igraph)

# Import data and prepare chart

df0 <- read.csv("data/reg_flow.csv", stringsAsFactors=FALSE)
df1 <- read.csv("data/reg_plot.csv", stringsAsFactors=FALSE)

net <- graph_from_data_frame(d=df0, vertices=df1, directed=T)
netm <- get.adjacency(net, attr="flow", sparse=F)
```

```
# Create chart

plot.new()
pheatmap(netm, col=brewer.pal(6,"RdPu"),
  cluster_rows=F,cluster_cols=F,cellwidth=35,cellheight=24,
  border_color="white",fontfamily="Lato Light", display_numbers=\
      T, number_color=matrix(ifelse(netm > 1.5, "white", "red"\
      ), nrow(netm)))

# Titling

mtext("Migration 2010-2015",3,line=1.5,adj=0,cex=1.75,family="\
    Lato Black",outer=T)
mtext("All figures in millions. Data Source: https://github.com/\
    cran/migest/tree/master/inst/vidwp",1,line=-1,adj=0,cex\
    =0.85,font=3,outer=T)

par(family="Lato Black")
mtext("▫",1,line=-7,adj=0.9,cex=7,col="grey80",outer=T)

dev.off()
```

In the script, we first load the package pheatmap, which was used in Sect. 6.3.4, RColorBrewer for the colour palette, and finally, igraph, which provides the functions for creating the relationship matrix netm. The matrix netm looks as in Fig. 6.7.

	Northern America	Africa	Europe	Eastern Europe & Central Asia	Western Asia	Southern Asia	Eastern Asia	Oceania	Latin America Caribbea
Northern America	0.112105	0.011244	0.213784	0.172959	0.026320	0.000000	0.044196	0.037692	0.20794
Africa	0.825409	3.498892	1.744666	0.036820	0.675984	0.000869	0.007214	0.107072	0.02251
Europe	0.449838	0.175242	2.006166	0.601772	0.223626	0.000014	0.010825	0.222712	0.16839
Eastern Europe & Central Asia	0.043438	0.001760	0.566555	1.354018	0.088158	0.000000	0.025689	0.002543	0.00143
Western Asia	0.323827	0.244112	0.431518	0.395673	4.519158	0.036058	0.007958	0.065140	0.00641
Southern Asia	1.420783	0.051368	1.168379	0.042952	3.106273	1.306389	0.505476	0.316133	0.00974
Eastern Asia	1.593311	0.021043	0.565937	0.085849	0.380948	0.002097	2.178887	0.359676	0.04099
Oceania	0.045753	0.006264	0.079224	0.015882	0.011272	0.000000	0.006773	0.184822	0.00412
Latin America & Caribbean	2.195385	0.004648	0.290048	0.009782	0.010966	0.000000	0.014653	0.022709	0.6824€

Fig. 6.7 Relationship matrix for world regions

We use the function pheatmap() to present the heat map. For the colour of the numbers we use the function ifelse(), so that in dark cells the font is light, and in light cells the font is dark. For the arrow that shows the direction of the migration, we select the font Lato Black and use the corresponding Unicode character magnified seven times.

6.4.6 Multiple Bar Chart

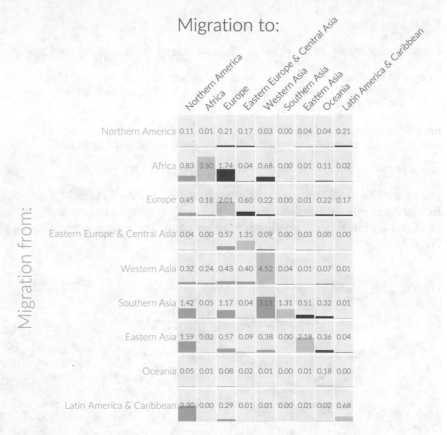

Migration 2010-2015

Migration to:

	Northern America	Africa	Europe	Eastern Europe & Central Asia	Western Asia	Southern Asia	Eastern Asia	Oceania	Latin America & Caribbean
Northern America	0.11	0.01	0.21	0.17	0.03	0.00	0.04	0.04	0.21
Africa	0.83	3.50	1.74	0.04	0.68	0.00	0.01	0.11	0.02
Europe	0.45	0.18	2.01	0.60	0.22	0.00	0.01	0.22	0.17
Eastern Europe & Central Asia	0.04	0.00	0.57	1.35	0.09	0.00	0.03	0.00	0.00
Western Asia	0.32	0.24	0.43	0.40	4.52	0.04	0.01	0.07	0.01
Southern Asia	1.42	0.05	1.17	0.04	3.11	1.31	0.51	0.32	0.01
Eastern Asia	1.59	0.02	0.57	0.09	0.38	0.00	2.18	0.36	0.04
Oceania	0.05	0.01	0.08	0.02	0.01	0.00	0.01	0.18	0.00
Latin America & Caribbean	2.20	0.00	0.29	0.01	0.01	0.00	0.01	0.02	0.68

Migration from:

All figures in millions. Data Source: https://github.com/cran/migest/tree/master/inst/vidwp

A variant that potentially shows the structure of the relationship matrix more clearly is a multiple bar chart.[10] Just like in the previous example, the relationship matrix is represented graphically. In this case, however, instead of different colour intensities, we select bar charts for the individual cells. The values are also written

---

[10]Here I follow a suggestion by Antony Unwin. See Unwin, Antony (2015): Graphical Data Analysis with R, Boca Raton, FL: CRC Press, p. 138f.

to the cells. Here, instead of an arrow, the direction of the migration is shown by headings for the rows and columns and by different colours.

```
pdf_file<-"pdf/networks_barchart_multiple.pdf"
cairo_pdf(bg="grey99", pdf_file,width=10,height=12)

library(igraph)
library(RColorBrewer)

# Import data and prepare chart

df0 <- read.csv("data/reg_flow.csv", stringsAsFactors=FALSE)
df1 <- read.csv("data/reg_plot.csv", stringsAsFactors=FALSE)

net <- graph_from_data_frame(d=df0, vertices=df1, directed=T)
netm <- get.adjacency(net, attr="flow", sparse=F)

maxvalue<-max(netm)
n<-nrow(netm)
m<-n
par(mfrow=c(n,m), omi=c(1,4,4,2), mai=c(0,0,0,0), family="Lato ⬎
      Light")

mycolor1<-rgb(255,0,210,maxColorValue=255)
mycolor2<-rgb(0,208,226,maxColorValue=255)

# Create chart

for(i in 1:n)
{
for(j in 1:m)
{
plot(1:1, xlim=c(0,1), ylim=c(0,1), type="n", axes=F)
if(i<j) mycolor<-mycolor1
if(i==j) mycolor<-"grey80"
if(i>j) mycolor<-mycolor2

if (i==1) text(0.5,1.2, df1$region[j], cex=2, xpd=NA, adj=0, srt⬎
      =45, col=mycolor1)
if (j==1) text(-0.1,0.5, df1$region[i], cex=2, xpd=NA, adj=1, ⬎
      col=mycolor2)

rect(0,0,1,1, col="grey95", border=NA)
rect(0,0,1,netm[i,j]/maxvalue, col=mycolor, border=NA)
text(0.5, 0.5, format(round(netm[i,j], 2), nsmall=2), cex=1.5, ⬎
      col="grey40")
}
}

# Titling

mtext("Migration to:", side=3, outer=T, cex=2.5, line=14, col=⬎
      mycolor1, adj=0)
```

```
mtext("Migration from:", side=2, outer=T, cex=2.5, line=25, col=⤸
      mycolor2, srt=90)
mtext("Migration 2010-2015", side=3, outer=T, cex=3, adj=1, at⤸
      =0.4, , line=22, col="grey50", family="Lato Black")
mtext("All figures in millions. Data Source: https://github.com/⤸
      cran/migest/tree/master/inst/vidwp",1,line=2.5, adj=1, at⤸
      =0.6, font=3, outer=T)
```

```
dev.off()
```

In the script, we again use get.adjacency() to create the relationship matrix. We save the maximum value under maxvalue so that all diagrams have the same scale. We define two colours for the upper and lower diagonals of the matrix. We then proceed through the relationship matrix by column and row in two loops, defining the corresponding colours for cells and labels in doing so. We draw the actual bar charts, which each consist of only one value, with the function rect(). This is followed by the usual labelling.

Chapter 7
Distributions

In addition to purely statistical presentation forms, such as histograms and box plots, visualisations of distributions include traditional presentation forms such as population pyramids or Lorenz curves. Obviously, not only populations can be shown in pyramids, and obviously an inequality, which is what a Lorenz curve commonly depicts, can be shown in a different form, too. We will therefore show alternative examples for both.

7.1 Histograms and Box Plots

7.1.1 Histograms Overlay

The figure shows the distribution of the women-to-men ratio in the German federal states of Brandenburg and Rhineland-Palatinate (Rheinland-Pfalz). With histograms, the choice of classification is vital. If we assume a normal or at least symmetrical distribution, then the mean should be plotted in the centre of the figure and the x-axis should span equal distances in both directions. If two distributions are stacked, then the mean of one of the distributions should be taken as reference point. The x-axis length then has to be chosen so it covers both distributions in their entirety. If the figures are stacked, then transparent colouring is most useful, as the overlap is then discernible as a third colour. Axis lines can be omitted. Since we are dealing with stable variables, graduation lines on the x-axis are not necessarily attached at the centre of the classes.

© Springer Nature Switzerland AG 2019
T. Rahlf, *Data Visualisation with R*, https://doi.org/10.1007/978-3-030-28444-2_7

Distribution of Women-Men-Ratio

Brandenburg and Rhineland-Palatinate

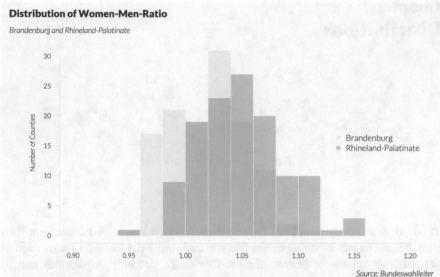

Source: Bundeswahlleiter

Data: see annex A, v_frauen_maenner.

```
pdf_file<-"pdf/histograms_overlay.pdf"
cairo_pdf(bg="grey98", pdf_file,width=11,height=7)

source("scripts/inc_datadesign_dbconnect.r")
par(omi=c(0.75,0.2,0.75,0.2),mai=c(0.25,1.25,0.25,0.25),family="↘
    Lato Light",las=1)

# Import data and prepare chart

sql<-"select * from v_women_men"
myDataset<-dbGetQuery(con,sql)
attach(myDataset)
myCol1<-rgb(191,239,255,180,maxColorValue=255)
myCol2<-rgb(255,0,210,80,maxColorValue=255)
brandenburg<-subset(myDataset,bundesland == 'Brandenburg')

# Create chart

hist(brandenburg$wm,col=myCol1,xlim=c(0.9,1.2),border=F,main='',↘
    xlab="Ratio",ylab="Number of Counties",axes=F)
axis(1,col=par("bg"),col.ticks="grey81",lwd.ticks=0.5,tck=-↘
    0.025)
axis(2,col=par("bg"),col.ticks="grey81",lwd.ticks=0.5,tck=-↘
    0.025)
rp<-subset(myDataset,bundesland == 'Rheinland-Pfalz')
hist(rp$wm,col=myCol2,xlim=c(0.9,1.2),border=F,add=T,main='')
legend("right",c("Brandenburg","Rhineland-Palatinate"),border=F,↘
    pch=15,col=c(myCol1,myCol2),bty="n",cex=1.25,xpd=T,ncol=1)
```

```
# Titling

mtext("Distribution of Women-Men-Ratio",3,line=1.8,adj=0,family=⟍
    "Lato Black",cex=1.5,outer=T)
mtext("Brandenburg and Rhineland-Palatinate",3,line=-0.2,adj=0,⟍
    font=3,cex=1.2,outer=T)
mtext("Source: Bundeswahlleiter",1,line=2,adj=1.0,font=3,cex⟍
    =1.2,outer=T)
dev.off()
```

In the script, data from a MySQL database are integrated by importing the view v_women_men. Since we want to stack two histograms, we have to define transparent colours; this will also make the overlap visible. We define two partial datasets: on the one hand all men/women ratios of the German federal state of Brandenburg, on the other hand all of the state Rhineland-Palatinate. A histogram is plotted for both databases using hist(), and the second one is added to the first using add=TRUE. It is important that we define the same display area on the x-axis for both figures. We do not explicitly specify the number of classes in the histograms; the default setting already provides an optimal result for our data.

7.1.2 Column Charts Coloured with ColorBrewer (Panel)

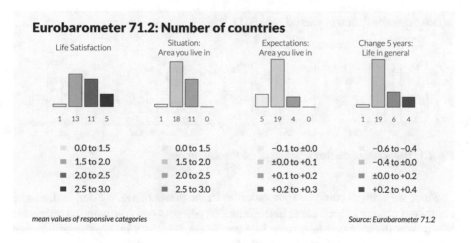

About the figure: the visualisation of frequency distributions can also be useful for rough classifications. In this example, we look at four variables from the Eurobarometer. The question we are asking ourselves is the distribution of four questions concerning life satisfaction. In 2009, the Eurobarometer 71.2 asked almost 30,000 people a series of questions relating to this topic. The possible responses to the two questions "On the whole, are you very satisfied, fairly satisfied, not very

satisfied or not at all satisfied with the life you lead?" and "How would you judge
the current situation in each of the following?" ranged between 1 ("Very satisfied"
or "very good") and 4 ("Not at all satisfied" or "very bad"). To the two questions
"What are your expectations for the next twelve months: will the next twelve months
be better, worse or the same, when it comes to…" and "Compared with 5 years ago,
would you say things have improved, gotten worse or stayed about the same when
it comes to …" the possible responses were 1 "better", 2 "worse" and 3 "same",
or 1 "improved", 2 "got worse" and 3 "stayed about the same". "Don't know" was
another optional response for all questions. If we are interested in the responses to
those questions in relation to countries, not individual persons, then we first have
to decide on an indicator. One option would be the proportion of people that have
chosen categories 1 and 2. Another would be to assume that the possible responses
correspond to points on a continuous scale and calculate averages. The latter is a
process commonly used for school grades, too. That means the arithmetic mean for
each of the 30 countries and each of the four questions is calculated first, and then
the distribution of those means is considered. One can arrive at useful classifications
by looking at the value distribution in histograms. The simplest way to do this is with

```
sql<-"select * from v_za4972_countries"
dataframe<-dbGetQuery(con,sql)
attach(dataframe)
par(mfcol=c(1,4))
for (i in 2:5) hist(dataframe[,i],main=names(dataframe[i]),xlab=\
      "")
```

which gives the following result (Fig. 7.1).

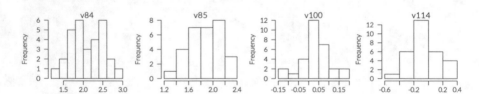

Fig. 7.1 Histogram of the four variables of the Eurobarometer

Since we want to compare four variables in the present case, we define the same
number of classes in each case. In the intended compact and comparative overview,
labels with the actual values would be confusing. This is why we use a different
form. First, we plot a bar chart instead of a histogram for each case. The difference
lies in the fact that the bars do not border each other. We choose the respective
frequency of the countries as labels for the bars, and for the colours a continuous
Brewer palette per diagram. This will also serve as our legend. Underneath the
individual figures, the range of values is listed. Since we are now able to plot them
one below the other, they are much easier to read. The classifications are the same

for the first two variables, and slightly different for the last two, which also do not measure the same things. We will get back to this figure in Sect. 10.3.4.

Data: See annex A, ZA4972: Eurobarometer 71.2 (May–Jun 2009).

```
pdf_file<-"pdf/columncharts_1x4.pdf"
cairo_pdf(bg="grey98", pdf_file,width=12,height=6)

library(RColorBrewer)
par(mfrow=c(2,4),omi=c(0.5,0.5,0.75,0.5),mai=c(0.5,0.5,0.5,0.5),\
    cex=1.1,family="Lato Light",las=1)

# Import data

source("scripts/inc_datadesign_dbconnect.r")
sql<-"select * from v_za4972_laender"
myDataset<-dbGetQuery(con,sql)
attach(myDataset)

# Create chart

myCuts<-c(0,1.5,2,2.5,3)
barplot(table(cut(v84,myCuts)),col=brewer.pal(4,"Reds"), ylim=c\
    (0,20), names.arg=table(cut(v84,myCuts)),axes=F,main="Life\
    Satisfaction")
myCuts<-c(0,1.5,2,2.5,3)
barplot(table(cut(v85,myCuts)),col=brewer.pal(4,"Greens"), ylim=\
    c(0,20),names.arg=table(cut(v85,myCuts)),axes=F,main="\
    Situation:\nArea you live in")
myCuts<-c(-0.1,0,0.1,0.2,0.3)
barplot(table(cut(v100,myCuts)),col=brewer.pal(4,"Blues"), ylim=\
    c(0,20),names.arg=table(cut(v100,myCuts)),axes=F,main="\
    Expectations:\nArea you live in")
myCuts<-c(-0.6,-0.4,0,0.2,0.4)
barplot(table(cut(v114,myCuts)),col=brewer.pal(4,"Purples"), \
    ylim=c(0,20),names.arg=table(cut(v114,myCuts)),axes=F,main\
    ="Change 5 years:\nLife in general")
par(family="Lato", cex=0.7)
plot.new()
bez<-c("0.0 to 1.5","1.5 to 2.0","2.0 to 2.5","2.5 to 3.0")
legend("center",bez,cex=2.05,border=F,bty="n",fill= brewer.pal\
    (4,"Reds"),y.intersp=1.3)
plot.new()
bez<-c("0.0 to 1.5","1.5 to 2.0","2.0 to 2.5","2.5 to 3.0")
legend("center",bez,cex=2.05,border=F,bty="n",fill= brewer.pal\
    (4,"Greens"),y.intersp=1.3)
plot.new()
bez<-c("−0.1 to ±0.0","±0.0 to +0.1","+0.1 to +0.2","+0.2 to \
    +0.3")
legend("center",bez,cex=2.05,border=F,bty="n",fill= brewer.pal\
    (4,"Blues"),y.intersp=1.3)
plot.new()
bez<-c("−0.6 to −0.4","−0.4 to ±0.0","±0.0 to +0.2","+0.2 to \
    +0.4")
```

```
legend("center",bez,cex=2.05,border=F,bty="n",fill= brewer.pal↘
      (4,"Purples"),y.intersp=1.3)

# Titling

mtext("Eurobarometer 71.2: Number of countries",3,line=1,adj=0,↘
      cex=2.15,family="Lato Black",outer=T)
mtext("mean values of responsive categories",1,line=1,adj=0,cex↘
      =1.25,font=3,outer=T)
mtext("Source: Eurobarometer 71.2",1,line=1,adj=1.0,cex=1.25,↘
      font=3,outer=T)
dev.off()
```

In the script, we first use mfrow=c(2,4) to split the figure into eight areas of equal size, two rows and four columns. The view v_za_4972 already contains the necessary aggregation of the variable means for the individual countries. This is followed by four calls of the barplot() function. To this end, we first define the previously determined classifications within the vector limits. With these classifications and the cut() function, the variables are organised into the respective classes; the class frequencies are simultaneously established using the table() function. The table call is utilised a second time to use the frequencies as axis labels. Four Brewer palettes are employed for the colours. Legends are then plotted into separate windows. Since legend() is a low-level function, we first need one plot.new() call, respectively, to create an empty figure window in which the legend can be written. The structure of the legend corresponds to the previous examples. With y.intersp=1.3, the line spacing is slightly increased to improve readability. Since we are using a panel illustration with multiple figures, we have to write headings and captions into the outer margin using outer=T. Please note that the 'minus' symbol is used for the labels, and not the dash shown on the keyboard.

7.1.3   Histograms (Panel)

The main question asked in the present and following two figures is what presentation options there are for distributions of a variable that are differentiated by a relatively large number of attributes of a factor or another categorical variable.

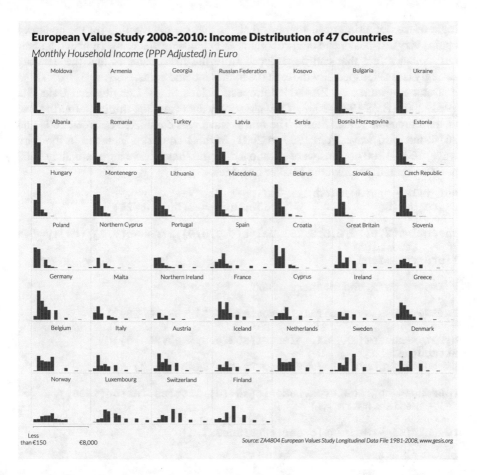

European Value Study 2008-2010: Income Distribution of 47 Countries

Monthly Household Income (PPP Adjusted) in Euro

About the figure: The figure shows the distribution of monthly net household income of respondents in selected countries. To allow a direct comparison between the countries, purchasing power parity (PPP) weighted values were used. Please note that the distributions are based on the results of a survey. Especially with statements concerning income it should be kept in mind that, without doubt, voluntary disclosures will not give a perfect representation. However, income values are usually a good reference point and are used for the analysis of socioeconomic questions in which an impact of the amount of income is suspected. The figure shows the distribution of the income disclosures for 46 countries. The same y-axis scaling was used for all countries, so frequencies are directly comparable. Similarly, x-axis span and class width are each identical. Data are sorted in descending order of their median, so that countries judged "poorest" by these criteria come first. The logic of the definition of histograms dictates that the distribution for the countries begins very left-heavy and then continuously extends further to the right and flattens out. Another fact that can be deduced from the histograms is that the data are strongly classed, especially those in the higher income areas.

Data: See annex A, ZA4804: European Values Study Longitudinal Data File 1981–2008 (EVS 1981–2008). Data are survey data from the third and fourth wave of the European Values Study. The fourth wave was conducted between 2008 and 2010, the third wave from 1999 to 2001. In most countries included in the third wave, 1000–1200 persons were questioned, in the fourth wave approximately 1500 persons, sometimes fewer in smaller countries.

```
pdf_file<-"pdf/histograms_7x7.pdf"
cairo_pdf(bg="grey98", pdf_file,width=25,height=25)

par(omi=c(2,0.75,2.5,0.25),mai=c(0,0,0,0),mfrow=c(7,7),family="↘
    Lato Light")
library(memisc)

# Import data and prepare chart

ZA4804<-spss.system.file("mydata/ZA4804_v3-0-0.sav")

myData<-subset(ZA4804,select=c(s002evs,s003,x047d))
attach(myData)
t<-subset(myData,x047d>0 & s002evs=="2008-2010")

tMedians<-aggregate(as.numeric(x047d),list(as.factor(s003)),↘
    median,na.rm =T)

tCountries<-tMedians[order(tMedians$x),1]

# Create chart

attach(t)
for (i in 1:(length(tCountries)-2))
{
Country<-subset(t,s003==tCountries[i])
hist(Country$x047d,main="",axes=F,xlab="",ylab="",xlim=c(0,8),↘
    ylim=c(0,1000),border="white",col="red",breaks=seq(from=-↘
    2,to=16,by=0.5))
text(4,900,tCountries[i],cex=3.0)
box(lty='dotdash',col='grey')
if (i==43) axis(1,cex.axis=3,at=c(0,8),labels=c("Less\nthan €150↘
    ","€8,000"),mgp=c(0,8,1))
}

# Titling

mtext("European Value Study 2008-2010: Income Distribution of 47↘
    Countries",3,line=10,adj=0,cex=3.8,family="Lato Black",↘
    outer=T)
```

```
mtext("Monthly Household Income (PPP Adjusted) in Euro",3,line↘
      =3,adj=0,cex=3.5,font=3,outer=T)
mtext("Source: ZA4804 European Values Study Longitudinal Data ↘
      File 1981-2008, www.gesis.org",1,line=7,adj=1.0,cex=2,font↘
      =3,outer=T)
dev.off()
```

In the script, we first define a very large window of 25 by 25 inches, a little larger than the short side of the A1 format. Data are read from an SPSS file using Martin Elff's memisc package, and limited to the required data. Here,

- s002evs is the EVS wave
- s003 the country
- x047d the purchasing power parity-weighted household income

After data import, we create a list of all featured countries and their respective medians. To do this, we use the aggregate() function. Aggregation functions have to be transferred as a list. The country list is sorted in ascending order by the median. In a loop, we now draw a histogram for each country, always using the same x- and y-axis areas and class widths. Once this is done, the heading is written above the histograms using text(), and a box of "dotdash" type is created. An x-axis label is attached in the lower left chart (the 43rd).

7.1.4   Box Plots for Groups: Sorted in Descending Order

Histograms are without doubt a very suitable form to present distributions. With limited space, the box plot is an alternative that can also explicitly plot a series of key quantities of a distribution. Here, the interquartile distance, the span of the first and third quartile, is depicted as a bar (a "box") in which the median (the second quartile) is added as a marker. In an extended variant, lines are joined on the left and right, so that it becomes a box-and-whisker plot. There are different definitions of these lines: the most frequently used is that their length is defined by the points that are less than 1.5 interquartile distances from the first or third quartile. This definition is also the default setting for R's boxplot function.

Income Distribution 2008-2010

European Values Study

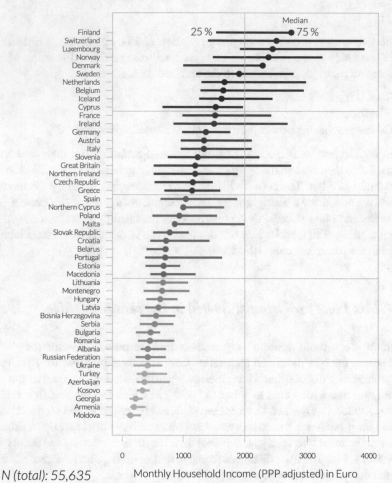

N (total): 55,635   Monthly Household Income (PPP adjusted) in Euro

Source: ZA4804 European Values Study Longitudinal Data File 1981-2008, www.gesis.org

The figure again shows a comparative income distribution in different countries, as in Sect. 7.1.3, but in the compact form of simplified box plots. In contrast to the complete box plot version, only the median and the interquartile distance (as span) are shown. Since we do not want to group anything, a continuous Brewer gamut in steps of ten is used as colour. Looking at the figure, we can derive that income disclosures from countries with a higher monthly net household income sometimes show a larger span than disclosures from countries with a lower income (each PPP weighted). However, when interpreting the results, it should be kept in mind that we present the quartile distance, and not the entire income. This means the figure shows that area around the median in which the inner 50% of the characteristic distribution lies. The selected illustration form is similar to the "dot charts" that we got to know in Sect. 6.1.11.

Data: see annex A, ZA4804: European Values Study Longitudinal Data File 1981–2008 (EVS 1981–2008).

```
pdf_file<-"pdf/boxplots_multiple.pdf"
cairo_pdf(bg="grey98", pdf_file,width=7,height=9)

par(omi=c(0.35,0.25,0.75,0.75),mai=c(0.95,1.75,0.25,0),family="\
      Lato Light",las=1)
library(RColorBrewer)
library(memisc)

# Import data and prepare chart

myDataFile<-"myData/ZA4804_v2-0-0.sav"
ZA4804<-spss.system.file(myDataFile)
myData<-subset(ZA4804,select=c(s002evs,s003,x047d))
attach(myData)
x<-subset(myData,x047d>0 & s002evs=="2008-2010")
attach(x)

tM<-aggregate(as.numeric(x047d),list(as.factor(s003)),median,na.\
      rm =T)

s003f<-factor(s003,levels=tM[order(tM$x),1])

myC1<-brewer.pal(6, "PuRd")[2]
myC2<-brewer.pal(6, "PuRd")[3]
myC3<-brewer.pal(6, "PuRd")[4]
myC4<-brewer.pal(6, "PuRd")[5]
myC5<-brewer.pal(6, "PuRd")[6]
myColour<-c(rep(myC1, 7), rep(myC2, 10), rep(myC3, 10), rep(myC4\
      , 10), rep(myC5,10))
par(fg="grey75")

# Create chart and other elements

boxplot(1000*x047d ~ s003f,horizontal=T,ylim=c(0,4000),border=NA\
      ,boxwex=0.25,las=1,col=myColour,outline=F,cex.axis=0.7)

points(sort(1000*tM$x,decreasing=T),length(unique(s003)):1,pch\
      =19,cex=1.15,col=rev(myColour))

abline(v=2000)
abline(h=seq(7.5,37.5,by=10))

par(fg="black")
mtext("25 %",3,at=1300,line=-2)
mtext("75 %",3,at=3000,line=-2)
```

```
mtext("Median",3,at=2800,line=-1, cex=0.75)

mtext("Monthly Household Income (PPP adjusted) in Euro",1,adj↘
      =0.5,line=2.5)

# Titling

mtext("Income Distribution 2008-2010",3,line=1.6,adj=0,cex=1.8,↘
      family="Lato Black",outer=T)
mtext("European Values Study",3,line=-0.2,adj=0,cex=1.5,font=3,↘
      outer=T)
mtext("Source: ZA4804 European Values Study Longitudinal Data ↘
      File 1981-2008, www.gesis.org",1,line=0,adj=1.0,cex=0.95,↘
      font=3,outer=T)
mtext("N (total): 55,635",1,line=-2,adj=0,cex=1.25,font=3,outer=↘
      T)
dev.off()
```

In the script, we compile a list of all featured countries and their respective medians as in the previous example. We want to aggregate by variable s003, with representation by the medians of x047d, sorted in descending order. To this end, we create the variable s003f as a factor with the country names from the median list and organised by levels (categories). To achieve the descending order by the median, we use a procedure that was suggested by Jim Porzak in an R mailing list.

That done, the box plots can be drawn. R expects a formula notation for this. Since the income is saved in thousands in the data frame, we multiply it by 1000 for a correct x-axis label. Using the boxwex parameter, the box plots are drawn narrower than usual. outline=F ensures that no rogue values are plotted. This means the representation of the box plot is limited to the span of the two quartiles. We choose a Brewer palette with six values as colour, and we use the second to sixth (the first is too light for our purposes). This means each of ten countries, at the end 7, are coloured. For ease of orientation, we also plot a vertical auxiliary line in the middle, and horizontal guide lines every ten rows.

The necessary labels come at the end of the script.

7.1.5   Box Plots for Groups: Sorted in Descending Order, Comparison of Two Polls

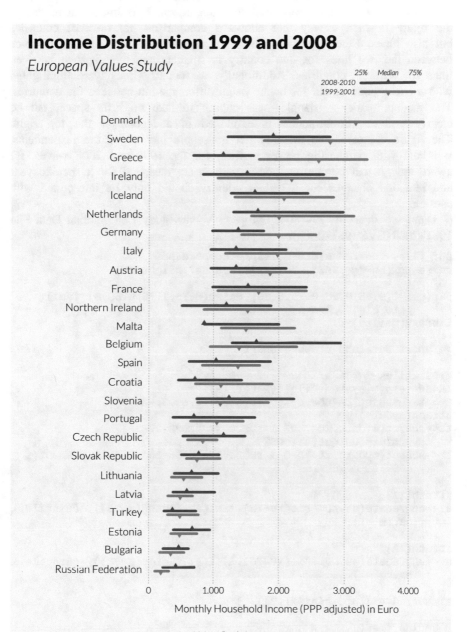

Income Distribution 1999 and 2008

European Values Study

Source: ZA4804 European Values Study Longitudinal Data File 1981-2008, www.gesis.org

About the figure: in contrast to the previous example, an additional temporal comparison is included here: How has the income distribution in the individual countries changed when compared with an earlier point in time? The survey wave from 1999 to 2001 was used for comparison. It is important to design the figure in such a way that allows a comparison not only of countries but also times. Here, quartile spans were represented such that the distance between the two lines for one country is smaller than the distance between lines for different countries. Additionally, the country lines were highlighted with a background colour for better visual differentiation between the countries. The median mark—a triangle—was chosen to take up little space, but be clearly visible. The structure is explained in a legend at the top right. The figure compares the monthly net household income of the respondents, weighted with purchasing power parities, for the 1999 and 2008 waves. Be aware though that purchasing power parities are designed for a specific year and temporal changes for a certain country should only be interpreted with caution.

Data: see annex A, ZA4804: European Values Study Longitudinal Data File 1981–2008 (EVS 1981–2008).

```
pdf_file<-"pdf/boxplots_multiple_compare.pdf"
cairo_pdf(bg="grey98", pdf_file,width=7,height=10)

par(omi=c(0.35,0.25,0.75,0.25),mai=c(0.75,1.75,0.55,0),family="\
    Lato Light",las=1)
library(memisc)

# Import data and prepare chart

myDatafile<-"myData/ZA4804_v2-0-0.sav"
ZA4804<-spss.system.file(myDatafile)
myData<-subset(ZA4804,select=c(s002evs,s003,x047d))
attach(myData)
t1<-subset(myData,x047d>0 & s002evs=="1999-2001")
t1_countries<-unique(t1$s003)
t2<-subset(myData,x047d>0 & s002evs=="2008-2010" & is.element(\
    s003,t1_countries))

attach(t1)
a1<-aggregate(as.numeric(x047d),list(as.factor(s003)), quantile,\
    na.rm =T)

attach(t2)
a2<-aggregate(as.numeric(x047d),list(as.factor(s003)), quantile,\
    na.rm =T)

a1.sorted<-a1[order(a1$x[,3]), ]

# Define chart

plot(1:1, type="n",xlim=c(0,4.25),ylim=c(0.5,51.5),axes=F,xlab="\
    ",ylab="",yaxs="i")
```

```
# Other elements

abline(v=c(0,1,2,3,4), lty="dotted", col="grey70")

myC1<-"gray55"
myC2<-"deeppink"
myBckgrnd<-rgb(191,239,255,70,maxColorValue=255)

for (i in 1:25)
{
rect(0,2*i-0.9,4.25,2*i+0.9, col=myBckgrnd, border=NA)
segments(a1.sorted$x[i,2],2*i-0.2,a1.sorted$x[i,4],2*i-0.2,lwd↘
     =4, col=myC1)
segments(a2$x[a2$Group.1==a1.sorted$Group.1[i],2],2*i+0.2,a2$x[↘
     a2$Group.1==a1.sorted$Group.1[i],4],2*i+0.2, col=myC2, lwd↘
     =4)
par(family="Symbola")
text(a1.sorted$x[i,3],2*i-0.4,"▼",col=myC1, cex=0.8)
text(a2$x[a2$Group.1==a1.sorted$Group.1[i],3],2*i+0.4,"▲",col=↘
     myC2,cex=0.8)

par(family="Lato Light")
text(-0.1,2*i,a1.sorted$Group.1[i],adj=1,xpd=T)
}
mtext(c(0, "1.000", "2.000", "3.000", "4.000"),1,at=c(0:4),cex↘
     =0.85)
mtext("Monthly Household Income (PPP adjusted) in Euro",1,adj↘
     =0.5,line=1.5)

# Titling

mtext("Income Distribution 1999 and 2008",3,line=1.6,adj=0,cex↘
     =1.8,family="Lato Black",outer=T)
mtext("European Values Study",3,line=-0.2,adj=0,cex=1.5,font=3,↘
     outer=T)
mtext("Source: ZA4804 European Values Study Longitudinal Data ↘
     File 1981-2008, www.gesis.org",1,linc=0,adj=1.0,cex=0.95,↘
     font=3,outer=T)

# Legend

par(new=T, omi=c(0,0,0,0), mai=c(8.5,5.5,0.5,0.55))
plot(0:1, xlim=c(0,1),ylim=c(0,1),type="n",axes=F,xlab="",ylab="↘
     ")
segments(0,0.42,1,0.42,col=myC1,xpd=T,lwd=4)
segments(0,0.57,1,0.58,col=myC2,xpd=T,lwd=4)
text(0,0.75,"25%",adj=0.5,cex=0.7,xpd=T,font=3)
text(1,0.75,"75%",adj=0.5,cex=0.7,xpd=T,font=3)
text(0.5, 0.75,"Median",adj=0.5,cex=0.7,font=3)
text(-0.1, 0.42,"1999-2001",adj=1,cex=0.65,xpd=T,font=3)
text(-0.1, 0.58,"2008-2010",adj=1,cex=0.65,xpd=T,font=3)
par(family="Symbola")
```

```
text(0.5,0.6,"▲",col=myC2,cex=0.8)
text(0.5,0.39,"▼",col=myC1,cex=0.8)
dev.off()
```

The script begins with the loading of the data as in the previous example. Here, we create two partial data frames. The first, t1, comprises the survey data from 1999 to 2001, the second, t2, those from 2008 to 2010. With is.element(s003, t1_countries), only values from countries that also feature in the first data frame are chosen for the second. The boxplot() function used in the previous example does not provide a parameter panel.first that allows for plotting elements of the figure, such as a coloured area or a grid, first. For this reason, we use a different approach in this example, drawing the box plots with the low-level functions segments() and points(). We first need aggregated data. This is easiest to achieve with the aggregate() function and the aggregation function quantile. The result of the first aggregation looks like this (the first five rows):

```
> a1[1:5, ]
          Group.1      x.0%       x.25%      x.50%      x.75%      x.100%
1         Austria 0.3726296 1.2669406 1.8631480 2.5338812 5.5149180
2         Belgium 0.6957128 0.9486993 1.3914256 2.2768783 3.2888241
3        Bulgaria 0.1688909 0.1688909 0.3377817 0.5477541 1.0133452
4         Croatia 0.3280854 0.7217878 1.1154903 1.9028952 3.4777050
5 Czech Republic 0.5065086 0.6014790 0.8389049 1.0605024 2.0102061
```

The output of the function is therefore a variable group.1 and a variable x that contains the minimum, the first quartile, the median, the second quartile, and the maximum in five columns, respectively. We create such an aggregated data frame for both survey periods, calling these a1 and a2. In the next step, we use a1 to create a sorted data frame a1.sorted, sorting by the third column of x (the median). We now define a common plot() that extends to the range of values of all quartiles in its x-axis. In the y-direction, we need twice the number of featured countries (25) and a "margin allowance" at the top and bottom, so that the y-axis range spans 0.5–51.5. To keep R from adding a standard 4%, yaxs="i" has to be set. This is followed by five vertical dotted guide lines and the specification of colours for the box plots and the background. Plotting of the actual data is done within a loop. First, a light-blue rectangle is laid behind each row. On this, the quartile spans are plotted using segments(). The first segment is processed row by row and contains the range of the value from the second column to the value of the fourth column of x within the data frame a1.sorted. The y-position is the numerator, shifted down by 0.2, respectively. The second segment has an almost identical structure. The difference is that the second data frame a2 is not processed row by row. Instead, the row, whose country corresponds to the current country of a1.sorted is picked out. Additionally, the y-position is shifted up by 0.2. Now we are only missing a marker for the median. Triangles with the tip pointing up and down are a good option. The plot symbol of filled triangles is available in R, but only with the tip pointing up. This is why we use the text() function for the image, and the Unicode characters "BLACK UP-POINTING TRIANGLE" and "BLACK DOWN-POINTING TRIANGLE" from the Unicode block "Geometric Shapes". Since the Lato font does not contain these characters, we choose the Symbola font. We shift the position of the triangles up or down by 0.4, so they protrude from the lines. Finally, we change the font back to Lato Light and then, in the last command within the loop, set the respective country's name left of the figure. After the loop come axes labels, title, and source details. At the end, we have to plot a legend at the top right; this is a bit more complex than before. We use an individual plot() call after calling par() with new=T and setting the margins so that the figure is plotted in the top left corner.

7.2 (Population) Pyramids

Population pyramids are a special form of distribution. They are very descriptive figures that are frequently found in publications, but are apparently not part of the standard repertoire of the many graphic software packages or spreadsheet analyses. The Web does offer a series of "workarounds", but with almost all of these, the misappropriation of standard figures is obviousi What constitutes a population pyramid? Generally, such figures illustrate the age distribution of a population differentiated by gender as bars arranged back-to-back. The expression "pyramid" is a relic from times in which populations presented in this way did actually resemble a pyramid: many young and few old people. However, this is no longer the case in advanced industrial nations. Here, the younger generations are much less represented than the older ones. Therefore, the distribution of modern populations resembles a mushroom rather than a pyramid. However, the term "population pyramid" is still valid today. With the help of such a population pyramid, a population's structure can be presented very clearly. With the appropriate illustration, its general structure, war-related losses of individual years, surplus of women or men, or portions of certain population groups are evident at a glance. The following examples show different aspects that can be described with population pyramids. The first example (Sect. 7.2.1) shows a variant in which different age classes are colour-coded; Sect. 7.2.2 shows two pyramids for comparison, with the surpluses of women and men mirrored on the opposing sides. In contrast, in Sect. 7.2.3, proportions of the population are suitably highlighted. The last two examples show the use of the pyramid for summarised data and for an estimation scale—since they can obviously also be useful for the representation of data other than age classes. In R, population pyramids can be created with individual packages, for example with the pyramid package by Minato Nakazawa. More flexible options are offered by the pyramid.plot() function from the plotrix package by Jim Lemon. The pyramids introduced here use the barplot() function, which—when extended with a few low-level functions—yields appealing results.

7.2.1   Pyramid with Multiple Colours

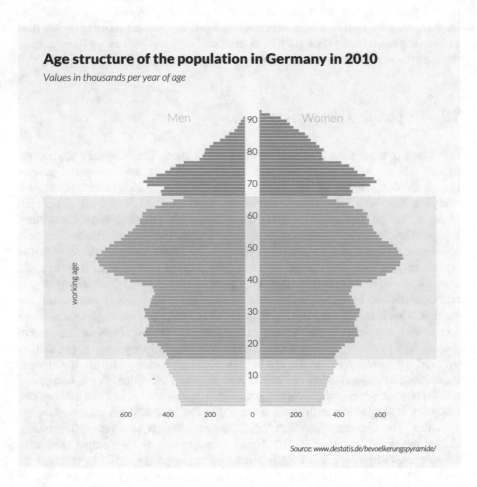

Age structure of the population in Germany in 2010
Values in thousands per year of age

Source: www.destatis.de/bevoelkerungspyramide/

About the figure: in the printed literature and on the Internet, there are numerous variants of population pyramids. In the present example, the age distribution is organised into three divisions to show the scope of the potentially employable population. To this end, the portion of people of working age (according to official definition, this includes everyone aged between 15 and 65 years) is shown in a different colour from those younger or older. The area of working age is additionally highlighted. In the present and the following figures, we omit axes and use only the age distribution and the frequencies as labels.

Data: the files men.text and women.txt were manually created. Data derive from the source text of the following website: https://www.destatis.de/bevoelkerungspyramide [Website in German]. In contrast to the files regularly offered by destatis, age classes are here continued to 99.

```
pdf_file<-"pdf/pyramids_multicoloured.pdf"
cairo_pdf(bg="grey98", pdf_file,width=9,height=9)

par(mai=c(0.5,1,0.5,0.5),omi=c(0.5,0.5,0.5,0.5),family="Lato ⬡
      Light",las=1)

# Import data and prepare chart

myWomen<-read.csv("myData/women.txt",header =F,sep=",")
for(i in 1:111) colnames(myWomen)[i]<-paste("x",i+1949,sep="")
myMen<-read.csv("myData/men.txt",header =F,sep=",")
for(i in 1:111) colnames(myMen)[i]<-paste("x",i+1949,sep="")

right<-myWomen$x2010
left<-myMen$x2010

myColour_right<-c(rep(rgb(210,210,210,maxColorValue=255),15),rep⬡
      (rgb(144,157,172,maxColorValue=255),50),rep(rgb⬡
      (225,152,105,maxColorValue=255),length(right)-65))
myColour_left<-myColour_right

# Create chart and other elements

barplot(right,axes=F,horiz=T,axis.lty=0,border=NA,col=myColour_⬡
      right,xlim=c(-750,750))
barplot(-left,axes=F,horiz=T,axis.lty=0,border=NA,col=myColour_⬡
      left,xlim=c(-750,750),add=T)

abline(v=0,lwd=28,col=par("bg"))
for (i in seq(10,90,by=10)) text(0,i+i*0.2,i,cex=1.1)
mtext(abs(seq(-600,600,by=200)),at=seq(-600,600,by=200),1,line=-⬡
      1,cex=0.80)

rect(-1000,15+15*0.2,1000,66+66*0.2,xpd=T,col=rgb⬡
      (210,210,210,90,maxColorValue=255), border=NA)

mtext("working age",2,line=1.5,las=3,adj=0.38)
mtext("Men",3,line=-5,adj=0.25,cex=1.5,col="grey")
mtext("Women",3,line=-5,adj=0.75,cex=1.5,col="grey")

# Titling

mtext("Age structure of the population in Germany in 2010",3,⬡
      line=-1.5,adj=0,cex=1.75,family="Lato Black",outer=T)
mtext("Values in thousands per year of age",3,line=-3.25,adj=0,⬡
      cex=1.25,font=3,outer=T)
mtext("Source: www.destatis.de/bevoelkerungspyramide/",1,line=0,⬡
      adj=1.0,cex=0.95,font=3,outer=T)
dev.off()
```

The script sorts the data in such a way that the years make up columns, and the age classes, rows. Once the data are imported, we use colnames() to create "talking" variable names. This means we can access the age distribution from

2010 with just x2010. The distribution for men and women is saved in the vectors right and left, respectively. For the colours, a vector is specified that contains a different hue for the first 15 values (years), the middle 50, and the remaining values.

Then we draw a bar chart for the right side using barplot(). However, we are already considering the left side as well by defining an x-axis scale from −750 to 750. The second barplot() call uses the negative values saved in left. Data are added to the existing diagram using add=T. Axes and axis labels are hidden. Now, abline() is used to put a vertical "corridor" in the background colour down the middle of the two bar charts; then seq(10,90,by=10) is used within a loop to plot the respective age in tens at the position i+i*0.2 (bar width plus width of the bar distance) every 10 years. The x-axis label is attached with mtext(), once more running through a sequence ranging from −600 to 600, with prefixes hidden in the negative range. The middle area, the one depicting the working-age population, is additionally indicated with a transparent grey rectangle.

7.2.2   Pyramids: Emphasis on the Outer Areas (Panel)

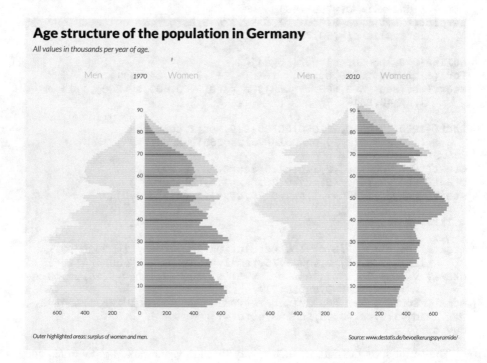

Age structure of the population in Germany

All values in thousands per year of age.

Outer highlighted areas: surplus of women and men.   *Source: www.destatis.de/bevoelkerungspyramide/*

About the figure: In this example, two pyramids are compared: on the one hand, the 1970 pyramid, on the other hand the one showing the age structure in

2010. Here, we add an effect that can sometimes be found in the literature: the surpluses of women or men are "mirrored" on the respective opposing side. So, when looking at e.g. the left side of the age structure of 1970, it can be seen that there was a slight surplus of men in the lower age classes, but a significant surplus of women in the age classes from the mid-50s onwards. The explanation for this is obvious: this is where the "war years" begin. In the 2010 pyramid, these years have "moved up". The war years are now 70 or older. The absence of the younger generation is also clearly visible. Here, we are quite clearly approaching a mushroom.

   Data: the files men.text and women.txt were manually created. Data derive from the source text of the following website: https://www.destatis.de/ bevoelkerungspyramide [website in German]. In contrast to the files regularly offered by destatis, age classes are here continued to 99.

```
pdf_file<-"pdf/pyramids_finely_outside_1x2.pdf"
cairo_pdf(bg="grey98", pdf_file,width=12,height=9)

#par(mai=c(0.2,0.1,0.2,0.1),omi=c(0.75,0.2,0.85,0.2),mfcol=c\
    (1,2),cex=0.75,
par(mai=c(0.2,0.1,0.8,0.1),omi=c(0.75,0.2,0.85,0.2),mfcol=c(1,2)\
    ,cex=0.75,family="Lato Light",las=1)

# Import data and prepare chart

myWomen<-read.csv("myData/women.txt",header =F,sep=",")
for(i in 1:111) colnames(myWomen)[i]<-paste("x",i+1949,sep="")
myMen<-read.csv("myData/men.txt",header =F,sep=",")
for(i in 1:111) colnames(myMen)[i]<-paste("x",i+1949,sep="")

myWSurplus<-(myWomen$x1970-myMen$x1970)
myWSurplus[myWSurplus < 0]<-0
myMSurplus<-(myMen$x1970-myWomen$x1970)
myMSurplus[myMSurplus < 0]<-0

right<-data.frame(myWomen$x1970-myWSurplus,myWSurplus)
left<-data.frame(myMen$x1970-myMSurplus,myMSurplus)

source("scripts/inc_pyramid.r")
mtext("1970",3,line=0,adj=0.5,cex=1,font=3)

myWSurplus<-(myWomen$x2010-myMen$x2010)
myWSurplus[myWSurplus < 0]<-0
myMSurplus<-(myMen$x2010-myWomen$x2010)
myMSurplus[myMSurplus < 0]<-0

right<-data.frame(myWomen$x2010-myWSurplus,myWSurplus)
left<-data.frame(myMen$x2010-myMSurplus,myMSurplus)

source("scripts/inc_pyramid.r")
mtext("2010",3,line=0,adj=0.5,cex=1,font=3)
```

```
# Titling

mtext("Age structure of the population in Germany",3,line=2,adj↘
      =0,cex=2.25,family="Lato Black",outer=T)
mtext("All values in thousands per year of age.",3,line=-0.5,adj↘
      =0,cex=1.25,font=3,outer=T)
mtext("Source: www.destatis.de/bevoelkerungspyramide/",1,line=2,↘
      adj=1.0,cex=0.95,font=3,outer=T)
mtext("Outer highlighted areas: surplus of women and men.",1,↘
      line=2,adj=0,cex=0.95,font=3,outer=T)
dev.off()
```

and

```
# inc_pyramid.r

myMark_right<-right
myMark_right[!(rownames(myMark_right) %in% seq(10,90,by=10)),]<-↘
      0
myMark_left<-left
myMark_left[!(rownames(myMark_left) %in% seq(10,90,by=10)),]<-0

myColour_right_outer<-rgb(255,0,210,50,maxColorValue=255)
myColour_right_inner<-rgb(255,0,210,120,maxColorValue=255)
myColour_left_inner<-rgb(191,239,255,220,maxColorValue=255)
myColour_left_outer<-rgb(191,239,255,100,maxColorValue=255)

myColour_right<-c(myColour_right_inner,myColour_right_outer)
myColour_left<-c(myColour_left_inner,myColour_left_outer)

myB1<-barplot(t(right),axes=F,horiz=T,axis.lty=0,border=NA,col=↘
      myColour_right,xlim=c(-750,750))
myB2<-barplot(-t(left),axes=F,horiz=T,axis.lty=0,border=NA,col=↘
      myColour_left,xlim=c(-750,750),add=T)

barplot(t(myMark_right),axes=F,horiz=T,axis.lty=0,border=NA,col=↘
      myColour_right,xlim=c(-750,750),add=T)
barplot(-t(myMark_left),axes=F,horiz=T,axis.lty=0,border=NA,col=↘
      myColour_left,xlim=c(-750,750),add=T)

abline(v=0,lwd=25,col=par("bg"))
mtext(abs(seq(-600,600,by=200)),at=seq(-600,600,by=200),1,line=-↘
      1,cex=0.80)
for (i in seq(10,90,by=10)) text(0,i+i*0.2,i,cex=1.1)

mtext("Men",3,line=0,adj=0.25,cex=1.5,col="grey")
mtext("Women",3,line=0,adj=0.75,cex=1.5,col="grey")
```

The script begins by splitting the illustration window into two equal parts using mfcol(). Data are imported as in the previous example. However, this time, we create a stacked bar chart. To this end, we first determine the surpluses of men and women by calculating the differences of men and women and only keeping the positive values, since we only want to add positive variations to the respective other side. right and left are then no longer vectors as in the previous example, but data frames with two columns each, containing the original values and the surpluses as columns. Since the bars are "stacked", that is, added, we have to subtract the surpluses from the original series. The plotting of the pyramids is done in a separate file inc_pyramid.r that is integrated with the source. We have transferred this part, because we need it twice in this script and another two times in the next example. In the integrated file, the two objects right and left are first duplicated and the copies are set to zero with the exception of the years 10, 20, 30, etc. These data make up a "scaffold" that can later serve as orientation lines to be put over the data. We proceed like this: my_Mark_right[condition,] sets all rows that fulfil the condition to zero. Since there is an empty space between the comma and the closing square class, all columns are carried over. The condition is !(rownames(my_Mark_right) %in% seq(10,90,by=10)). This means: row numbers of my_Mark_right should not (exclamation mark) be included in the sequence 10, 20, 30, ... Then the colours for right and left, both inside (inner) and outside (outer), are defined and combined. Now follow four barplot() calls. The first two match the previous example (definition of the figure by the right part, addition of the left part to the existing one). Then come another two calls that put the "scaffold" over the existing figure. This is done in the same colours. Since the colours are transparent, the colouring becomes more intense in the years that are a multiple of ten, which results in internal orientation lines with every tenth value. The remainder of the script matches the procedure of the previous example: drawing of a wide centre strip and individual labels. After the call of the integrated file, the section heading in the main script is set to "1970" and the procedure is repeated for 2010. As usual, headings and captions come last.

7.2.3   Pyramids: Emphasis on the Inner Areas (Panel)

About the figure: another frequent variant—albeit not as frequent as surplus pyramids—is the depiction of proportions. There are proportions of e.g. foreigners or, as in our example, the proportion of married men and women. As is clear from the figure, the age of marriage for women begins in their early 20s in 1970, but only in their mid-20s in 2010; for men, it is their mid- and late 20s. This difference is particularly obvious in old age: here, the proportion of married men is markedly higher than that of married women—a sure sign that men are marrying younger women, but not the other way around.

Age structure and married proportion of the population in Germany
All values in thousands per year of age

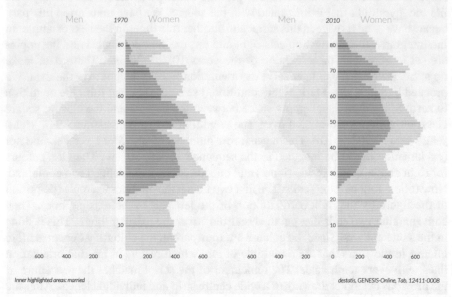

Inner highlighted areas: married   destatis, GENESIS-Online, Tab. 12411-0008

Data: Go to http://www-genesis.destatis.de and enter "12411-0008" as a search string. The result is the German population table "Bevölkerung: Deutschland, Stichtag, Altersjahre, Nationalität, Geschlecht/Familienstand" (Population: Germany, reference date, age in years, nationality, sex/marital status), differentiated according to the stated dimensions. Without further selection/limitation, you will get a table differentiated by Germans and foreigners and within nationality by sex, but only for the year 2010. Nationality, sex, and the age classes are provided in such a way that you will receive all attributes if you do not make a specific selection. We need two tables: one for the year 2010 and one for the year 1970. 2010 is the default selection, so that you can immediately click on Werteabruf (value retrieval). The table will be displayed, and with the first icon, you can save it in MS Excel format. Save the file as 12411-0007-2010.xls in your working directory. In the breadcrumb navigation above the row Ergebnis—12411-0007 (Result—12411-0007), click on Tabellenaufbau (table structure) to return to the selection, and then click on Zeit auswählen (select time). Place a tick mark in the last row at 31.12.1970 (31/12/1970) and click on übernehmen (apply) at the bottom, then again on Werteabruf. You will receive the respective table for the reference day 31/12/1970 and can now save that as 12411-0007-1970.xls in your working directory. Before data import, we deleted the headers and saved the files under a new name (with "reformatted" appended). These are:

```
X1   Male unmarried
X2   Male married
X3   Male widowed
```

```
X4   Male divorced
X5   Female unmarried
X6   Female married
X7   Female widowed
X8   Female divorced

pdf_file<-"pdf/pyramids_finely_inside_1x2.pdf"
cairo_pdf(bg="grey98", pdf_file,width=12,height=9)

par(mai=c(0.2,0.1,0.8,0.1),omi=c(0.75,0.2,0.85,0.2),mfcol=c(1,2)↘
     ,cex=0.75,family="Lato Light",las=1)

# Import data and prepare chart
library(gdata)
x1991<-read.xls("myData/12411-0008_1991_reformatted.xlsx", ↘
     encoding="latin1")
x1991<-x1991[1:nrow(x1991)-1,]

attach(x1991)

myF_married<-X6/1000
myF_unmarried<-(X5+X7+X8)/1000
myM_married<-X2/1000
myM_unmarried<-(X1+X3+X4)/1000

right<-data.frame(myF_married,myF_unmarried)
left<-data.frame(myM_married,myM_unmarried)

source("scripts/inc_pyramid.r")
mtext("1970",3,line=0,adj=0.5,cex=1,font=3)

x2010<-read.xls("myData/12411-0008_2010_reformatted.xlsx", ↘
     encoding="latin1")
x2010<-x2010[1:nrow(x2010)-1,]

attach(x2010)

myF_married<-X6/1000
myF_unmarried<-(X5+X7+X8)/1000
myM_married<-X2/1000
myM_unmarried<-(X1+X3+X4)/1000

left<-data.frame(myM_married,myM_unmarried)
right<-data.frame(myF_married,myF_unmarried)

source("scripts/inc_pyramid.r")

mtext("2010",3,line=0,adj=0.5,cex=1,font=3)

# Titling

mtext("Age structure and married proportion of the population in↘
     Germany",3,line=2,adj=0,cex=2.25,family="Lato Black",↘
     outer=T)
```

```
mtext("All values in thousands per year of age",3,line=-0.5,adj↖
    =0,cex=1.25,font=3,outer=T)
mtext("destatis, GENESIS-Online, Tab. 12411-0008",1,line=2,adj↖
    =1.0,cex=0.95,font=3,outer=T)
mtext("Inner highlighted areas: married",1,line=2,adj=0,cex↖
    =0.95,font=3,outer=T)
dev.off()
```

The script effectively proceeds just as in Sect. 7.2.2. Only the data are made up differently: we create each married and unmarried group, which we just add up in this case though. The last row ("85 years and older") is ignored each time.

7.2.4   Pyramids with Added Line (Panel)

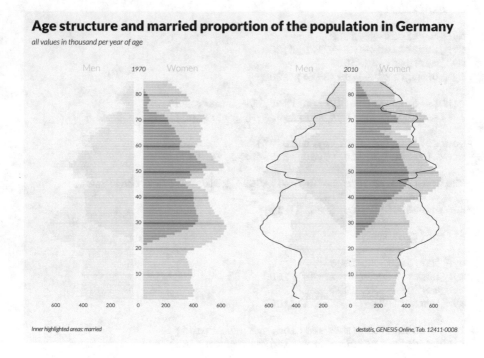

A useful addition for reasons of comparison is sketched-in lines of other pyramids. For illustration purposes, we overlay the right pyramid with the outer margins of the last example as lines. To this end, we have to add the two rows

```
right_line<-f_married+f_unmarried
left_line<--(m_married+m_unmarried)
```

before the attach(x2010) row, and these two rows

```
lines(right_line,b1,type="l",col="black")
```

```
lines(left_line,b2,type="l", col="black")
```

after integration of inc_pyramid.r.

7.2.5   Aggregated Pyramids

If the number of data is limited, such as with surveys, then pyramids with rougher classification than an age class per year should be used.

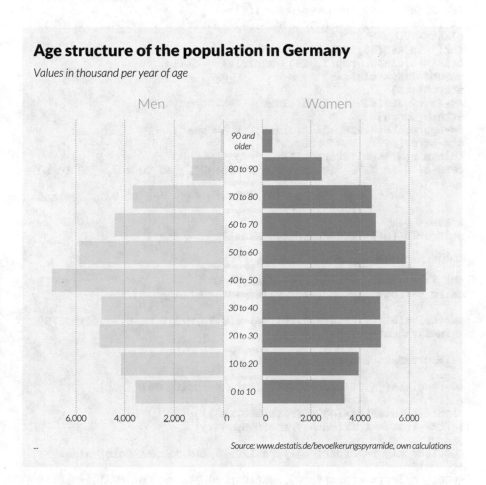

Age structure of the population in Germany

Values in thousand per year of age

Men   Women

	90 and older
	80 to 90
	70 to 80
	60 to 70
	50 to 60
	40 to 50
	30 to 40
	20 to 30
	10 to 20
	0 to 10

6.000   4.000   2.000   0   0   2.000   4.000   6.000

Source: www.destatis.de/bevoelkerungspyramide, own calculations

In this case, it might be advisable to use alternating light and dark colours for the successive classes; it will make reading individual classes easier. However, this does not apply if stacked bars are used. We will now show a simple example with single-coloured parts.

About the figure: If data are not available for all years, but only for a few classes, then the appearance has to be slightly adjusted. On the one hand, longer labels between bars have to be considered; on the other hand, guide lines can be attached since more importance is attributed to the individual values rather than the overall impression as was the case in the previous examples. For this example, data were created by aggregation from the two txt files men.txt and women.txt.

```
men<-read.csv("data/men.txt", header = FALSE, sep = ",")
for(i in 1:100) rownames(men)[i]<-i
for(i in 1:111) colnames(men)[i]<-paste("x", i+1949, sep="")
women<-read.csv("data/women.txt", header = FALSE, sep = ",")
for(i in 1:100) rownames(women)[i]<-i
for(i in 1:111) colnames(women)[i]<-paste("x", i+1949, sep="")
class<-NULL
for(i in seq(10,100, by=10))
class<-c(class, rep(i, 10)) men$class<-class
women$class<-class
attach(men)
m<-aggregate(x2010, list(class), FUN="sum")
attach(women)
f<-aggregate(x2010, list(class), FUN="sum")
mf<-merge(m, f, by="Group.1")
colnames(mf)<-c("Group", "M", "F")
bez<-c("0 to 10", "10 to 20", "20 to 30", "30 to 40", "40 to 50"\
    ,
        "50 to 60", "60 to 70", "70 to 80", "80 to 90", "90 and\
            nolder")
mf$bez<-bez
write.xlsx(mf, "data/popclass.xlsx")
```

The result was exported as XLS file.

```
pdf_file<-"pdf/pyramids_aggregated.pdf"
cairo_pdf(bg="grey98", pdf_file,width=8,height=8)

par(mai=c(0.2,0.25,0.8,0.25),omi=c(0.75,0.2,0.85,0.2),cex=0.75,
  family="Lato Light",las=1)

# Import data and prepare chart

x<-read.xls("myData/popclass.xlsx", encoding="latin1")

right<-t(as.matrix(data.frame(800,x$F)))
left<--t(as.matrix(data.frame(800,x$M)))

myColour_right<-c(par("bg"),rgb(255,0,210,150,maxColorValue=255)\
    )
myColour_left<-c(par("bg"),rgb(191,239,255,maxColorValue=255))

# Create charts and other elements

b1<-barplot(right,axes=F,horiz=T,axis.lty=0,border=NA,col=\
    myColour_right,xlim=c(-8000,8000))
```

```
barplot(left,axes=F,horiz=T,axis.lty=0,border=NA,col=myColour_↘
      left,xlim=c(-7500,7500),add=T)

abline(v=seq(0,6000,by=2000)+800,col="darkgrey",lty=3)
abline(v=seq(-6000,0,by=2000)-800,col="darkgrey",lty=3)

mtext(format(seq(0,6000,by=2000),big.mark="."),at=seq(0,6000,by↘
      =2000)+800,1,line=0,cex=0.95)
mtext(format(abs(seq(-6000,0,by=2000)),big.mark="."),at=seq(-↘
      6000,0,by=2000)-800,1,line=0,cex=0.95)
text(0,b1,x$des,cex=1.25,font=3)

mtext("Men",3,line=1,adj=0.25,cex=1.5,col="darkgrey")
mtext("Women",3,line=1,adj=0.75,cex=1.5,col="darkgrey")

# Titling

mtext("Age structure of the population in Germany",3,line=2,adj↘
      =0,cex=1.75,family="Lato Black",outer=T)
mtext("Values in thousand per year of age",3,line=-0.5,adj=0,cex↘
      =1.25,font=3,outer=T)
mtext("Source: www.destatis.de/bevoelkerungspyramide, own ↘
      calculations",1,line=2,adj=1.0,cex=0.95,font=3,outer=T)
mtext("...",1,line=2,adj=0,cex=0.95,font=3,outer=T)
dev.off()
```

   In the script, we first read the previously aggregated data right and left and then specify these so they each contain an inner column with the constant 800. This is necessary because the space between the left and right bar has to be larger than in the previous examples: there has to be enough space for labels such as "0 to 10" etc. The inner areas with a width of 800 contain the background colour. Then the bars are plotted. Since we still require the y position of the individual bars for the labels, the first diagram object is saved as b1. We then draw orientation lines using abline() and attach the x-axis labels with mtext(). Y-axis labelling is done using text() at the b1 positions. At the end come the usual titles.

7.2.6   Bar Charts as Pyramids (Panel)

International Social Survey Programme: Social Inequality IV

Q10a: Groups tending towards top and bottom. Where would you put yourself now on this scale?

AR-Argentina   AU-Australia   AT-Austria   BE-Belgium   BG-Bulgaria

CL-Chile   CN-China   TW-Taiwan   HR-Croatia   CY-Cyprus

CZ-Czech Republic   DK-Denmark   EE-Estonia   FI-Finland   FR-France

DE-Germany   HU-Hungary   IS-Iceland   IL-Israel   IT-Italy

JP-Japan   KR-South Korea   LV-Latvia   NZ-New Zealand   NO-Norway

PH-Philippines   PL-Poland   PT-Portugal   RU-Russia   SK-Slovak Republic

SI-Slovenia   ZA-South Africa   ES-Spain   SE-Sweden   CH-Switzerland

TR-Turkey   UA-Ukraine   GB-GBN-Great Britain and/or United Kingdom   US-United States

Top,Highest,10

Bottom,Lowest,01

50%

Source: ZA5400: ISSP 2009

About the figure: The representation form of a population pyramid is not only suited for paired frequencies, but also for survey data. The present example from the International Social Survey Programme shows the response to the question "In our society there are groups which tend to be towards the top and groups which tend to be towards the bottom. Below is a scale that runs from top to bottom. Where would you put yourself now on this scale?". The respondents were asked to rate themselves on a scale from "1 bottom, lowest 01" to "10 top, highest, 10". If the frequency values are concentrated in the middle, then the result is an ostensive distribution pattern. The result is not an age pyramid, but a "self-perception pyramid" of the respondents in their respective countries. The figure delivers the respective distribution of self-perception (as percentages) in the society for each of the 38 surveyed countries. For such a large number of individual graphics, the elements of each individual graphic should be limited to the bare minimum. If we use the same x-axis width for all charts, then we don't need individual x-axis labels. We can also do without y-axis labels since the number of response options is always the same. The data derive from the SPSS data frame ZA5400: International Social Survey Programme: Social Inequality IV-ISSP 2009. They can be downloaded from http://gesis.org by registered users (doi:10.4232/1.10736). The ISSP is a social science survey program that was founded in 1984 and conducts yearly surveys on varying social science topics. Each year, focus topics are selected that are then repeated in specific intervals. In 2009, the topic was "social inequality", that means questions addressing social background, income, discrimination, attitude towards corruption etc. Respondents were aged 18 or older (Finland: 15–74, Italy, Japan: 16 and older, Norway: 19–80, Sweden 17–79). The data frame contains 350 variables (questions, socio-demographic characteristics of respondents, interviewer data). A total of 55,238 people were included in the survey. We are using the following variables:

V5: "Country" and V44: "Q10a Groups tending towards top and bottom. Where would you put yourself now on this scale?"

```
pdf_file<-"pdf/barcharts_pyramids_8x5.pdf"
cairo_pdf(bg="grey98", pdf_file,width=8.27,height=11.7)

par(omi=c(0.25,0.1,1.0,0.1),mai=c(0.1,0.1,0.55,0.1),mfrow=c(8,5),\
    family="Lato Light",las=1)
library(Hmisc)

# Import data and prepare chart

ISSP<-spss.get("myData/ZA5400_v3-0-0.sav",use.value.labels=T)
attach(ISSP)
countries<-as.data.frame(table(V5))
country_list<-countries$V5[1:45]

for (i in 1:length(country_list))
[
land<-subset(ISSP,ISSP$V5==country_list[i])
attach(land)
```

```
y<-as.data.frame(prop.table(table(V44))*100)
if (!(is.na(y[1,2])))
{
left<--(y$Freq/2)
right<-y$Freq/2

# Create charts

barplot(left,horiz=T,xlim=c(-30,30),border="orange",col="orange",
   main=country_list[i],cex.axis=0.6,axes=F)
barplot(right,horiz=T,add=T,border="orange",col="orange",cex.axis↘
      =0.6,axes=F)
segments(-25,-0, 25,0)
}
}

# Legend

myDes<-c("Bottom,Lowest,01","",".","",".","",".","","",","" Top,↘
      Highest,10")
n<-length(myDes)
plot(0:n,type="n",axes=F,xlab="",ylab="", xlim=c(-25,25),ylim=c↘
      (0,11)) #
for (i in 1:n) text(0,i+0.5,myDes[i],cex=0.9,xpd=T)
segments(-25,-0, 25,0)
text(0,-1.5, "50%", xpd=T)

# Titling

mtext("International Social Survey Programme: Social Inequality ↘
      IV",3,line=2.2,adj=0,cex=1.4,family="Lato Black",outer=T)
mtext("Q10a: Groups tending towards top and bottom. Where would ↘
      you put yourself now on this scale?",3,line=0,adj=0,cex↘
      =1.1,outer=T)
mtext("Source: ZA5400: ISSP 2009",1,line=0,adj=1,cex=0.7,outer=T,↘
      font=3)
dev.off()
```

The script uses the spss.get() function from Frank Harrells Hmisc package to read the data, working with "value labels" of the data rather than their actual values. A list of countries is created using the table() function (using e.g. unique() would have been an alternative). Then a partial data frame containing all data for a single country from the list is generated in a loop. For the data of this country, we use prop.table() to create a distribution of the V44 variable as percentages and to save these as a data frame. Now we generate centred bars for all non-missing values by dividing the respective frequency by two, with one half plotting the left bars of a bar chart as negative value, while the other half is added in a second barplot() call using the add=T function. The country name output happens during the first call. In total, the distributions of 38 countries are plotted. In an 8×5 graphic layout, we can use the second to last for a legend in which we name the lowest and highest values. At the end, there come the usual labels.

7.3 Inequality

A graphic illustration form for the representation of inequality is the Lorenz curve, introduced by Max Otto Lorenz more than 100 years ago. It is typically used to visualise income distributions. On the x-axis, the illustration shows the cumulative percentage of the statistical unit, such as the "population" or the respondents. The cumulative percentages of income are plotted on the y-axis. Since both criterions share the same dimension and span (from 0 to 100%), the figures should be square. In a completely uniform distribution, the result would be a diagonal line; with identical axis lengths, the line would be at a 45° angle. The more inequality in the distribution, the greater the deviation of the actual distribution line from this straight line. The area between the straight line and the curve of the actual distribution expresses the power of the inequality: the larger the area, the greater the inequality of the criterion's distribution. In the literature, there are variants for both individual as well as classified data. In this paragraph, we describe three examples for the representation of Lorenz curves, followed by three variants for the visualisation of income distributions with common bar charts.

7.3.1 Simple Lorenz Curve

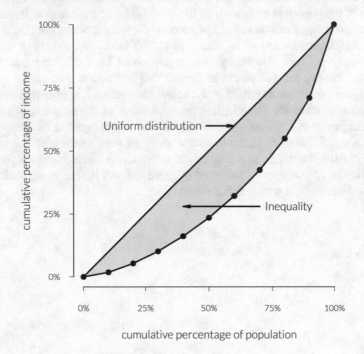

Income Distribution in the USA in 2000

(10 classes)

Source: United Nations University, UNU-WIDER World Income Inequality Database

About the figure: This example shows the income distribution in the USA in 2000, arranged in ten categories. In the illustration, the category-arrangements are also identified with dots. The area created by the difference to the uniform distribution is highlighted in colour. A frame or further auxiliary lines are not necessary. Data were taken from the World Income Inequality Database V2.0c May 2008, which is available from http://www.wider.unu.edu/research/database/. I filtered data from the WIID2C.xls file, which is provided for download on that site, and typed the filtered data into a different XLS file.

```
pdf_file<-"pdf/lorenzcurves_10.pdf"
cairo_pdf(bg="grey98", pdf_file,width=6.5,height=7.25)

par(mai=c(0,0,0,0),omi=c(0.75,0.5,0.85,0.2),pin=c(4,4),family="\
      Lato Light",las=1)

# Read data and prepare chart

library(gdata)
myData<-read.xls("myData/income_ten_classes.xlsx",head=T,skip=1,\
      dec=".",encoding="latin1")
attach(myData)
U<-rep(10,10)
U_cum<-c(0,cumsum(U/100))
U2_cum<-c(0,cumsum(U2/100))

# Define chart and other elements

plot(U_cum,U2_cum,type="n",axes=F,xlab="cumulative percentage of\
      population",ylab="cumulative percentage of income",xlim=c\
      (0,1),ylim=c(0,1))
lines(U_cum,U2_cum,lwd=2)
points(U_cum,U2_cum,pch=19)
x<-array(c(0,1,0,1),dim=c(2,2))
lines(x,lwd=2,col="black")
text(0.08,0.585,"Uniform distribution",adj=c(0,0))
text(0.72,0.265,"Inequality",adj=c(0,0))
arrows(0.4,0.28,0.7,0.28,length=0.10,angle=10,code=1,lwd=2,col="\
      black")
arrows(0.49,0.6,0.6,0.60,length=0.10,angle=10,code=2,lwd=2,col="\
      black")
xx<-c(U_cum,rev(U_cum))
yy<-c(U2_cum,rev(U_cum))
polygon(xx,yy,col=rgb(255,97,0,50,maxColorValue=255),border=F)
source("scripts/inc_axes_with_lines_lorenz.r")

# Titling

mtext("Income Distribution in the USA in 2000",side=3,line=1,cex\
      =1.5,family="Lato Black",adj=0,outer=T)
mtext("(10 classes)",side=3,line=-1.5,cex=1.25,font=3,adj=0,\
      outer=T)
mtext("Source: United Nations University, UNU-WIDER World Income\
      Inequality Database",1,line=1,adj=1,cex=0.85,font=3,outer\
      =T)
dev.off()
```

Axis generation is integrated separately:

```
axis(1,col=rgb(24,24,24,maxColorValue=255),
        col.ticks=rgb(24,24,24,maxColorValue=255),
lwd.ticks=0.5,cex.axis=0.75,xlim=c(0,1),tck=-0.015,
        at=c(0,0.25,0.5,0.75,1),labels=c("0%","25%","50%","75%",\
              "100%"))
axis(2,col=rgb(24,24,24,maxColorValue=255),
```

```
col.ticks=rgb(24,24,24,maxColorValue=255),
lwd.ticks=0.5,cex.axis=0.75,xlim=c(0,1),tck=-0.015,
at=c(0,0.25,0.5,0.75,1),labels=c("0%","25%","50%","75%",\
     "100%"))
```

In the script, we first use cumsum() to create cumulative values from the data of the 10 categories' XLS file, and then use plot() and lines() to plot these values. Here, U is the universal distribution forming the 45° line in the Lorenz chart. To emphasise the class basis, the individual points are added using points(). End points are added using array() and then follow the individual labels. Now we still need to colour the area between the 45° line and the curve. For this, we use the polygon() function and hand over the xx and yy vectors. These vectors have previously been generated using c(U_cum, rev(U_cum)) or c(U2_cum,rev(U_cum)). The result is the value pairs that, starting at bottom left and continuing counter-clockwise, form the coordinates for the corner points of the shape that we want to colour. It looks like this:

```
> xx[1:15]
 [1] 0.0 0.1 0.2 0.3 0.4 0.5 0.6 0.7 0.8 0.9 1.0 1.0 0.9 0.8 0.7
> as.numeric(format(round(yy[1:15],1), nsmall=1))
 [1] 0.0 0.0 0.1 0.1 0.2 0.2 0.3 0.4 0.5 0.7 1.0 1.0 0.9 0.8 0.7
```

Finally, prior to the usual labels, we plot the axes with the percentages. Since we will use this type of axis in the subsequent examples as well, we save the two commands and embed the file using source().

7.3.2 Lorenz Curves Overlay

About the figure: When comparing two distributions, it is a good idea to super-impose one over the other. The type of figure matches the previous example. Here, we present two income distributions of the German General Social Survey ("Allgemeine Bevölkerungsumfrage der Sozialwissenschaften—ALLBUS"), which show that the inequality is greater in the survey done in 2008 than in the one done in 1988.

Lorenz Curve of Income Distribution in 1988 and 2008

German General Social Survey

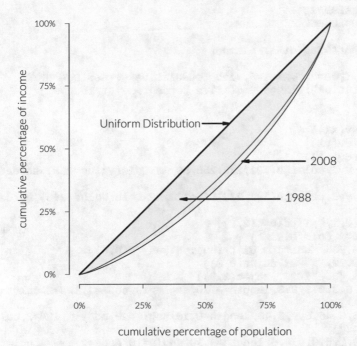

Source: GESIS ZA 4600, ZA 1670

Data are taken from the survey project Allgemeine Bevölkerungsumfrage der Sozialwissenschaften (ALLBUS) [German General Social Survey] and were kindly provided by Michael Terwey.

```
pdf_file<-"pdf/lorenzcurves_lc_overlay.pdf"
cairo_pdf(bg="grey98", pdf_file,width=6.5,height=6.5)

par(mai=c(0.25,0,0,0),omi=c(0.4,0.2,0.5,0.2),pin=c(4,4),family="
    Lato Light",las=1)
library(Hmisc)
library(ineq)

# Read data and prepare chart

myData<-spss.get("myData/AEQU+.por",use.value.labels=T)
x2008<-subset(myData$EQINCOM1,myData$EQINCOM1 > 0 & myData$V2 ==
    "STUDIEN-NR. 4000")
Lc.Eplus2008<-Lc(x2008)
```

```
x1988<-subset(myData$EQINCOM1,myData$EQINCOM1 > 0 & myData$V2 ==↘
      "STUDIEN-NR. 1670")
x1988<-x1988[1:length(x2008)]
Lc.Eplus1988<-Lc(x1988)
x1<-Lc.Eplus1988$p
y1<-Lc.Eplus1988$L
x2<-Lc.Eplus2008$p
y2<-Lc.Eplus2008$L

# Define chart and other elements

plot(x1,y1,type="n",axes=F,xlab="cumulative percentage of ↘
      population",ylab="cumulative percentage of income")
lines(x1,y1)
lines(x2,y2)
xx<-c(x1,rev(x1))
yy1<-c(y1,rev(y2))
yy2<-c(y1,rev(x1))
polygon(xx,yy1,col=rgb(191,239,255,80,maxColorValue=255),border=↘
      F)
polygon(xx,yy2,col=rgb(191,239,255,120,maxColorValue=255),border↘
      =F)
x<-array(c(0,1,0,1),dim=c(2,2))
lines(x,lwd=2,col="black")
text(0.08,0.585,"Uniform Distribution",adj=c(0,0))
text(0.82,0.29,"1988",adj=c(0,0))
text(0.92,0.435,"2008",adj=c(0,0))
arrows(0.4,0.3,0.8,0.3,length=0.10,angle=10,code=1,lwd=2,col="↘
      black")
arrows(0.65,0.45,0.9,0.45,length=0.10,angle=10,code=1,lwd=2,col=↘
      "black")
arrows(0.49,0.6,0.6,0.6,length=0.10,angle=10,code=2,lwd=2,col="↘
      black")
source("scripts/inc_axes_with_lines_lorenz.r")

# Titling

mtext("Lorenz Curve of Income Distribution in 1988 and 2008",↘
      side=3,line=0.25,cex=1.45,family="Lato Black",outer=T,adj↘
      =0)
mtext("German General Social Survey",3,line=-2,adj=0,outer=T,cex↘
      =1.05,font=3)
mtext("Source: GESIS ZA 4600, ZA 1670",1,line=0.6,adj=1,outer=T,↘
      cex=0.85,font=3)
dev.off()
```

Similar to the previous example, the script starts with calculating and plotting the two Lorenz curves. Only in this case, we do not plot them next to each other, but on top of each other—in the same figure. That also means that the areas created with the polygon() function are superimposed onto each other in transparent colours. Labels and titles follow as in Sect. 7.3.1.

7.3.3 Lorenz Curves (Panel)

Lorenz Curve of Income Distribution in 1988 and 2008

German General Social Survey

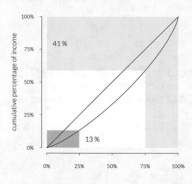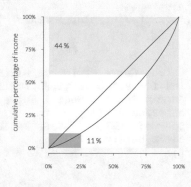

Source: GESIS ZA 4600, ZA 1670

About the figure: Another possible variant for the comparison of two Lorenz curves is the representation as a panel in which both curves are arranged next to each other. In these cases, it is recommended to separately mark the percentages of the total income that fall on the lower and upper quartile, respectively. Using the same ALLBUS data as in the previous example, we can show that 13% of the total income of all respondents went on the lower 25% of respondents in 1988, and 41% on the upper 25%; in 2008, it was 11 and 44%. Data are taken from the survey project Allgemeine Bevölkerungsumfrage der Sozialwissenschaften (ALLBUS) [German General Social Survey] and were kindly provided by Mike Terwey.

```
pdf_file<-"pdf/lorenzcurves_lc_1x2.pdf"
cairo_pdf(bg="grey98", pdf_file,width=14,height=8)

par(mfcol=c(1,2),mai=c(0.25,0.25,0.25,0.25),omi=c↘
     (1.25,0.5,1.25,0.5),pin=c(4.5,4.5),cex=1.3,family="Lato ↘
     Light",las=1,family="Lato Light")
library(Hmisc)
library(ineq)

# Read data and prepare chart

myData<-spss.get("myData/AEQU+.por",use.value.labels=T)
x2008<-subset(myData$EQINCOM1,myData$EQINCOM1 > 0 & myData$V2 ==↘
     "STUDIEN-NR. 4600")
Lc.Eplus2008<-Lc(x2008)
x1988<-subset(myData$EQINCOM1,myData$EQINCOM1 > 0 & myData$V2 ==↘
     "STUDIEN-NR. 1670")
Lc.Eplus1988<-Lc(x1988)
```

```
x<-Lc.Eplus1988$p
y<-Lc.Eplus1988$L

# Create chart

source("scripts/inc_plot_lorenz.r")
x<-Lc.Eplus2008$p
y<-Lc.Eplus2008$L
source("scripts/inc_plot_lorenz.r")

# Titling

mtext("Lorenz Curve of Income Distribution in 1988 and 2008",3,↘
      line=1.5,adj=0,cex=1.85,family="Lato Black",outer=T)
mtext("German General Social Survey",3,line=-0.5,adj=0,cex=1.85,↘
      font=3,outer=T)
mtext("Source: GESIS ZA 4600, ZA 1670",1,line=2,adj=1,cex=1.05,↘
      font=3,outer=T)
dev.off()
```

and

```
axis(1,col=rgb(24,24,24,maxColorValue=255),col.ticks=rgb↘
      (24,24,24,maxColorValue=255),lwd.ticks=0.5,cex.axis=0.75,↘
      xlim=c(0,1),tck=-0.015,at=c(0,0.25,0.5,0.75,1),labels=c("↘
      0%","25%","50%","75%","100%"))
axis(2,col=rgb(24,24,24,maxColorValue=255),col.ticks=rgb↘
      (24,24,24,maxColorValue=255),lwd.ticks=0.5,cex.axis=0.75,↘
      xlim=c(0,1),tck=-0.015,at=c(0,0.25,0.5,0.75,1),labels=c("↘
      0%","25%","50%","75%","100%"))
```

In this case, the script first reads individual data from an SPSS file; from this data, the Lorenz curve data are then calculated using the Lc() function from the ineq package. The function returns the values of the Lorenz curve with p and L. The number of calculated values corresponds to the number of input data, which means no classes are formed. Since the Lorenz curve is plotted twice in the same way, we are saving the necessary steps into a file inc_plot_lorenz.r. After specifying the graphic using plot() of the n type, the 25 and 75% positions of the data are determined. For x1988, there are 2149 cases, which means that the position for 25% of the data would be 537 (rounded), and the position for 75% of the data, 1611. Now we plot three rectangles: First a large one spanning the entire data area. Second comes a rectangle spanning from 0 to the 75% position and the associated y value, and last a third one over the other two that reaches the 25% position and the associated y value. The result is the desired area colouring of the 25 and 75% areas. After plotting the Lorenz curve and the diagonal x, we label the areas with the y values of the 25 or 75% x values. Finally we call the file for plotting of the axes. At the end, there come the usual labels.

7.3.4   Comparison of Income Proportions with Bar Chart (Quintile)

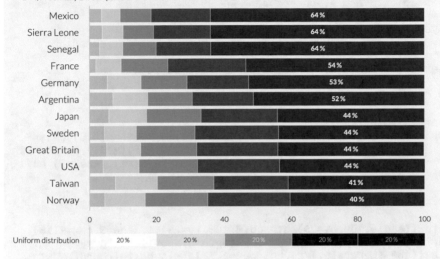

Income Distribution over five Classes in different Countries

In Mexico the richest 10% of income recipients held over 45% of the overall income in 2000, in the USA it was 29%, in Germany 24. Compared to 1984 the share did increase.

Source: World Income Inequality Database V2.0c 2008

About the figure: Another obvious representation of income distributions is a 100% bar chart. This is also a good choice for easy comparison of income distributions in multiple countries. In this example, we compare the income distribution (split into five classes) of 12 countries. Countries are sorted in descending order by the proportion that falls on the highest class. For all countries and classes, we use only one shade of grey and a colour in the shades of a Brewer palette. For the highest income class, the proportion of the total income is written into the respective part of the bar. This would not be an optimal choice for all classes. Additionally, the lowest class has often such a small proportion in the distributions that it leaves no room for a label. For ease of orientation, country names are separated from the bars by a vertical line; an x-axis is omitted. We use a theoretical uniform distribution as legend, which is offset from the actual data with grey-scale values. By way of illustration, the number of the uniform distribution (20%) can be written into each class. Data were taken from the World Income Inequality Database V2.0c May 2008, which is available from http://www.wider.unu.edu/research/database/. I filtered data from the WIID2C.xls file, which is provided for download on that site, and wrote the filtered data into a different XLS file.

```
pdf_file<-"pdf/lorenzcurves_barcharts_05.pdf"
cairo_pdf(bg="grey98", pdf_file,width=12,height=9)

par(omi=c(0.5,0.5,1.1,0.5),mai=c(0,2,0,0.5),family="Lato Light",\
    las=1)
library(fBasics)
library(gdata)

# Read data and prepare chart

myDataFile<-"myData/income_five_classes.xlsx"
myData<-read.xls(myDataFile,head=T,skip=1,dec=".",encoding="\
    latin1")
layout(matrix(c(1,2),ncol=1),heights=c(80,20))

# Create chart

par(mai=c(0,1.75,1,0))
bp1<-barplot(as.matrix(myData),ylim=c(0,6),width=c(0.5),axes=F,\
    horiz=T,col=c("grey",seqPalette(5,"OrRd")[2:5]),border=par\
    ("bg"),names.arg=gsub("."," ",names(myData),fixed=T),cex.\
    names=1.55)

# Other elements

mtext(seq(0,100,by=20),at=seq(0,100,by=20),1,line=0,cex=1.15)
arrows(0,-0.03,0,7.30,lwd=1.5,length=0,xpd=T,col="grey")
text(100-(myData[5,]/2),bp1,cex=1.1,labels=paste(round(myData\
    [5,],digits=0),"%",sep=" "),col="white",family="Lato Black\
    ",xpd=T)

# Create chart

par(mai=c(0.55,1.75,0,0))
bp2<-barplot(as.matrix(rep(20,5)),ylim=c(0,0.5),width=c(0.20),\
    horiz=T,col=seqPalette(5,"Greys"),border=par("bg"),names.\
    arg=c("Uniform distribution"),axes=F,cex.names=1.25)

# Other elements

arrows(0,-0.01,0,0.35,lwd=1.5,length=0,xpd=T,col="grey")
text(c(10,30,50,70,90),bp2,labels=c("20 %","20 %","20 %","20 %",\
    "20 %"),col=c("black","black","white","white","white"),xpd\
    =T)

# Titling

title(main="Income Distribution over five Classes in different \
    Countries",line=3,adj=0,cex.main=2.25,family="Lato Black",\
    outer=T)
```

```
myBreak<-strsplit( strwrap("In Mexico the richest 10% of income ⬎
       recipients held over 45% of the overall income in 2000, in⬎
        the USA it was 29%, in Germany 24. Compared to 1984 the ⬎
        share did increase.",width=110),"\n")
for(i in seq(along=myBreak))
{
mtext(myBreak[[i]],line=(1.8-i)*1.5,adj=0,side=3,cex=1.25,outer=⬎
      T)
}
mtext("Source: World Income Inequality Database V2.0c 2008",side⬎
      =1,adj=1,cex=0.95,font=3,outer=T)
dev.off()
```

In the script, we load the fBasics package for this figure, as we can use it to create sequential colour palettes. Data are read from an XLS file. Since we design the legend as a separate bar, we define a layout that splits the figure horizontally into two areas. The upper area makes up 80% of the height, the lower area 20%. In the upper one, we use barplot() to draw a bar chart and the seqPalette function to define the colours. The first colour is too bright, which is why we replace it with grey. We also use gsub(".", " ", names(data), fixed=T) to remove dots in the variable names (which are in there if the variable names of the original data contained spaces). X-axis labelling is done using mtext() and the seq() function. The arrows() function is used to add a vertical orientation line between the y-axis label and the bar, and we then add the percent values for the last part into the bars using text(). We can once more add the y position using bpl, the object previously created by the barplot() function. The second bar chart bp2, which makes up our legend, consists of five repetitions of the value 20, and contains a grey scale. Here, we add text to each segment. Since the subheading is too long for a row, we wrap the text using the functions strsplit() and strwrap().

7.3.5   Comparison of Income Proportions with Bar Chart (Decile)

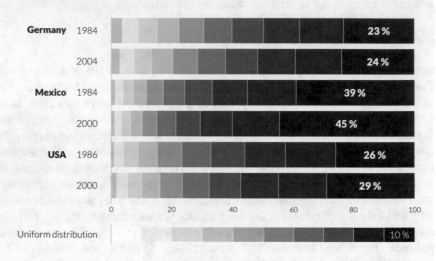

Income Distribution over ten Classes in three Countries

In Mexico the richest 10% of income recipients held over 45% of the overall income in 2000, in the USA it was 29%, in Germany 24. Compared to 1984 the share did increase.

Source: World Income Inequality Database V2.0c 2008

About the figure: Even if we have ten classes instead of five, we can still limit ourselves to one colour with appropriate gradations. In the current example, two distinct points of time are compared for three countries. In this case, the two levels should also be typographically distinguishable. The first level (here: the countries) should not be repeated in the labels but should be highlighted with a bolder typeface cut. In this case, we limit the legend to the labels of one segment.

Data are again taken from the World Income Inequality Database V2.0c May 2008.

```
pdf_file<-"pdf/lorenzcurves_barcharts_10.pdf"
cairo_pdf(bg="grey98", pdf_file,width=12,height=9)

library(fBasics) # for seqPalette
library(gdata)

layout(matrix(c(1,2,1,2),2,2),heights=c(6,1))
par(omi=c(1,0.5,1.25,0.25),mai=c(0,2.65,0.75,0.25),cex=1.5,\
    family="Lato Light",las=1)
```

```
# Read data

myData<-read.xls("myData/income_ten_classes.xlsx",head=T,skip=1,⬎
      dec=".", encoding="latin1")

# Create chart and other elements

bp1<-barplot(as.matrix(myData),ylim=c(0,3),width=c(0.45),axes=F,⬎
      horiz=T,col=c("grey",seqPalette(10,"OrRd")[2:10]),border=⬎
      par("bg"),names.arg=c("2000","1986","2000","1984","2004","⬎
      1984"))
arrows(0,-0.01,0,3.25,lwd=1.5,length=0,xpd=T,col="grey")
text(100-(myData[10,]/2),bp1,col="white",cex=1.1,family="Lato ⬎
      Black",labels=paste(round(myData[10,],digits=0),"%",sep=" ⬎
      "),xpd=T)
text(-15,bp1[2],"USA",family="Lato Black",adj=1,xpd=T)
text(-15,bp1[4],"Mexico",family="Lato Black",adj=1,xpd=T)
text(-15,bp1[6],"Germany",family="Lato Black",adj=1,xpd=T)

# Create chart and other elements

par(mai=c(0,2.65,0.1,0.25))
bp2<-barplot(as.matrix(rep(10,10)),ylim=c(0,0.5),width=c(0.25),⬎
      axes=F,horiz=T,col=seqPalette(10,"Greys"),border=par("bg")⬎
      ,names.arg=c("Uniform distribution"))
arrows(0,-0.01,0,0.35,lwd=1.5,length=0,xpd=T,col="grey")
text(95,bp2,labels="10 %",col="white",xpd=T)
mtext(seq(0,100,by=20),at=seq(0,100,by=20),3,line=0,cex=1.15)

# Titling

mtext("Income Distribution over ten Classes in three Countries",⬎
      line=2,adj=0,cex=2.25,family="Lato Black",outer=T)
myBreak<-strsplit( strwrap("In Mexico the richest 10% of income ⬎
      recipients held over 45% of the overall income in 2000, in⬎
       the USA it was 29%, in Germany 24. Compared to 1984 the ⬎
      share did increase.",width=110),"\n")
for(i in seq(along=myBreak))
{
mtext(myBreak[[i]],line=1.8-i,adj=0,side=3,cex=1.25,outer=T)
}
mtext("Source: World Income Inequality Database V2.0c 2008",1,⬎
      line=1.5,adj=1,cex=0.95,font=3,outer=T)
dev.off()
```

In the script, we again load the fBasics package for this figure. As before, data are read from an XLS file and the figure is split into two parts. However, since we intend to use hierarchical bar labelling, we first plot the first level, the years, as a name argument of the barplot() function. The rest of the labelling matches the previous example, but we also add the countries in front of the first years using text(). The rest of the example matches the previous one.

7.3.6 Comparison of Income Proportion with Panel-Bar Chart (Quintile)

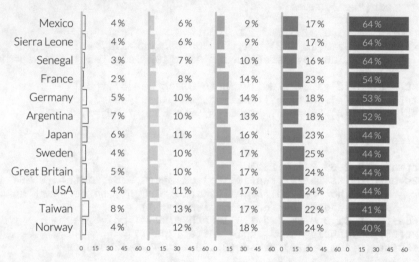

Income Distribution over five Classes in different Countries
In Mexico the richest 20% of income recipients hold over 64% of the overall income, in Norway the figure is 40%. Compared interntionally Germany is in the upper half.

Mexico	4%	6%	9%	17%	64%
Sierra Leone	4%	6%	9%	17%	64%
Senegal	3%	7%	10%	16%	64%
France	2%	8%	14%	23%	54%
Germany	5%	10%	14%	18%	53%
Argentina	7%	10%	13%	18%	52%
Japan	6%	11%	16%	23%	44%
Sweden	4%	10%	17%	25%	44%
Great Britain	5%	10%	17%	24%	44%
USA	4%	11%	17%	24%	44%
Taiwan	8%	13%	17%	22%	41%
Norway	4%	12%	18%	24%	40%

0 15 30 45 60 0 15 30 45 60 0 15 30 45 60 0 15 30 45 60 0 15 30 45 60

Source: World Income Inequality Database V2.0c 2008

About the figure: This is a variant of the second to last example, in which the individual segments of the bar do not come together horizontally, but in which each income class makes up an individual bar chart. This structure was explained in Sect. 6.1.5.

The advantage of this variant lies in the fact that the differences of the individual attributes are easier to compare. In this variant, we can also write the percent values for each class into the figure, as the space is certainly available. A legend is not necessary in this example.

Data once more derive from the World Income Inequality Database V2.0c May 2008.

```
pdf_file<-"pdf/lorenzcurves_panel_05.pdf"
cairo_pdf(bg="grey98", pdf_file,width=11,height=8)

par(omi=c(0.5,0.5,1.1,0.5),family="Lato Light",las=1)
layout(matrix(data=c(1,2,3,4,5),nrow=1,ncol=5),widths=c↘
    (2.0,1,1,1,1),heights=c(1,1))

library(gdata)
```

```
# Read data and prepare chart

myData<-read.xls("myData/income_five_classes.xlsx",skip=1,dec="."\
      ", encoding="latin1")
tmyData<-t(myData)
transparency<-c(0,50,100,150,200)
number_colour<-c("black","black","black","black","white")
pos<-c(45,45,45,45,35)
par(cex=1.05)

# Create chart and other elements

for (i in 1:5) {
if (i == 1)
{
par(mai=c(0.25,1.75,0.25,0.15))
bp1<-barplot(tmyData[ ,i],horiz=T,cex.names=1.6,axes=F,names.arg\
      =gsub("."," ",names(myData),fixed=T),xlim=c(0,60),col=rgb\
      (43,15,52,0,maxColorValue=255))
} else
{
par(mai=c(0.25,0.1,0.25,0.15))
bp2<-barplot(tmyData[ ,i],horiz=T,axisnames=F,axes=F,xlim=c\
      (0,60),col=rgb(200,0,0,transparency[i],maxColorValue=255),\
      border=par("bg"))
}
text(pos[i],bp1,adj=1,labels=paste(round(myData[i ,],digits=0),"\
      %",sep=" "),col=number_colour[i],xpd=T,cex=1.3)
mtext(seq(0,60,by=15),at=seq(0,60,by=15),1,line=0,cex=0.85)
arrows(0,-0.1,0,14.6,lwd=2.5,length=0,xpd=T,col="grey")
}

# Titling

title(main="Income Distribution over five Classes in different \
      Countries",line=3,adj=0,cex.main=1.75,family="Lato Black",\
      outer=T)
myBreak<-strsplit( strwrap("In Mexico the richest 20% of income \
      recipients hold over 64% of the overall income, in Norway \
      the figure is 40%. Compared interntionally Germany is in \
      the upper half.",width=110),"\n")
for(i in seq(along=myBreak))
{
mtext(myBreak[[i]],line=(1.8-i)*1.7,adj=0,side=3,cex=1.25,outer=\
      T)
}
mtext("Source: World Income Inequality Database V2.0c 2008",1,\
      line=2,adj=1,font=3)
dev.off()
```

In the script, we need our data in two different formats in this instance: on the one hand, as before in the format in which they are ordered in the XLS table, on the other hand in transposed format. We use a loop to go through each class and create a bar chart for the i-th column of the transposed data. The output of y-axis labels happens during the first run, but not in the subsequent ones. Within the loop, the text() function writes the value of the bar, mtext() the x-axis labels and arrows() again vertical orientation lines at 0 for each individual bar chart. The rest matches the previous examples.

Chapter 8
Time Series

This chapter is dedicated to time series. As before, there are typical application forms to be differentiated. First, we will look at how to present "short" time series. For this, we will also use columns. Then, we show how to plot areas below, between or above time series. In our experience, the presentation of daily, weekly or monthly values always proves a bit tricky; so we deal with those, too. The chapter concludes with special cases that cannot be allocated to the above groups.

8.1 Short Time Series

8.1.1 Column Chart for Developments

About the figure: Time series should generally be shown as lines. In exceptional cases, especially with short series, a representation with columns may be possible. However, the time dimension should always be horizontal. Instead of using horizontal grid lines, the columns are broken in the background colour. This makes orientation easier and spares us from "chart junk". Especially in business reports, a variant is often used in which the previous year is highlighted in a different colour and there are additional labels with the value. If the topic is a growing development, then the y-axis can be attached on the right side to achieve a harmonic overall impression. If it is clear which years the figure concerns (e.g. because it is stated in the subheading), then x-axis labelling can be limited to two numbers for the years. Obviously, "01" has to be used instead of "1" for years prior to 2010. The y-scale should start at 0 and can then finish just below the maximal value. If the units are stated in the subheading, then they do not have to be repeated on the y-axis.

© Springer Nature Switzerland AG 2019   239
T. Rahlf, *Data Visualisation with R*, https://doi.org/10.1007/978-3-030-28444-2_8

Sales Development Microsoft

2002–2011, figures in Bill. US-Dollars

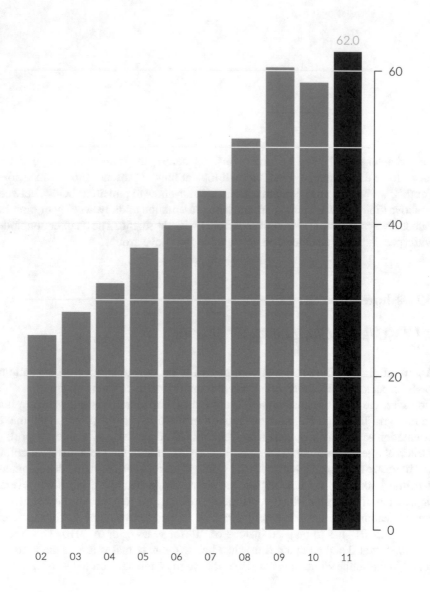

Source: money.cnn.com

Data were taken from the website http://money.cnn.com and manually entered into the script.

```
pdf_file<-"pdf/timeseries_columns_progress.pdf"
cairo_pdf(bg="grey98", pdf_file,width=6,height=9)

par(las=1,cex=0.9,omi=c(0.75,0.25,1.25,0.25),mai=c\
      (0.5,0.25,0.5,0.75),family="Lato Light",las=1)

#  Read data and prepare chart

myData<-c(25296,28365,32187,36835,39788,44282,51122,60420,58437,
          62484)/1000
myLabels<-c(2002:2011)
myColours<-c(rep("olivedrab",length(myData)-1),"darkred")

# Create chart and other elements

barplot(myData,border=NA,col=myColours,names.arg=substr(myLabels\
      ,3,4),axes=F,cex.names=0.8)
abline(h=c(10,20,30,40,50,60,70,80),col=par("bg"),lwd=1.5)
axis(4,at=c(0,20,40,60))
text(11.5,myData[10]+0.025*myData[10],format(round(myData[10]),\
      nsmall=1),adj=0.5,xpd=T,col="darkgrey")

# Titling

mtext("Sales Development Microsoft",3,line=4,adj=0,family="Lato \
      Black",outer=T,cex=2)
mtext("2002-2011, figures in Bill. US-Dollars",3,line=1,adj=0,\
      cex=1.35,font=3,outer=T)
mtext("Source: money.cnn.com",1,line=2,adj=1.0,cex=1.1,font=3,\
      outer=T)
dev.off()
```

The script begins with the specification of borders, followed by definition of the data directly within the data vector. Labelling is defined as the sequence from 2002 to 2011. Colour is also a vector that is filled with the colour olivedrab for every column aside from the last one. The last column becomes dark red. The last two numbers of the years are used for the labels. The axis is suppressed and is later plotted separately at position 4 (=right). The horizontal lines created with abline() in the background colour are drawn on top of the columns to give them the "broken" impression. Finally, the respective value is plotted over the last bar at the position myData[10]+0.025*myData[10].

8.1.2   Column Chart with Percentages for Growth Developments

Sales Development Microsoft

Figures in Bill. US-Dollars

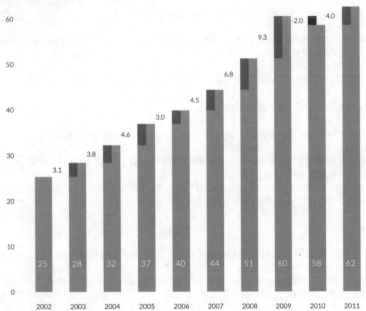

Source: money.cnn.com

In contrast to the previous one, the focus of this figure's message is the change. However, this is not shown in isolation (such as a growth rate), but highlighted in the absolute numbers. Since the width or area of the columns is irrelevant, and only height is important, it is not necessary to mark the entire higher area. A very descriptive effect is achieved by only emphasising half the width of the change. This has the additional advantage that negative changes can also be shown. In such cases (here, 2010), a thin red column is positioned on top of the main column rather than a green one within. The exact dimension of all changes is written next to the column, and the overall magnitude of the value is plotted inside the column, in this case in white numbers. If the y-axis is attached on the left, as in this example, and the height of the columns is reflected by the number within each column, a reduced axis representation without a vertical line suffices. Data were taken from the website http://money.cnn.com and manually entered into the script.

```
pdf_file<-"pdf/timeseries_columns_share_growth.pdf"
cairo_pdf(bg="grey98", pdf_file,width=11,height=9)
```

```r
par(las=1,cex=0.9,omi=c(0.75,0.5,1.25,0.5),mai=c(0.5,1,0,1),↘
      family="Lato Light",las=1)

# Define data

myData<-c(25296,28365,32187,36835,39788,44282,51122,60420,58437,
            62484)/1000
myLabels<-c(2002:2011)
myGrowth<-0
for (i in 2:length(myData)) myGrowth<-c(myGrowth,myData[i]-↘
      myData[i-1])
myValueLeft<-myData-myGrowth

x<-rbind(t(myData),t(myData))
y<-rbind(t(myValueLeft),rep(0,length(myData)))
f1<-"darkgreen"; f2<-"grey60"
myColours<-c(f1,f2)
for (i in 1:length(myData)-1) myColours<-c(myColours,f1,f2)

for (i in 1:length(myData))
{
if (y[1,i]>x[1,i])
{
  tmp<-x[1,i]; x[1,i]<-y[1,i]; y[1,i]<-tmp
  myColours[(2*i)-1]<-"darkred"
}
}

# Create chart and other elements

barplot(x,beside=T,border=NA,col=myColours,space=c(0,2),axes=F)
barplot(y,beside=T,border=NA,col=rep("grey60",2*length(myData)),↘
      add=T,names.arg=myLabels,space=c(0,2),axes=F)
axis(2,col=par("bg"),col.ticks="grey81",lwd.ticks=0.5,tck=-↘
      0.025)

hoehe<-0.1*max(myData)
j<-1
k<-j
for (i in 1:length(myData))
{
if (j > 1) k<-k+4
text(k+1.3,hoehe,format(round(x[2,i]),nsmall=0),cex=1.25,adj=0,↘
      xpd=T,col="white")
j<-j+3
if (i<length(myData)) text(k+3.1,y[1,i+1]+((x[1,i+1]-y[1,i+1])/↘
      2),
  format(round(myGrowth[i+1],1),cex=0.75,nsmall=1),adj=0)
}

# Titling

mtext("Sales Development Microsoft",3,line=4,adj=0,family="Lato ↘
      Black",outer=T,cex=2)
```

```
mtext("Figures in Bill. US-Dollars",3,line=1,adj=0,cex=1.35,font↘
     =3,outer=T)
mtext("Source: money.cnn.com",1,line=2,adj=1.0,cex=1.1,font=3,↘
     outer=T)
dev.off()
```

In the script, data are defined as in the previous example; we then assign the value 0 to a variable growth. The subsequent loop then creates a vector containing the growth as the difference from the previous value. myValueleft contains the original data minus the growths. Next, a vector x is created, which contains the transposed data twice as rows. The vector y also comprises two rows: the first one with the transposed data from myValueleft, the second with zeros.

```
> x
        [,1]    [,2]    [,3]    [,4]    [,5]    [,6]    [,7]   [,8]    [,9]   [,10]
[1,] 25.296 28.365 32.187 36.835 39.788 44.282 51.122 60.42 60.420 62.484
[2,] 25.296 28.365 32.187 36.835 39.788 44.282 51.122 60.42 58.437 62.484
>
> y
        [,1]    [,2]    [,3]    [,4]    [,5]    [,6]    [,7]    [,8]    [,9]   [,10]
[1,] 25.296 25.296 28.365 32.187 36.835 39.788 44.282 51.122 58.437 58.437
[2,]  0.000  0.000  0.000  0.000  0.000  0.000  0.000  0.000  0.000  0.000
>
```

These are going to be our columns, for which we still have to specify colours. First, the "normal" bars are set to grey and the growths to green. Now we still have to consider the possibility that the difference from the previous year may be negative. To do this, we go over all the years. If y is greater than x, then x and y have to be swapped and the colour is changed to dark red. Now the data are in the required format. First, we draw a barplot() with x. Here the parameter beside=T is important, as it plots the two barplots next to each other. Both bars are of equal height with the exception of those years with a negative difference from the previous year. The left half is green (when the difference is positive) or red (if the difference is negative); the right half is always grey. On top of these, we now plot y, also always in grey. This means we cover the green or red columns from bottom up, so only the green and red areas of the difference remain visible at the top. Then we draw the reduced axis without a line. Now come two labels. On the one hand, we write the respective values into the columns at 0.1*max(myData), that is at 10% of the height of the highest value. k has to be increased by 4 with every pass of the loop, since we put two bars next to each other and defined a gap of 2. If again transposed by 1.3, the label appears within the bar. The second label, the value of the difference, is set to appear at half the height next to the respective difference. The x-position is 3.1 steps next to k. At the end come the usual titles.

8.1.3   Quarterly Values as Columns

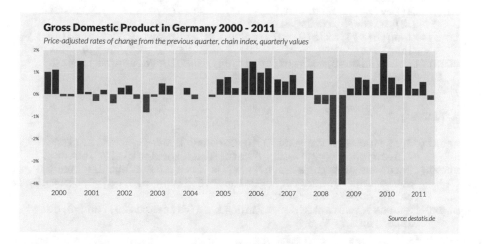

About the figure: With the representation of growth rates, there are often examples in which the negative values are shown in a different colour than the positive values. If quarterly values are displayed, then a period of one entire year should be visually identified: here, this is done by highlighting each 1-year period with a slightly darker colour, which makes them stand out from the background. In the present case, the background was tinted yellow. For x-axis labelling, the years are sufficient. The four values per year can easily be distinguished. Data can be downloaded as an XLS table from http://www.destatis.de. Please be aware that data are sorted in reversed order.

```
pdf_file<-"pdf/timeseries_quarterly_columns.pdf"
cairo_pdf(bg="grey98", pdf_file,width=14,height=7)

library(gplots)
library(gdata)
par(omi=c(0.65,0.75,0.95,0.75),mai=c(0.9,0,0.25,0.02),fg="
    cornsilk",bg="cornsilk",family="Lato Light",las=1)

# Read data and prepare chart

gdp<-read.xls("myData/GDP_germany_quarter.xlsx",sheet=2)
x<-rev(gdp$priceadjusted)
t<-unique(gdp$year)

# Create chart and other elements

par(mfcol=c(1,length(t)))
for (i in length(t):1)
{
```

```
xt<-subset(gdp$priceadjusted,gdp$year == t[i])
myColours<-rep("blue4",length(xt))
for (j in 1:length(xt)) if(xt[j]<0) myColours[j]<-"coral4"
barplot2(rev(xt),border=NA,bty="n",col=rev(myColours),ylim=c(-\
    4,2),axes=F,prcol="bisque1")
if (i==length(t)) axis(2,col="cornsilk",cex.axis=1.25,at=c(-4:2)\
    ,labels=c("-4%","-3%","-2%","-1%","0%","1%","2%"))
mtext(t[i],1,line=2,col=rgb(64,64,64,maxColorValue=255),cex\
    =1.25)
}

# Titling

mtext("Gross Domestic Product in Germany 2000 - 2011",3,line\
    =2.5,adj=0,cex=2,family="Lato Black",col="Black",outer=T)
mtext("Price-adjusted rates of change from the previous quarter,\
    chain index, quarterly values",3,line=-0.5,adj=0,cex=1.5,\
    font=3,col="Black",outer=T)
mtext("Source: destatis.de",1,line=1,adj=1,cex=1.25,font=3,col="\
    Black",outer=T)
dev.off()
```

In the script, we require the gplots package for the barplot2() function. This function provides the prcol parameter, which allows us to separately colour the background of the data area. For the figure's background, however, we first define "cornsilk". The fg parameter has to be defined with this colour as well, as not doing so would result in a border around the individual figures. Data are then read and inverted (the Statistisches Bundesamt provides them in reversed order). Then years are extracted from the data using the unique() function. This is what we use to define the number of graphic windows: one per year. Since data are sorted in reversed order, we perform the loop backwards by using for (i in length(t):1). Within the loop, the data frame is filtered year by year and one bar chart is plotted, respectively. The colour is set to blue; for negative values, red. Using the prcol parameter of the barplot2() function, the background colour can be selected to contrast with the overall background. Years are written under the respective graphic using mtext(t[i],...). During the first pass, when the condition i==length(t) is true (we are counting backwards after all), a y-axis is drawn and percent labels are attached. At the end come two titles and two captions.

8.1.4   *Quarterly Values as Lines with Value Labels*

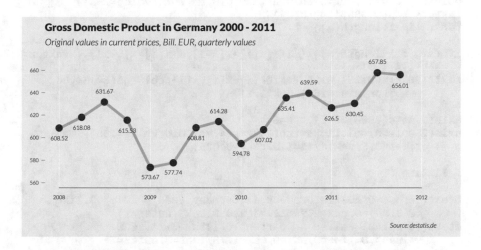

In this figure, the individual values are not only connected by a line, but also highlighted with dots of a different colour. Every dot is labelled, either above or below, with its respective value. The position of the label should be dependent on readability: if the values to the left and right are smaller than the value to be labelled, then the label should be attached above the value; if both are greater, it should be below. If one value is smaller and the other greater, then a case-by-case decision has to be made. The background was coloured darker in this instance, with white vertical auxiliary lines at turn of each year (i.e., the January values). For the x-axis, the years alone are sufficient; tick marks are not needed. A title for the y-axis can also be omitted as the unit is stated in the subtitle. Data can be downloaded as an XLS table from http://www.destatis. de.

```
pdf_file<-"pdf/timeseries_quarterly_lines.pdf"
cairo_pdf(bg="grey98", pdf_file,width=14,height=7)

par(omi=c(0.65,0.75,0.95,0.75),mai=c(0.9,0,0.25,0.02),fg=rgb↘
    (64,64,64,maxColorValue=255),bg="azure2",family="Lato ↘
    Light",las=1)

# Read data and prepare chart

gdp<-read.xls("myData/GDP_germany_quarter.xlsx",sheet=1)
gdp<-subset(gdp,gdp$year > 2007)
x<-ts(rev(gdp$jeworiginal),start=2008,frequency=4)

# Create chart and other elements

plot(x,type="n",axes=F,xlim=c(2008,2012),ylim=c(560,670),xlab=""↘
    ,ylab="")
```

```
abline(v=c(2008:2012),col="white",lty=1,lwd=1)
lines(x,lwd=8,type="b",col=rgb(0,0,139,80,maxColorValue=255))
points(x,pch=19,cex=3,col=rgb(139,0,0,maxColorValue=255))
faktor<-rep(0.985,length(x))
for (i in 1:length(x))
{
if (i>1 & i<length(x)) { if (x[i]>x[i-1] & x[i]>x[i+1]) { faktor⬎
    [i]<-1.015 } }
text((2008+i*0.25)-0.25,faktor[i]*x[i],x[i],col=rgb(64,64,64,⬎
    maxColorValue=255),cex=1.1)
}
axis(1,at=c(2008:2012),tck=0)
axis(2,col=NA,col.ticks=rgb(24,24,24,maxColorValue=255),lwd.⬎
    ticks=0.5,cex.axis=1.0,tck=-0.025)

# Titling

mtext("Gross Domestic Product in Germany 2000 - 2011",3,line⬎
    =2.3,adj=0,cex=2,family="Lato Black",outer=T)
mtext("Original values in current prices, Bill. EUR, quarterly ⬎
    values",3,line=0,adj=0,cex=1.75,font=3,outer=T)
mtext("Source: destatis.de",1,line=1,adj=1,cex=1.25,font=3,outer⬎
    =T)
dev.off()
```

In the script, both the general and the data background are set to the colour "cornsilk". The same data as in the previous example are read, but we only retain data from 2008 and later, and use ts() with the frequency=4 parameter to construct a quarterly time series. The chart is then specified with the plot() function, but the type="n" parameter keeps it from being plotted. That only happens with lines() and points() two rows later. This separation is important, because we use abline() to draw a white vertical line for orientation in between. Now comes a loop that passes every point of the time series and performs a case analysis: if the current value is greater than the neighbouring values, then the y-position of the label is multiplied by a factor of 1.015 and effectively moved upwards; in any other case, the factor remains at 0.985 so that the label is attached below the value. At the end come the axes and titles. Inconveniently located labels, which have not been considered in the case analysis, can be manually corrected in Inkscape. Alternatively, the respective position can be individually set for each point.

8.1.5 *Short Time Series Overlayed*

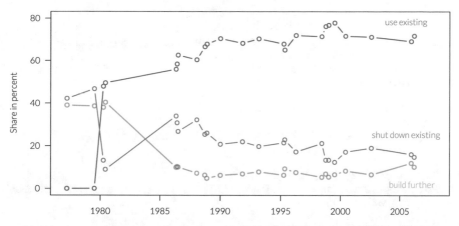

x

V39 — Attitude Towards Nuclear Energy

Seit einigen Jahren werden zur Energieversorgung Kernkraftwerke gebaut. Wie ist Ihre Meinung dazu: Sollte man auch weiterhin Kernkraftwerke bauen oder sollte man die Energieversorgung auf andere Weise sicherstellen? (1980 bis 1995 lautete die Frage:) (1980:) Über Kernkraftwerke wird ja viel diskutiert. **(1986-1988)** Denken Sie nun bitte einmal an die KKW in der Bundesrepublik. Was meinen Sie dazu: **(1990-1991:)** Wenn

Sie an die Kernkraftwerke in der Bundesrepublik denken. Was meinen Sie: **(1993-2006:)** Wenn Sie an die Kernkraftwerke hier in Deutschland denken. Was meinen Sie: Sollen weitere Kernkraftwerke gebaut werden, sollen nur die vorhandenen genutzt werden, ohne neue Kernkraftwerke zu bauen oder sollen die vorhandenen Kernkraftwerke stillgelegt werden?

The figure shows three variables of the Politbarometer, a survey program conducted by the Forschungsgruppe Wahlen (Research Group Elections). The survey has been conducted at predominantly monthly intervals since 1977, there is now a reasonably large data frame of time series. Many questions have multiple answer options, such as the question concerning the attitude towards nuclear power. This type of representation with the variable name as title could be used in documentation of a data frame (that is why no sources are stated, assuming that they can be found elsewhere). In those circumstances, it is useful to list all answer options. As can be seen from the figure, there is an abrupt drop of the proportion of the response option "decommission existing ones". Simultaneously, it is obvious that this has no substantial reason, but is based on the inclusion of a third response option "use existing ones", which maintains the largest proportion throughout the depicted time frame. With survey programs spanning such a long time frame, questions are sometimes modified over time. In such cases, the variants should be listed even if this results in an extensive legend. In the present figure, the explanatory text is placed in two columns underneath the figure.

Data: see appendix A, ZA2391: Politbarometer 1977–2011 (partial cumulation).

```
pdf_file<-"pdf/timeseries_short_inc.pdf"
cairo_pdf(bg="grey98", pdf_file,width=9,height=4.2)

source("scripts/inc_datadesign_dbconnect.r")
par(mar=c(3,5,0.5,2),omi=c(0,0,0,0),family="Lato Light",las=1)

# Read data

sql<-"select dezjahr, v39_weitere, v39_nutzen, v39_stilllegen ⤸
      from t_za2391_zeitreihen"
myDataset<-dbGetQuery(con,sql)
attach(myDataset)

# Create chart and other elements

plot(type="n",xlab="",ylab="Share in percent",dezjahr,v39_⤸
      weitere,ylim=c(0,80))

myVars1<-c("dezjahr","v39_weitere")
myPoints1<-subset(myDataset[myVars1],!is.na(myDataset[myVars1]$⤸
      v39_weitere))

myVars2<-c("dezjahr","v39_nutzen")
myPoints2<-subset(myDataset[myVars2],!is.na(myDataset[myVars2]$⤸
      v39_nutzen))

myVars3<-c("dezjahr","v39_stilllegen")
myPoints3<-subset(myDataset[myVars3],!is.na(myDataset[myVars3]$⤸
      v39_stilllegen))

myColour1<-rgb(200,97,0,150,maxColorValue=255)
myColour2<-rgb(100,97,0,maxColorValue=255)
myColour3<-rgb(130,130,130,maxColorValue=255)

points(myPoints1,col=myColour1,lwd=2,type="b")
points(myPoints2,col=myColour2,lwd=2,type="b")
points(myPoints3,col=myColour3,lwd=2,type="b")

text(2006,2,"build further",col=myColour1)
text(2005.5,78,"use existing",col=myColour2)
text(2005.5,25,"shut down existing",col=myColour3)

dev.off()
```

as well as in LaTeX

```
\documentclass{article}
\usepackage[paperheight=21cm, paperwidth=29.7cm, top=1.25cm, ⤸
      left=0cm, right=0cm, bottom=0cm]{geometry}
\usepackage{multicol}
\usepackage[german]{babel}
\usepackage{graphicx,color}
```

```
\usepackage{fontspec}
\setmainfont[Mapping=text-tex]{Lato Light}
\setlength{\columnsep}{1.5pc}
\definecolor{myBackground}{rgb}{0.99,0.99,0.99}
\pagecolor{myBackground}
\linespread{1.2}
\begin{document}
\pagestyle{empty}
\fontsize{24pt}{18pt}\selectfont
\hspace*{2.0cm}
\textbf{V39 — Attitude Towards Nuclear Energy}
\begin{center}
\fontsize{12pt}{14pt}\selectfont
\vspace*{0.15cm}
\includegraphics[width=1.00\textwidth]{../pdf/timeseries_short_\
    inc.pdf}
\vspace*{0.25cm}
\begin{minipage}[t]{26.5cm}
\begin{multicols}{2}
Seit einigen Jahren werden zur Energieversorgung Kernkraftwerke \
    gebaut. Wie ist Ihre Meinung dazu: Sollte man auch \
    weiterhin Kernkraftwerke bauen oder sollte man die \
    Energieversorgung auf andere Weise sicherstellen?
(1980 bis 1995 lautete die Frage:) (1980:) Über Kernkraftwerke \
    wird ja viel diskutiert. \textbf{(1986-1988)}
Denken Sie nun bitte eimmal an die KKW in der Bundesrepublik. \
    Was meinen Sie dazu: \textbf{(1990-1991:)} Wenn Sie an die\
    Kernkraftwerke in der Bundesrepublik denken. Was meinen \
    Sie: \textbf{(1993-2006:)} Wenn Sie an die Kernkraftwerke \
    hier in Deutschland denken. Was meinen Sie: Sollen weitere\
    Kernkraftwerke gebaut werden, sollen nur die vorhandenen \
    genutzt werden, ohne neue Kernkraftwerke zu bauen oder \
    sollen die vorhandenen Kernkraftwerke stillgelegt werden?
\end{multicols}
\end{minipage}
\end{center}
\end{document}
```

In the script, data are read from a MySQL table. The three variables are first defined as respective point pairs; during this procedure, we first exclude the missing values. Then, the series are plotted one after the other using points() and selecting the "b" type. With this type, points and lines are plotted, but lines are not drawn all the way to the points.

8.2   Areas Underneath and Between Time Series

8.2.1   Areas Between Two Time Series

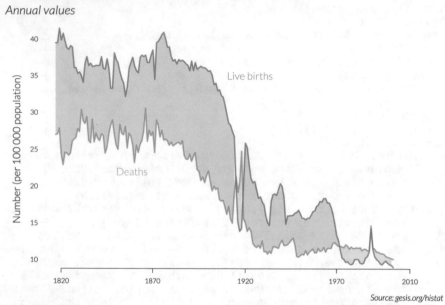

Live births and deaths in Germany 1820-2001
Annual values

Source: gesis.org/histat

About the figure: When presenting two time series in one figure, it can be useful to colour highlight the area between the time series. Any overlap of the series has to be considered when choosing a colour for the area. In the present case, the area reflects the difference between the number of live births and deaths in Germany between 1820 and 2001. There are two organge-coloured phases in which the number of deaths exceeds the number of live births.

Data are available from http://www.gesis.org/histat/ for registered users.

```
pdf_file<-"pdf/timeseries_areas_between.pdf"
cairo_pdf(bg="grey98", pdf_file,width=11.69,height=8.27)

par(mai=c(1,1,0.5,0.5),omi=c(0,0.5,1,0),family="Lato Light",las↘
    =1)

# Import data and prepare chart

library(gdata)
rs<-read.xls("myData/B1_01.xls",1,header=F,encoding="latin1")
myColour1_150<-rgb(68,90,111,150,maxColorValue=255)
myColour1_50<-rgb(68,90,111,50,maxColorValue=255)
myColour2_150<-rgb(255,97,0,150,maxColorValue=255)
```

```
myColour2_50<-rgb(255,97,0,50,maxColorValue=255)
attach(rs)

# Define graphic and other elements

plot(V1,V11,axes=F,type="n",xlab="",ylab="Number (per 100 000 ⬐
      population)",cex.lab=1.5,xlim=c(1820,2020),ylim=c(10,40),⬐
      xpd=T)
axis(1,at=c(1820,1870,1920,1970,2010))
axis(2,at=c(10,15,20,25,30,35,40),col=par("bg"),col.ticks="⬐
      grey81",lwd.ticks=0.5,tck=-0.025)
lines(V1,V11,type="l",col=myColour1_150,lwd=3,xpd=T)
lines(V1,V12,type="l",col=myColour2_150,lwd=3)
text(1910,35,"Live births",adj=0,cex=1.5,col=myColour1_150)
text(1850,22,"Deaths",adj=0,cex=1.5,col=myColour2_150)
myBegin<-c(1817,1915,1919,1972); ende<-c(1914,1918,1971,2000)
myColour<-c(myColour1_50,myColour2_50,myColour1_50,myColour2_50)
for (i in 1:length(myBegin))
{
mySubset<-subset(rs,V1 >= myBegin[i] & V1 <= ende[i])
attach(mySubset)
xx<-c(mySubset$V1,rev(mySubset$V1)); yy<-c(mySubset$V11,rev(⬐
      mySubset$V12))
polygon(xx,yy,col=myColour[i],border=F)
}

# Titling

mtext("Live births and deaths in Germany 1820-2001",3,line=1.5,⬐
      adj=0,family="Lato Black",cex=2.2,outer=T)
mtext("Annual values",3,line=-0.75,adj=0,font=3,cex=1.8,outer=T)
mtext("Source: gesis.org/histat",1,line=3,adj=1,cex=1.2,font=3)
dev.off()
```

The script begins with margin settings and then specifies two colour gradations. The first one contains a transparency value of 150, the second one of 50. Both axes and lines are suppressed, and the axis then plotted separately with the axis() function. Then, the time series are drawn and labelled using text(). The areas between the time series are filled using the polygon() function. Since we want the areas to be coloured differently—depending on which series is above and which below—we have to distinguish several segments. To do this, we specify four starting points and four end points. The loop now passes through these four time segments. A partial data frame is constructed for each segment and used to create two new vectors. The first one contains the x-axis values, initially from left to right and then, using the rev() function, from right to left. Accordingly, the second one contains first the y-values of the first row from left to right, and then the y-values of the second row from right to left. Using these data, the polygon() function can now fill in the areas.

8.2.2   Areas as Corridor with Time Series (Panel)

About the figure: Another use of areas in time series is the presentation of a "corridor"—an area within which the time series move. In the present case, we compare the development of eight time series. In this example, the corridor is the respective smallest and greatest annual value of each of the eight time series. So as not to overload the image, the figure is composed of four graphics arranged as a 2 × 2 panel; aside from the corridor made up of all eight time series, each graphic only contains two matching series. With this type of panel, the x-axis can be omitted in the top two graphics, and labels and scale lines suffice for the y-axes. The names of the time series should not be stated in the legend, but written directly on the time series or at least in relatable proximity.

Prices for wheat and rye in Europe 1200-1960

in g silver/100 kg, decennial means

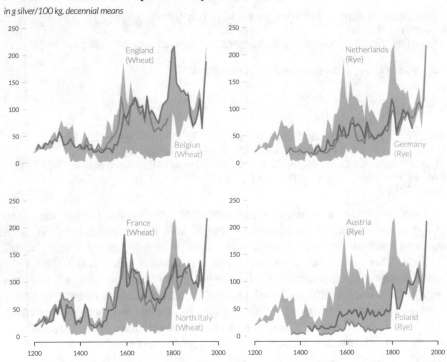

Data derive from Wilhelm Abel's book "Agrarkrisen und Agrarkonjunktur. Eine Geschichte der Land- und Ernährungswirtschaft Mitteleuropas seit dem hohen Mittelalter." [Publication in German]. Data are available as a data frame ZA8082 from http://gesis.org/histat.

```
pdf_file<-"pdf/timeseries_areas_corridor_2x2.pdf"
cairo_pdf(bg="grey98", pdf_file,width=10.65,height=9.2)

source("scripts/inc_datadesign_dbconnect.r")
par(mai=c(0.6,0.5,0,0),omi=c(0.2,0.5,1.25,0.25),mfcol=c(2,2),\
     family="Lato Light",las=1)

# Import data and prepare chart

myColour1_150<-rgb(68,90,111,150,maxColorValue=255)
myColour2_150<-rgb(255,97,0,150,maxColorValue=255)

y1variable<-c("England_Weizen",
        "Frankreich_Weizen",
        "Niederlande_Roggen",
        "Oesterreich_Roggen")
y2variable<-c("Belgien_Weizen",
        "Oberitalien_Weizen",
        "Deutschland_Roggen",
        "Polen_Roggen")

y1label<-c("England\n(Wheat)",
         "France\n(Wheat)",
         "Netherlands\n(Rye)",
         "Austria\n(Rye)")
y2label<-c("Belgiun\n(Wheat)",
         "North Italy\n(Wheat)",
         "Germany\n(Rye)",
         "Poland\n(Rye)")

for (i in 1:length(y1variable))
{
sql<-paste("select jahr x,",y1variable[i]," y1,",y2variable[i],"\
       y2,Min_Getreide,Max_Getreide from z8082 where jahr > 0",\
       sep="")
rs<-dbGetQuery(con,sql)

attach(rs)


# Create chart and other elements

plot(x,y1,axes=F,type="n",xlab="",ylab="",cex.lab=0.8,xlim=c\
     (1200,2000),ylim=c(0,250),xpd=T)
xx<-c(x,rev(x))
yy<-c(Max_Getreide,rev(Min_Getreide))
polygon(xx,yy,col=rgb(68,90,111,80,maxColorValue=255),border=F)
lines(x,y1,type="l",col=myColour1_150,lwd=3,xpd=T)
lines(x,y2,type="l",col=myColour2_150,lwd=3)
text(1600,200,y1label[i],adj=0,cex=1.3,col=myColour1_150)
text(1810,25,y2label[i],adj=0,cex=1.3,col=myColour2_150)
if (i==2 | i==4) axis(1,at=c(1200,1400,1600,1800,2000))
axis(2,col=par("bg"),col.ticks="grey81",lwd.ticks=0.5,tck=-\
     0.025)
```

```
}

# Titling

mtext("Prices for wheat and rye in Europe 1200-1960",3,line=3,↘
      adj=0,family="Lato Black",cex=1.8,outer=T)
mtext("in g silver/100 kg, decennial means",3,line=1,adj=0,font↘
      =3,cex=1.2,outer=T)
dev.off()
```

In the script, colours are specified first, followed by variable names and labels. Then a loop is used to build an SQL command with these variables, and to generate one respective data frame containing two time series, and the minimal and maximal values that are also stored in the database. The area is plotted with the polygon() function as in the previous example, and so are the series. With this panel representation, we can omit the x-axes in the top two charts, so we use one condition to limit the axis output to passes 2 and 4.

8.2.3   Forecast Intervals (Panel)

UN Population Forecasts
values in millions, 5 year values

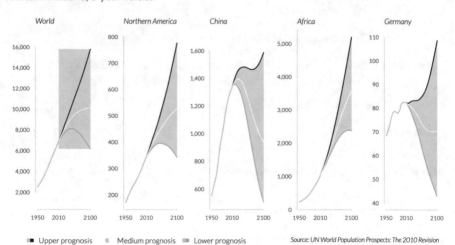

The figure shows UN population prognoses for the world, the USA, China, Africa, and Germany. Prognoses are based on the development from 1950 to 2010, and due to different hypotheses, they build one divergent corridor each. A maximal, a minimal, and a medium development was calculated. The individual charts are arranged next to each other such that each spans the entire vertical extent. This results in different scaling in each graphic, so we require an individual y-axis for each chart. The unit explanation "values in millions" comes with the first graphic. x-axis labels are limited to the start and end values and the start of the forecast (2010). In this case, the series' colours are explained in a legend, since they are consistent throughout all five graphics. The legend has been designed to also show the colour of the area. That way, the white line is also visible. Data are provided by the UN as a CSV file on http://esa.un.org/unpd/wpp. They look like this:

```
"Country","Variable","Variant","Year","Value"
"China","Population (thousands)","Medium variant","1950",550771
"China","Population (thousands)","High variant","1950",550771
"China","Population (thousands)","Low variant","1950",550771
"China","Population (thousands)","Constant-fertility variant","1950",550771
"China","Population (thousands)","Medium variant","1955",608360
"China","Population (thousands)","High variant","1955",608360
"China","Population (thousands)","Low variant","1955",608360
"China","Population (thousands)","Constant-fertility variant","1955",608360
"China","Population (thousands)","Medium variant","1960",658270
.
.
.
```

As you can see, they are arranged below each other, which means that forecast variants are not variables but factors.

```
pdf_file<-"pdf/timeseries_forecast_interval_1x5.pdf"
cairo_pdf(bg="grey98", pdf_file,width=11,height=7)

par(mfcol=c(1,5),omi=c(1.0,0.25,1.45,0.25),mai=c(0,0.75,0.25,0),\
     family="Lato Light",las=1)

# Import data and prepare chart

UNPop<-read.csv("myData/UNPop.csv")
mySelection<-c("World","Northern America","China","Africa","\
     Germany")
ymin<-c(1000,170,500,200,40)
ymax<-c(17000,800,1700,5200,110)
myTitle<-c("World","Northern America","China","Africa","Germany"\
     )

# Create charts and other elements

for (i in 1:length(mySelection)) {
source("scripts/inc_forecast_interval_05.r")
mtext(myTitle[[i]],side=3,adj=0,line=1,cex=1.1,font=3)

if (myTitle[[i]] == "World")
{
```

```
legend(1900,-1750,c("Upper prognosis ","Medium prognosis","Lower⤸
      prognosis"),fill=c("grey","grey","grey"),border=F,xpd=NA,⤸
      pch=15,col=c("black","white","orange"),bty="n",cex=1.6,⤸
      ncol=3)
}
}

# Titling

mtext("UN Population Forecasts",3,line=7,adj=0,cex=2.25,family="⤸
      Lato Black",outer=T)
mtext("values in millions, 5 year values",3,line=3.5,adj=0,cex⤸
      =1.75,font=3,outer=T)
mtext("Source: UN World Population Prospects: The 2010 Revision"⤸
      ,1,line=5,adj=1.0,cex=0.95,font=3,outer=T)
dev.off()
```

and

```
# inc_forecast_interval_05.r
myCountry<-subset(UNPop,UNPop$Country==mySelection[i] & UNPop$⤸
      Variant=="Medium variant")
myPrognoses<-subset(UNPop,UNPop$Country == mySelection[i] & Year⤸
       >= 2010)
myPrognosis_L<-subset(myPrognoses,myPrognoses$Variant=="Low ⤸
      variant")$Value/1000
myPrognosis_M<-subset(myPrognoses,myPrognoses$Variant=="Medium ⤸
      variant")$Value/1000
myPrognosis_H<-subset(myPrognoses,myPrognoses$Variant=="High ⤸
      variant")$Value/1000
myYears<-seq(2010,2100,by=5)
attach(myCountry)

plot(axes=F,type="n",xlab="",ylab="",Year,Value/1000,ylim=c(ymin⤸
      [[i]],ymax[[i]]))
axis(1,tck=-0.01,col="grey",at=c(1950,2010,2100),cex.axis=1.2)
axis(2,tck=-0.01,col="grey",at=py<-pretty(c(myPrognosis_L,Value/⤸
      1000,myPrognosis_H)),labels=format(py,big.mark=","),cex.⤸
      axis=1.2)

xx<-c(myYears,rev(myYears))
yy<-c(myPrognosis_H,rev(myPrognosis_L))
polygon(xx,yy,col=rgb(68,90,111,50,maxColorValue=255),border=F)

lines(Year,Value/1000,col="grey",lwd=2)
lines(myYears,myPrognosis_H,col="black",lwd=2)
lines(myYears,myPrognosis_L,col="orange",lwd=2)
lines(myYears,myPrognosis_M,col="white",lwd=2)
```

In the script, we split the figure into five windows using mfcol, then read the data from a CSV file. We begin by defining the selection, the minimal and maximal values for the y-axes for the respective selections, and the individual headings. For the number of elements, a loop is executed that embeds a file inc_forecast_interval_05.r. That file first reads those rows of the data frame in which the variable MyCountry matches the desired value. Since the desired values "Low variant", "Medium variant", and "High variant" are not ordered by columns but by rows, we create three subsets. We then specify an empty plot, followed by the axes. With the pretty() function, which is called within axis() and immediately assigned to a variable py, we achieve "pretty" y-axis scaling. The axis() function would attempt this anyway, but since we are using the labels parameter to show thousands separators, we also specify the position using at. The procedure for creation of areas matches the two previous examples. We use lines() to draw three lines on each of these areas: a black one at the top, an orange one at the bottom, and a white line in the middle. After embedding the inc_forecast_interval_05.r file, the title is also written within the loop; during the first pass, output also includes an explanatory text "values in millions" and the legend. We design the legend so that grey-highlighted area of the forecast corridors also becomes visible with the fill parameter.

8.2.4 Forecast Intervals Index (Panel)

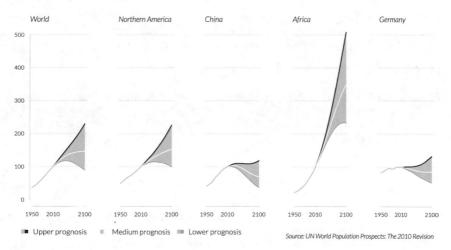

UN Population Forecasts

2010=100, 5 year values

   The figure again shows UN population prognoses for the world, the USA,
China, Africa, and Germany; in this instance though, they are shown as indexes
with the 2010 value set to 100. In this variant, it is immediately obvious
that the prognoses for Africa assume a markedly more dynamic development.
Since the unit is the same throughout (the index), we only need a single y-
axis in the first graphic. However, horizontal auxillary lines should be added
to all charts for ease of orientation. Occasionally, there is criticism on the
practice of equating a year with the value 100 ("2010=100"), which is also
common practice with the Statistische Bundesamt (German Statistical Federal
Office); however, we do not follow this criticism here. Data are provided by
the UN and can be downloaded as a CSV file from http://esa.un.org/unpd/
wpp.

```
pdf_file<-"pdf/timeseries_forecast_interval_1x5_index.pdf"
cairo_pdf(bg="grey98", pdf_file,width=11,height=7)

par(mfcol=c(1,5),omi=c(1.0,0.25,1.45,0.25),mai=c(0,0.75,0.25,0),↘
      family="Lato Light",las=1)

# Import data and prepare chart

UNPop<-read.csv("myData/UNPop.csv")

mySelection<-c("World","Northern America","China","Africa","↘
      Germany")
ymin<-rep(0, 5)
ymax<-rep(500, 5)
myTitle<-c("World","Northern America","China","Africa","Germany"↘
      )

for (i in 1:length(mySelection)) {
source("scripts/inc_forecast_interval_05_index.r")
mtext(myTitle[i],side=3,adj=0,line=1,cex=1.1,font=3)

if (i==1)
{
legend(1900,-70,c("Upper prognosis ","Medium prognosis","Lower ↘
      prognosis"),fill=c("grey","grey","grey"),border=F,pch=15,↘
      xpd=NA,col=c("black","white","orange"),bty="n",cex=1.6,↘
      ncol=3)
}
}

# Titling

mtext("UN Population Forecasts",3,line=7,adj=0,cex=2.25,family="↘
      Lato Black",outer=T)
mtext("2010=100, 5 year values",3,line=3.5,adj=0,cex=1.75,font↘
      =3,outer=T)
mtext("Source: UN World Population Prospects: The 2010 Revision"↘
      ,1,line=5,adj=1.0,cex=0.95,font=3,outer=T)
dev.off()
```

and

```
# inc_forecast_interval_05_index.r
myLand<-subset(UNPop,UNPop$Country==mySelection[i] & UNPop$
    Variant=="Medium variant")
myPrognoses<-subset(UNPop,UNPop$Country == mySelection[i] & Year
    >= 2010)
myPrognosis_L<-subset(myPrognoses,myPrognoses$Variant=="Low
    variant")$Value/1000
myPrognosis_M<-subset(myPrognoses,myPrognoses$Variant=="Medium
    variant")$Value/1000
myPrognosis_H<-subset(myPrognoses,myPrognoses$Variant=="High
    variant")$Value/1000
myYears<-seq(2010,2100,by=5)
attach(myLand)
myBase<-(Value[13]/1000)

plot(axes=F,type="n",xlab="",ylab="",Year,Value/1000,ylim=c(ymin
    [i],ymax[i]))
py<-c(0,100,200,300,400,500)
abline(h=py[2:6], col="lightgray",lty="dotted")
axis(1,tck=-0.01,col="grey",at=c(1950,2010,2100),cex.axis=1.2)
py<-c(0,100,200,300,400,500)
if (mySelection[i]=="World")
{
axis(2,tck=-0.01,col="grey",at=py,labels=format(py,big.mark=".")
    ,cex.axis=1.2)
}
xx<-c(myYears,rev(myYears))
yy<-c(100*myPrognosis_H/basis,rev(100*myPrognosis_L/basis))

polygon(xx,yy,col=rgb(192,192,192,maxColorValue=255),border=F)

lines(Year,100*(Value/1000)/basis,col="grey",lwd=2)
lines(myYears,100*myPrognosis_H/basis,col="black",lwd=2)
lines(myYears,100*myPrognosis_L/basis,col="orange",lwd=2)
lines(myYears,100*myPrognosis_M/basis,col="white",lwd=2)
```

The script largely works as in the previous example. The differences lie on the one hand in identical scaling of all y-axes, and on the other hand in the presentation of an index value that uses value[13], the year 2010 value, as base value.

8.2.5   Time Series with Stacked Areas

Gross electricity generation in Bavaria 1990-2011
All values in mil. kWh, annual figures

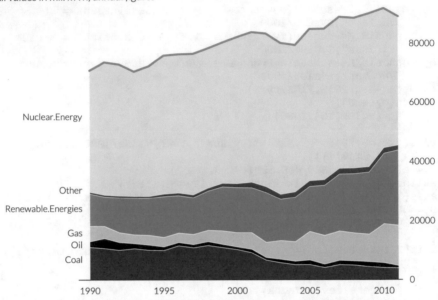

Source: www.statistik.bayern.de

About the figure: A frequently encountered variant when presenting multiple
time series is that of stacked areas. In these, the individual time series' values are
added and the space between them is filled with areas. The present example shows
the gross power generation in the German federal state of Bavaria between 1990 and
2011. Colours are chosen to be associated with the respective types of energy. For
ease of viewing, a white line is added between the different areas, and the upper edge
is emphasised with a slightly darker line than the uppermost area. Depending on the
available data, such figures should be taken with a pinch of salt, since an ascent or
descent of the uppermost series can give the impression that the series varied in the
respective direction. However, this could just as easily have been caused by one or
more of the lower series. Data can be downloaded from the website of the Bavarian
State Statistical Office (http://www.statistik.bayern.de/).

```
pdf_file<-"pdf/timeseries_stacked_areas.pdf"
cairo_pdf(bg="grey98", pdf_file,width=9,height=7)

library(plotrix)
library(gdata)
par(mai=c(0.5,1.75,0,0.5),omi=c(0.5,0.5,0.8,0.5),family="Lato ⤸
    Light",las=1)
```

```
# Import data and prepare chart

myData<-read.xls("myData/Power_generation_Bavaria.xlsx", \
     encoding="latin1")

myC1<-"brown"
myC2<-"black"
myC3<-"grey"
myC4<-"forestgreen"
myC5<-"blue"
myC6<-"lightgoldenrod"

myYears<-myData$Year
myData$Year<-NULL
Complete<-myData$Complete
myData$Complete<-NULL

fg_org<-par("fg")
par(fg=par("bg"))


# Create chart and other elements

stackpoly(myData,main="",xaxlab=rep("", nrow(myData)),border="\
     white",stack=TRUE,col=c(myC1,myC2,myC3,myC4,myC5,myC6), \
     axis2=F, ylim=c(0,95000))
lines(Complete, lwd=4, col="lightgoldenrod4")
par(fg=fg_org)
mtext(seq(1990,2010,by=5), side=1, at=seq(1,21,by=5), line=0.5)
segments(0.25,0,22.25,0,xpd=T)
ypos<-c(7000,12000,16000,24000,30500,55000)
myDes<-names(myData)
text(rep(0.5,6), ypos, myDes, xpd=T, adj=1)

# Titling

mtext("Gross electricity generation in Bavaria 1990-2011",3,line\
     =1.5,adj=0,family="Lato Black",cex=1.75,outer=T)
mtext("All values in mil. kWh, annual figures",3,line=-0.2,adj\
     =0,font=3,cex=1.25,outer=T)
mtext("Source: www.statistik.bayern.de",1,line=1,adj=1,cex=0.9,\
     font=3,outer=T)
dev.off()
```

In the script, we use the stackpoly() function from Jim Lemon's plotrix package for the presentation of stacked time series. We first specify suitable colours for the individual series, read the data, save the "complete" series in a separate vector, and then delete it from the data frame. The axes created by the stackpoly() function are suppressed by setting the parameter of fg to the background colour. However, we first save the original value fg, so we can restore it after creation of the chart. The stackpoly() call is then done with "empty" x-axis labels. By default, the function creates a y-axis on both sides, left and right; axis2=F suppresses the left one. After the call, we use lines() to enhance the upper border, and then restore the original settings of the parameter value of fg. The x-axis is created with the mtext() function, and the line is added with segments(). At the end, the labels for the areas are attached at the positions ypos on the left side. We can simply use the variable names to do this. Finally, the usual titles and sources were added.

8.2.6 Areas Under a Time Series

Gross national product of Chile

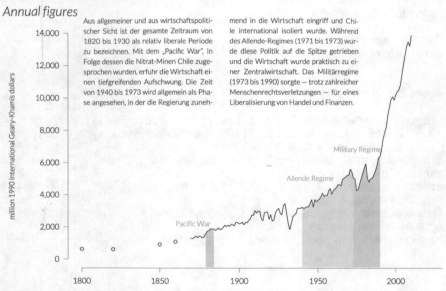

Annual figures

Aus allgemeiner und aus wirtschaftspolitischer Sicht ist der gesamte Zeitraum von 1820 bis 1930 als relativ liberale Periode zu bezeichnen. Mit dem „Pacific War", In Folge dessen die Nitrat-Minen Chile zugesprochen wurden, erfuhr die Wirtschaft einen tiefgreifenden Aufschwung. Die Zeit von 1940 bis 1973 wird allgemein als Phase angesehen, in der die Regierung zunehmend in die Wirtschaft eingriff und Chile international isoliert wurde. Während des Allende-Regimes (1971 bis 1973) wurde diese Politik auf die Spitze getrieben und die Wirtschaft wurde praktisch zu einer Zentralwirtschaft. Das Militärregime (1973 bis 1990) sorgte – trotz zahlreicher Menschenrechtsverletzungen – für eines Liberalisierung von Handel und Finanzen.

Source: Rolf Lüders, The Comparative Economic Performance of Chile 1810-1995, www.ggdc.net/maddison

About the figure: Areas under time series can be especially useful for emphasis of individual periods in time series or "phases". These phases could just be shown as a rectangular area over the entire height of the chart. The highlighted time periods stand out more though, if the upper part of the area stops under the time series. Different phases can have different colours (they obviously have to if they immediately follow each other). In this example, we choose different graduations of a single colour. Phases should be labelled in their immediate vicinity; a legend is not necessary. Since the series is missing values at the start, both points and lines are represented. Given the specific form of the data (a strong growth trend), it pays to use the free space for an explanatory text. In this case, text is formatted in two columns and justified. Data are taken from Angus Maddison's database and can be downloaded from http://www.ggdc.net/maddison. I transferred them into a separate XLS table.

```
pdf_file<-"pdf/timeseries_areas_below_inc.pdf"
cairo_pdf(bg="grey98", pdf_file,width=11.69,height=8.26)

library(gdata)
par(cex.axis=1.1,mai=c(0.75,1.5,0.25,0.5),omi=c(0.5,0.5,1.1,0.5)\
     , mgp=c(6,1,0),family="Lato Light",las=1)

# Import data and prepare chart

colour<-rgb(68,90,111,150,maxColorValue=255)
myData<-read.xls("myData/chile.xlsx", encoding="latin1")
attach(myData)

# Define chart and other elements

plot(x,y,axes=F,type="n",xlab="",xlim=c(1800,2020),ylim=c\
     (0,14000),xpd=T,ylab="million 1990 International Geary-\
     Khamis dollars")
axis(1,at=pretty(x),col=colour)
axis(2,at=py<-pretty(y),col=colour,cex.lab=1.2,labels=format(py,\
     big.mark=","))

y<-ts(y,start=1800,frequency=1)
points(window(y, end=1869))
lines(window(y, start=1870))

myShapeColour1<-rgb(0,128,128,50,maxColorValue=255)
myShapeColour2<-rgb(0,128,128,80,maxColorValue=255)
mySelection<-subset(myData,x >= 1879 & x <= 1884)
attach(mySelection)
polygon(c(min(mySelection$x),mySelection$x,max(mySelection$x)),c\
     (-500,mySelection$y,-500),col=myShapeColour2,border=NA)
text(1860,2200,adj=0,col=colour,"Pacific War")
mySelection<-subset(myData,x >= 1940 & x <= 1973)
attach(mySelection)
polygon(c(min(mySelection$x),mySelection$x,max(mySelection$x)),c\
     (-500,mySelection$y,-500),col=myShapeColour1,border=NA)
text(1930,5000,adj=0,col=colour,"Allende Regime")
mySelection<-subset(myData,x >= 1973 & x <= 1990)
attach(mySelection)
polygon(c(min(mySelection$x),mySelection$x,max(mySelection$x)),c\
     (-500,mySelection$y,-500),col=myShapeColour2,border=NA)
text(1960,6800,adj=0,col=colour,"Military Regime")

# Titling

mtext("Gross national product of Chile",3,line=2,adj=0,cex=2.4,\
     family="Lato Black", outer=T)
mtext("Annual figures",3,line=-0.5,adj=0,cex=1.8,font=3, outer=T\
     )
mtext("Source: Rolf Lüders, The Comparative Economic Performance\
      of Chile 1810-1995, www.ggdc.net/maddison",1,line=3,adj\
     =1.0,cex=0.95,font=3)
dev.off()
```

Here the integration into LaTeX:

```
\documentclass{article}
\usepackage[paperheight=21cm, paperwidth=29.7cm,
  top=0cm, left=0cm, right=0cm, bottom=0cm]{geometry}
\usepackage{multicol}
\usepackage[german]{babel}
\usepackage{color}
\usepackage{fontspec}
\usepackage[abs]{overpic}
\setmainfont[Mapping=text-tex]{Lato Light}
\setlength{\columnsep}{1cm}
\definecolor{hintergrund}{rgb}{0.9412,0.9412,0.9412}
\pagecolor{hintergrund}
\begin{document}
\pagestyle{empty}
\begin{center}
\fontsize{12pt}{17pt}\selectfont
\begin{overpic}[scale=1,unit=1mm]{../pdf/timeseries_areas_below_\
    inc.pdf}
\put(60,128){\begin{minipage}[t]{16.25cm}
\begin{multicols}{2}
Aus allgemeiner und aus wirtschaftspolitischer Sicht ist der \
    gesamte Zeitraum von 1820 bis 1930 als relativ liberale \
    Periode zu bezeichnen. Mit dem „Pacific War", in Folge \
    dessen die Nitrat-Minen Chile zugesprochen wurden,  erfuhr\
     die Wirtschaft einen tiefgreifenden Aufschwung. Die Zeit \
    von 1940 bis 1973 wird allgemein als Phase angesehen, in \
    der die Regierung zunehmend in die Wirtschaft eingriff und\
     Chile international isoliert wurde. Während des Allende-\
    Regimes (1971 bis 1973)  wurde diese Politik auf die \
    Spitze getrieben und die Wirtschaft wurde praktisch zu \
    einer Zentralwirtschaft. Das Militärregime (1973 bis 1990)\
     sorgte — trotz zahlreicher Menschenrechtsverletzungen — f\
    ür eines Liberalisierung von Handel und Finanzen.
\end{multicols}
\end{minipage}}
\end{overpic}
\end{center}
\end{document}
```

In the script, we first specify the figure for the data in the usual way using plot(). Data are only available for isolated years prior to 1870. If we drew the time series with the lines() function or with the type="l" parameter, then these individual values would not be visible, as no line would be drawn between them. For this reason, we distinguish two sections, which can easily be specified using the window() function: data up to 1860 are plotted as points, data from 1870 onwards as lines. Now we have to add the areas under the time series. We can do this with the polygon() function that was introduced in Sect. 7.3.1. Complementary labels within the area are then attached by embedding the PDF file created by R into a LaTeX file (as described in Sect. 4.1).

8.2.7   Time Series with Trend (Panel)

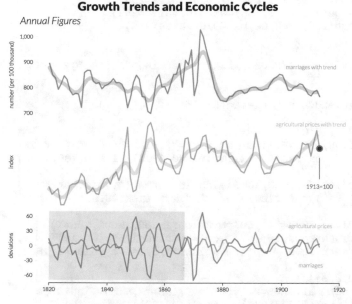

About the figure: In this variant, three charts with two time series each are arranged above each other. The top two charts show both the original values as well as smoothed values that eliminate short-term fluctuations. Such smoothed values can be expressed with a marginally thicker line type and transparent form of the original colour. The chart in the middle shows index values with the base year 1913, which should be marked accordingly: unlike the example in Sect. 8.2.4, this fact is not obvious from the heading. Any labels that are attached to the series (instead of being listed in a legend) can summarise both series in this instance. The lower chart shows the deviation of the original from the smoothed values of the upper two charts. The colours match the upper series. Additionally, an area from 1820 to 1867 is highlighted with a coloured rectangle. Instead of cross headings, the y-axis title serves as an explanatory statement. Only the lowest chart gets an x-axis. On the middle one, the y-axis label has been plotted on the right. In this case, the explanatory text is added in the form of a marginal note. Data derive from a study by Reinhard Spree entitled "Wachstumstrends und Konjunkturzyklen von 1820 bis 1913" ("Growth trends and economical cycles from 1820 to 1913") [in German]. Data can be downloaded from the time series database http://www.gesis.org/histat.

```
pdf_file<-"pdf/timeseries_with_trend_3x1_inc.pdf"
cairo_pdf(bg="grey98", pdf_file,width=11,height=9.5)

par(mfcol=c(3,1),cex.axis=1.4,mgp=c(5,1,0),family="Lato Light",
    las=1)
par(omi=c(0.5,0.5,1.1,0.5),mai=c(0,2,0,0.5))
```

```
# Prepare chart and import data

myColour1_150<-rgb(68,90,111,150,maxColorValue=255)
myColour1_50<-rgb(68,90,111,50,maxColorValue=255)
myColour2_150<-rgb(255,97,0,150,maxColorValue=255)
myColour2_50<-rgb(255,97,0,50,maxColorValue=255)

library(gdata)
myData<-read.xls("myData/z8053.xlsx", encoding="latin1")
attach(myData)

# Define graphic and other elements

par(mai=c(0,1.0,0.25,0))
plot(year,marriage,axes=F,type="n",xlab="",ylab="number (per 100↘
        thousand)",cex.lab=1.5,xlim=c(1820,1920),ylim=c(700,1000)↘
        ,xpd=T)
axis(2,at=py<-c(700,800,900,1000),labels=format(py,big.mark=",")↘
        ,col=par("bg"),col.ticks="grey81",lwd.ticks=0.5,tck=-↘
        0.025)
lines(year,marriage,type="l",col=myColour1_150,lwd=3,xpd=T)
lines(year,marriagetrend,type="l",col=myColour1_50,lwd=10)
text(1910,880,"marriages with trend",cex=1.5,col=myColour1_150)

par(mai=c(0,1.0,0,0))
plot(year,agricultural,axes=F,type="n",xlab="",ylab="index",cex.↘
        lab=1.5,xlim=c(1820,1920),ylim=c(40,130))
axis(4,at=c(40,70,100,130),col=par("bg"),col.ticks="grey81",lwd.↘
        ticks=0.5,tck=-0.025)
lines(year,agricultural,type="l",col=myColour2_150,lwd=3)
lines(year,agriculturaltrend,type="l",col=myColour2_50,lwd=10)
text(1910,125,"agricultural prices with trend",cex=1.5,col=↘
        myColour2_150,xpd=T,)
text(1913,60,"1913=100",cex=1.5,col=rgb(100,100,100,↘
        maxColorValue=255))

arrows(1913,68,1913,90,length=0.10,angle=10,code=0,lwd=2,col=rgb↘
        (100,100,100,maxColorValue=255))
points(1913,100,pch=19,col="white",cex=3.5)
points(1913,100,pch=1,col=rgb(25,25,25,200,maxColorValue=255),↘
        cex=3.5)
points(1913,100,pch=19,col=rgb(25,25,25,200,maxColorValue=255),↘
        cex=2.5)

par(mai=c(0.5,1.0,0,0))
plot(year,marriagez,axes=F,type="n",xlab="",ylab="deviations",↘
        cex.lab=1.5,xlim=c(1820,1920),ylim=c(-70,70))
axis(1,at=pretty(year))
axis(2,at=c(-60,-30,0,30,60),col=par("bg"),col.ticks="grey81",↘
        lwd.ticks=0.5,tck=-0.025)
rect(1820,-70,1867,70,border=F,col="grey90")
lines(year,marriagez,type="l",col=myColour1_150,lwd=3)
lines(year,agriculturalz,type="l",col=myColour2_150,lwd=3)
text(1910,-40,"marriages",col=myColour1_150,cex=1.5)
```

```
text(1910,40,"agricultural prices ",col=myColour2_150,cex=1.5)

# Titling

mtext("Growth Trends and Economic Cycles",3,adj=0.5,line=3,cex⤸
    =2.1,outer=T,family="Lato Black")
mtext("Annual Figures",3,adj=0.06,line=0,cex=1.75,outer=T,font⤸
    =3)
dev.off()
```

In the script, the regular pattern of first defining the chart using plot(), then plotting the axes and data, and finally labelling the latter is followed for a total of three blocks. The margins before the individual charts have to be individually set. In the second chart, the index base value is highlighted with a line drawn with the arrows() function and three superposed dots. In the third chart, the area from 1820 to 1867 is highlighted with a grey area using the rect() function before drawing the lines. The integration into LaTeX essentially follows the previous example. We can omit the overpic environment in this case, as the text is set to the right of the charts. The integration into LaTeX is as follows:

```
\documentclass[a4paper,landscape]{article}
\usepackage[german]{babel}
\usepackage[top=0.1in, left=0.1in, right=2.4in, bottom=0.1in]{⤸
    geometry}
\usepackage{graphicx,color}
\usepackage{ragged2e}
\definecolor{orange}{RGB}{255,97,0}
\definecolor{blue}{RGB}{68,90,111}
\definecolor{hintergrund}{rgb}{0.99,0.99,0.99}
\pagecolor{hintergrund}
\renewcommand{\baselinestretch}{1.2}
\usepackage{fontspec}
\setmainfont[Mapping=text-tex]{Lato Regular}
\begin{document}
\thispagestyle{empty}
\vspace*{2cm}
\marginpar{\RaggedRight
\vspace*{-1cm}
Spree, Reinhard: \textit{Wachstumstrends und Konjunkturzyklen in⤸
    der deutschen Wirtschaft von 1820 bis 1913}. Vandenhoeck ⤸
    \& Ruprecht, Göttingen 1978. GESIS Köln, Deutschland ⤸
    ZA8053 Datenfile Version 1.0.0.\\~\\Spree unternimmt den ⤸
    Versuch, Wachstum und Konjunktur anhand von 18 ausgewä⤸
    hlten Indikatoren zu bestimmen. Die hier exemplarisch ⤸
    dargestellten Reihen zeigen die Eheschließungen und den ⤸
    Index der Nahrungsmittelpreise. Wie man deutlich sieht, ⤸
    verläuft die Konjunktur bei den Eheschließungen und den ⤸
    Nahrungsmittelpreisen bis etwa 1867 gegenläufig, erst ⤸
    danach löst sich dieser Zusammenhang auf. Steigende ⤸
    Nahrungsmittelpreise gehen bis 1867 offensichtlich mit ⤸
    einer Verschlechterung der Versorgungslage einher, was ⤸
    viele Menschen dazu veranlasste, nicht zu heiraten. Diese ⤸
    Gegenläufigkeit zwischen Heiratsverhalten und ökonomischer⤸
```

```
    Versorgungslage ist typisch für agrarisch geprägte ⟍
    Volkswirtschaften. Erst im Laufe der ⟍
    Hochindustrialisierung hat sich dieser Zusammenhang aufgel⟍
    öst.}\begin{picture}(0,485)
\put(-50,-55)
{\includegraphics[scale=0.89]{../pdf/timeseries_with_trend_3x1_⟍
    inc.pdf}}\end{picture}
\end{document}
```

8.3   Presentation of Daily, Weekly and Monthly Values

8.3.1   Daily Values with Labels

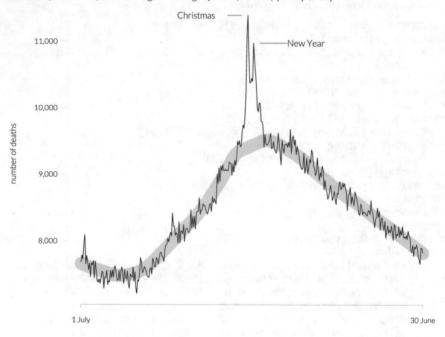

Death risk on Christmas and New Year 1979-2004 (USA)

Number of deaths before reaching the emergeny room, sums of years per day

Source: David Phillips, Gwendolyn E. Barker, Kimberly E. Brewer, Christmas and New Year as risk factors for death, Social Science & Medicine 71 (2010) 1463-1471

The figure shows the number of deaths before arrival at an emergency department in the USA from 1979 to 2004. Each individual value represents the sum of all deaths on that day of the year over 26 years. It is obvious that there is a continuous increase towards the turn of the year followed again by a decline. The risk of death is greatly increased on Christmas and New Year. The orange line shows the trend throughout the year without consideration for those outliers as a so-called

LOWESS (locally weighted scatterplot smoothing) estimation, a robust weighted rolling median. Since the main focus of this figure is on the extreme values on Christmas and New Year, and aside from that mainly the trend pattern is of interest, axis labelling can be kept to a minimum. We do not need auxillary lines for orientation, nor do we need a line as y-axis; for the x-axis, the start and end values suffice.

Data: The time series was kindly provided by its creator, David Phillips, as a CSV file.

```
pdf_file<-"pdf/timeseries_daily.pdf"
cairo_pdf(bg="grey98", pdf_file,width=10,height=8.27)

par(cex.axis=1.1,omi=c(1,0.5,0.95,0.5),mai=c(0.1,1.25,0.1,0.2),\
      mgp=c(5,1,0),family="Lato Light",las=1)

# Import data

christmas<-read.csv(file="myData/allyears.calendar.byday.dat.a",\
      head=F,sep=" ",dec=".")
attach(christmas)

# Create chart

plot(axes=F,type="n",xlab="",ylab="number of deaths",V1,V2)

# other elements

axis(1,tck=-0.01,col="grey",cex.axis=0.9,at=V1[c(1,length(V1))],\
      labels=c("1 July","30 June"))
axis(2,at=py<-pretty(V2),labels=format(py,big.mark=","),cex.axis\
      =0.9,col=par("bg"),col.ticks="grey81",lwd.ticks=0.5,tck=-\
      0.025)
points(V1,V2,type="l")
points(lowess(V2,f=1/5),type="l",lwd=25,col=rgb(255,97,0,70,\
      maxColorValue=255))
text(123,V2[179],"Christmas",cex=1.1)
arrows(157,V2[179],172,V2[179],length=0.10,angle=10,code=0,lwd\
      =2,col=rgb(100,100,100,100,maxColorValue=255))
arrows(192,V2[185],220,V2[185],length=0.10,angle=10,code=0,lwd\
      =2,col=rgb(100,100,100,100,maxColorValue=255))
text(240,V2[185],"New Year",cex=1.1)

# Titling

mtext("Death risk on Christmas and New Year 1979-2004 (USA)",3,\
      line=1.5,adj=0,cex=2,family="Lato Black",outer=T)
mtext("Number of deaths before reaching the emergeny room, sums \
      of years per day",3,line=-0.2,adj=0,cex=1.35,font=3,col="\
      black",outer=T)
mtext("Source: David Phillips, Gwendolyn E. Barker, Kimberly E. \
      Brewer, Christmas and New Year as risk factors for death, \
      Social Science & Medicine 71 (2010) 1463-1471",1,line=3,\
      adj=1,cex=0.75,font=3,outer=T)
dev.off()
```

In the script, we first increase the distance of the axis titles a little using mgp=c(5,1,0), so the y-axis title is not too close to the axis increment labels. We only want the first and last value to show on the x-axis. This can be done by specification of at=V1[c(1,length(V1))]; the labels are given as text. As in Sect. 8.2.3, we use the assignment to the variable py within the function for axis creation; that way, we can format the value labels with a comma as thousand separator. In addition to the actual time series, we also add a trend that smooths out short-term fluctuations as well as outliers. The trend is calculated as a LOWESS curve. This is broadly speaking a weighted robust rolling median. Please check the R help of the function for further details. As a general rule, it delivers good visual results for the presentation of trends. The "smoothness" is controlled via the parameter f. The greater this value, the smoother the course of the trend line. In the present case, it is used to illustrate the "normal" course, which does not consider outliers. Labels are added at suitable spots using arrows() and text().

8.3.2 Daily Values with Labels and Week Symbols (Panel)

Facebook Gross National Happiness Index 2010-2012

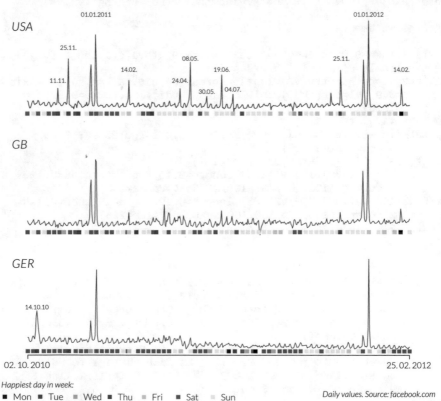

Daily values. Source: facebook.com

The figure shows the Facebook Gross National Happiness (GNH) Index for the USA, Great Britain, and Germany from 2 October 2010 to 25 February 2012. This index is a value developed by the Facebook Data Team, which is based on the evaluation of words from status updates of Facebook users. In the USA, the greatest amplitudes are turns of the year, 2010/11 New Year's Eve and 2010/2011 New Year. Immediately before that, there are strong positive values for the Christmas days. The same can be seen for Great Britain and Germany, while Christmas days are hardly pronounced in Germany. For the USA, there are further positive peaks on the public holidays Veterans Day (11 November), Thanksgiving (25 November), Memorial Day (30 May), and Easter Sunday (24 April). Other happy days are Valentine's Day (14 February) and Mother's Day (8 May). In Great Britain, Christmas and New Year's Eve/New Year show a similar pattern to the USA. However, we can also see a negative amplitude on the 9th of May, which is obviously related to the violent riots in London's suburbs during those days. Aside from the high values around the turn of the year and markedly smaller ones for the Christmas holidays, the index for Germany only shows a distinct peak around 14 October 2010: maybe the rescue of Chilean miners? Once more we can keep the labelling of the figure very minimal. For the x-axis, it suffices to state the start and end values on the lowest chart. Since the absolute height of the index value does not matter, we can omit y scaling completely. In addition to the index values, the figure also shows the day(s) with the highest index value within a week. For this, markings in seven colours are used in weekly intervals. Colours are explained in a legend at the bottom left. Data are available from Facebook's data team.

```
pdf_file<-"pdf/timeseries_daily_weekly_symbols_3x1.pdf"
cairo_pdf(bg="grey98", pdf_file,width=9,height=8.27)

par(mfcol=c(3,1),cex.axis=1.4,omi=c(1,0.5,0.75,0.5),mai=c⤸
    (0.1,0.2,0.1,0.2),family="Lato Light",las=1)

# Prepare chart and import data

library(gdata)

myCountry<-c("USA","GB","GER")
myCountryName<-c("USA","GB","GER")
for (i in 1:length(myCountry))
{
myFile<-paste("myData/Facebook Happiness Index-",myCountry[i],".⤸
    csv",sep="")
myData<-read.csv(myFile,skip=1)
myData$x<-as.Date(myData$Date)
myData<-subset(myData,myData$x>"2010-10-01")

# Define chart and other elements

plot(myData$x,myData$GNH,type="n",axes=F,ylab="",ylim=c(-⤸
    0.05,0.4))
points(myData$x,myData$GNH,type="l",col="darkblue")
```

```
myData$myYear<-as.numeric(format(myData$x,"%Y"))
myData$myCw<-as.numeric(format(myData$x,"%V"))
myNewData<-myData[order(myData$myYear,myData$myCw,myData$GNH),]
myDay<-NULL
myColour<-NULL
n<-nrow(myNewData)-1
for (j in 1:n)
{
if(myNewData$myCw[j+1] != myNewData$myCw[j])
{
  myDay<-c(myDay,as.character(myNewData$x[j]))
  myColour<-c(myColour,weekdays(myNewData$x[j]))
}
}
myColour<-as.numeric(as.factor(myColour))
points(as.Date(myDay),rep(-0.05,length(myDay)),pch=15,cex=1.5, ⤸
    col=myColour)

mtext(myCountryName[i],3,line=-3,adj=0,cex=1.3,font=3)
source("scripts/inc_daily_weeksymbols_datelabels.r")
}
axis(1,at=myData$x[c(1,length(myData$x))],labels=format(myData$x⤸
    [c(1,length(myData$x))],"%d. %m. %Y "))
mtext("Happiest day in week:",1,line=3,adj=0,cex=0.9,font=3,⤸
    outer=T)
legend(as.Date("2010-08-22"),-0.21,c("Mon","Tue","Wed","Thu","⤸
    Fri","Sat","Sun"),pch=15,col=c(1,2,3,4,5,6,7),ncol=7,bty="⤸
    n",xpd=NA,cex=1.5)

# Titling

mtext("Facebook Gross National Happiness Index 2010-2012",3,line⤸
    =1,adj=0,cex=2,family="Lato Black",outer=T)
mtext("Daily values. Source: facebook.com",1,line=4.25,adj=1,cex⤸
    =0.9,font=3,outer=T)
dev.off()
```

The script starts with the specification of three countries in one vector. A CSV file is read for each country, and the first line is ignored. We then generate a date variable x and filter the data frame for all data after 1 October 2010; then, the charts are created using plot() and points(). To display the week symbols, two new columns are generated with format(). %Y outputs the year, %V the week number within the year. Finally, a second data frame myNewdata is generated, which sorts the original data by year, calendar week, and by GNH value. Now, two variables myDay and myColour are initialised. The loop passes the new data frame from start to finish. Changes in calendar weeks are also the positions of the greatest GNH values, since this is how data were sorted. The day's date and number of the weekday are attached to day and colour, respectively. After completion of the loop, we convert the vector colour to a numerical vector, so that Monday becomes a 1, Tuesday a 2, etc. Points are then plotted in the colour of the weekday at the respective day's position and in y-direction at position −0.05. We use R's internal colour numbers to do this, which have, e.g., a 6 for purple and a 7 for yellow. In our next step, we insert the myCountryName into the three charts using mtext(), and then add the individual date labels via an embedded file. At the end come an x-axis, which is limited to the start and end value as in Sect. 8.3.1, a legend that explains the colours for the week symbols, and title and sources. A content-related variant of the figure is one that displays the unhappiest days of the week instead of the happiest ones (Fig. 8.1). The sole differences lie in the row

```
newdata<-data[order(data$year, data$cw, -data$GNH), ]
```

where a minus symbol is inserted before data$GNH, which means that we do not sort in ascending, but descending order, and in the fact that "happiest day" is replaced with "unhappiest day" in the legend.

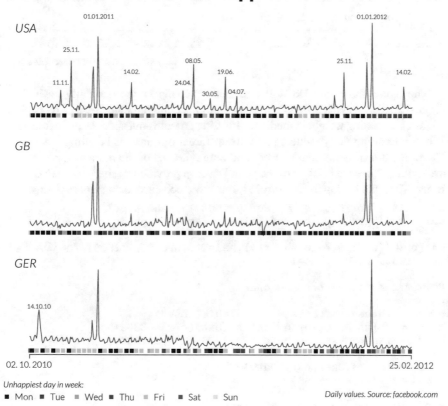

Fig. 8.1 Daily values with labels and week symbols (negative)

8.3.3 Daily Values with Monthly Labels

Exchange rate US Dollar/Euro 2010-2012
Daily values

Source: www.ecb.stats/exchange/eurofxref/html/index/html

The figure shows the US dollar to Euro exchange rate on a daily basis from August 2010 to August 2012. The daily values are complemented by a LOWESS filter, a robust rolling weighted median. This filter smooths out short-term fluctuations. When presenting daily values, there are different options for labelling the x-axis. Here, the turn of the month is used, and abbreviations of the names of the months are written above the number of the year. Data are provided by the European Central Bank. They can be downloaded from http://www.ecb.europa.eu/stats/exchange/.

```
pdf_file<-"pdf/timeseries_daily_monthly_labels.pdf"
cairo_pdf(bg="grey98", pdf_file,width=14,height=8)

par(omi=c(0.65,0.75,0.95,0.75),mai=c(0.9,0.75,0.25,0),family="\
    Lato Light",las=1)

# Import data and prepare chart

euro<-read.csv("myData/eurofxref-hist.csv")
euro<-euro[as.Date(euro$Date)>as.Date("2010-08-01"),]
myBeginOfMonth<-seq(as.Date("2010-08-01"),as.Date("2012-08-01"),\
    by="1 months")
myDays<-rev(as.Date(euro$Date))
myValues<-rev(euro$USD)

# Define chart and other elements

plot(myDays,myValues,axes=F,type="n",xlab="",ylab="")
lines(myDays,myValues,col="grey")
myLColour<-rgb(255,97,0,100,maxColorValue=255)
lines(lowess(myDays,myValues,f=1/25),col=myLColour,lwd=5)
par(mgp=c(3,1.5,0))
```

```
axis(1,at=myBeginOfMonth,labels=format(myBeginOfMonth,"%b\n%Y"),
    cex.axis=0.75)
axis(2)
legend("bottomleft","Lowess-filter f=1/25",pch=15,col=myLColour,
    bty="n",cex=2)

# Titling

mtext("Exchange rate US Dollar/Euro 2010-2012",3,line=0,adj=0,
    cex=2.5,family="Lato Black",outer=T)
mtext("Daily values",3,line=-2,adj=0,cex=1.75,font=3,outer=T)
mtext("Source: www.ecb.stats/exchange/eurofxref/html/index/html"
    ,1,line=4,adj=1.0,cex=1.25,font=3)
dev.off()
```

The script reads the data from a CSV file and is initially limited to those values that are greater than 01/08/2010. We then generate a sequence of monthly values from 01/08/2010 to 01/08/2012 using seq. Date and USD are read in reverse as days and values, since the data are again provided in reverse order as they were in Sect. 8.1.3. In addition to the actual time series, we once more plot a LOWESS filter as trend, as we did in Sect. 8.3.1. We use the sequence myBeginOfMonth for our x-axis labels; the format determines that the output is the abbreviated name of the month, and the year in four-digit format, each separated with a line break.

8.3.4   Time Series from Weekly Values (Panel)

Google Trends: Number of search queries 2004-2012

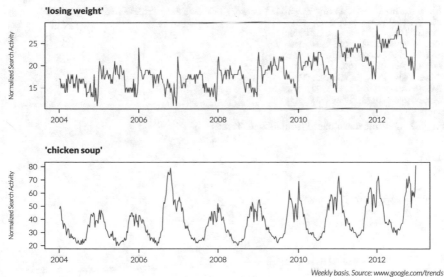

Weekly basis. Source: www.google.com/trends

The figure shows the frequency of search queries for the two terms "losing weight" and "chicken soup" on a weekly basis from 2004 to 2012. Google offers the frequency of a series of search terms on a weekly basis in the form of a "normalized search activity" index. Strong periodicity is obvious for both terms: while interest in losing weight decreases between the start and the end of the year, and rapidly increases again on Christmas and New Year's Eve, interest in chicken soup fluctuates more evenly. One could say that interest decreases towards summer and increases towards winter. For this topic, we have chosen a suitable, more technical appearing representation form. It largely corresponds to R's default settings. We should choose an aspect ratio that makes the cyclical patterns easy to recognise. William S. Cleveland clearly explained how the aspect ratio for a graphic should be chosen: so that the absolute values of all lines' inclines are 45°. Although this rule cannot claim universal validity, it is almost always a good starting point for a solution suitable for the data and the desired message. Even though these are weekly values, the x-axis is only labelled with years at the turn of the year. For the subtitles, the colour of the respective series was selected.

Data can be downloaded as a text file from http:/www.google.com/trends by anyone with a google account. For the present example, data were read into a SQL table google_trends. This table is available as an SQL dump at https://link.springer.com/chapter/10.1007/978-3-030-28444-2_1. Data deliver values on a weekly basis, each assigned to a concrete date (Figs. 8.2 and 8.3).

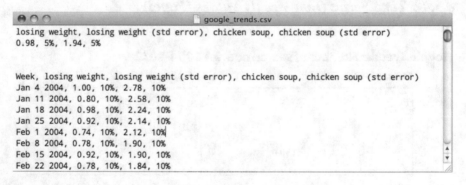

```
●○○                              google_trends.csv
losing weight, losing weight (std error), chicken soup, chicken soup (std error)
0.98, 5%, 1.94, 5%


Week, losing weight, losing weight (std error), chicken soup, chicken soup (std error)
Jan 4 2004, 1.00, 10%, 2.78, 10%
Jan 11 2004, 0.80, 10%, 2.58, 10%
Jan 18 2004, 0.98, 10%, 2.24, 10%
Jan 25 2004, 0.92, 10%, 2.14, 10%
Feb 1 2004, 0.74, 10%, 2.12, 10%
Feb 8 2004, 0.78, 10%, 1.90, 10%
Feb 15 2004, 0.92, 10%, 1.90, 10%
Feb 22 2004, 0.78, 10%, 1.84, 10%
```

Fig. 8.2 CSV file downloaded from Google Trends

Fig. 8.3 Import of the Google Trends CSV file into MySQL

```
pdf_file<-"pdf/timeseries_weekly_2x1.pdf"
cairo_pdf(bg="grey98", pdf_file,width=9,height=6)

source("scripts/inc_datadesign_dbconnect.r")
par(omi=c(0.5,0.25,0.25,0.5),mai=c(0.25,0.75,0.75,0.25),mfcol=c↘
    (2,1),
  cex=0.7,mgp=c(4,1,0),family="Lato Light",las=1)

# Import data

sql<-"select left(week,10) week, losing_weight, chicken_soup ↘
    from google_trends"
myData<-dbGetQuery(con,sql)
myData$x<-as.Date(myData$week)
attach(myData)

# Create chart

plot(x,losing_weight,type="l",col="darkblue",ylab="Normalized ↘
    Search Activity",cex.axis=1.2)
mtext("'losing weight'",3,line=1,adj=0,cex=1,col="darkblue",↘
    family="Lato Black")

plot(x,chicken_soup,type="l",col="darkred",ylab="Normalized ↘
    Search Activity",cex.axis=1.2)
mtext("'chicken soup'",3,line=1,adj=0,cex=1,col="darkred",family↘
    ="Lato Black")
```

```
# Titling

mtext("Google Trends: Number of search queries 2004-2012",3,line↘
     =-2,adj=0,cex=1.5,family="Lato Black",outer=T)
mtext("Weekly basis. Source: www.google.com/trends",1,line=1,adj↘
     =1.0,cex=0.85,font=3,outer=T)
dev.off()
```

In the script, we first split the figure into two horizontal areas using mfcol=c(2,1) and then read the data from the SQL database. With the SQL command STR_TO_Date(), the dates from the original's form of, e.g., "Jan 4 2004" are read into R in a form that can be easily converted into a date column with as.Date(). That done, we use the plot() function to create two graphs and largely keep the default settings.

8.3.5   Monthly Values (Panel)

Skalometer SPD, CDU, FDP, Grüne: 1977-2007

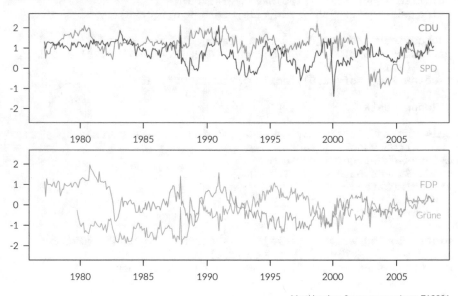

Monthly values. Source: www.gesis.org, ZA2391

The figure is another example of the representation of throughout-the-year data with a year label. The presentation shows the so-called Skalometer from the Politbarometer, a regular survey conducted in Germany. The documentation states the following [original in German]: "In the years from 1977 until June 1987, as well as for the split group 1 of the months 08-12/87 and 01-12/88, the question was: and what do you think—in general terms—of the political parties? Please answer using this scale. +5 means that you rate the party highly. −5 means that you think nothing of it. The values in between allow you to grade your response. (...) (For the respondents of the split group 2 of the months 09-10/87, as well as for all respondents of the years 1989–2007, the question was:) And now more specifically about the parties: Imagine a thermometer ranging from plus 5 to minus 5, with a zero point in between. Use this thermometer to tell me what you think about the individual parties. +5 means that you rate the party high. −5 means that you think nothing of it. With the values in between, you can grade your opinion. (...)" The results of this survey are shown on a monthly basis for a period of 30 years; therefore, there are more than 300 points in time. The layout largely matches the previous example. However, we show monthly values here, not weekly ones. The y-axis scaling in the upper and lower chart should be the same. Since the expression "Skalometer" is already in the title, we can omit a y-axis label. We can accommodate two time series per chart. The colours should resemble the colours of the political parties; titles appear directly next to the series, which is preferred over a legend in this case. The aspect ratio is chosen to make the long-term development of the series easily visible.

Data: see appendix A, ZA2391:Politbarometer 1977–2011 (partial accumulation).

```
pdf_file<-"pdf/timeseries_monthly_2x1.pdf"
cairo_pdf(bg="grey98", pdf_file,width=9,height=6.2)

source("scripts/inc_datadesign_dbconnect.r")
par(omi=c(0.65,0.55,0.55,0.25),mai=c(0.25,0.25,0.25,0.25),mfcol=\
    c(2,1),family="Lato Light",las=1)

# Read data

sql<-"select dezjahr decyear,v8,v9,v11,v12 from t_za2391_\
    zeitreihen"
myData<-dbGetQuery(con,sql)
attach(myData)

# Create chart and other elements

plot(type="n",xlab="",ylab="Mittelwert",decyear, v8,col=rgb\
    (255,97,0,150,maxColorValue=255),lwd=2,ylim=c(-2.5,2.5))
myVars1<-c("decyear","v8")
myVars2<-c("decyear","v9")
myColour1<-rgb(255,0,0,150,maxColorValue=255)
myColour2<-rgb(0,0,0,150,maxColorValue=255)
points(myData[myVars1],col=myColour1,lwd=1,type="l")
points(myData[myVars2],col=myColour2,lwd=1,type="l")
text(2007.5,0,"SPD",col=myColour1)
text(2007.5,2,"CDU",col=myColour2)

plot(type="n",xlab="",ylab="Mittelwert",decyear, v8,col=rgb\
    (255,97,0,150,maxColorValue=255),lwd=2,ylim=c(-2.5,2.5))
myVars3<-c("decyear","v11")
```

```
myVars4<-c("decyear","v12")
myColour3<-"orange"
myColour4<-rgb(0,128,0,150,maxColorValue=255)
points(myData[myVars3],col=myColour3,lwd=1,type="l")
points(myData[myVars4],col=myColour4,lwd=1,type="l")
text(2007.5,1,"FDP",col=myColour3)
text(2007.5,-0.5,"Grüne",col=myColour4)

# Titling

mtext("Skalometer SPD, CDU, FDP, Grüne: 1977-2007",3,line=0,adj↘
      =0,cex=1.5,family="Lato Black",outer=T)
mtext("Monthly values. Source: www.gesis.org, ZA2391",1,line=3,↘
      adj=1.0,cex=0.95,font=3)
dev.off()
```

In the script, data are again read via SQL from a database. In the data table, we define a variable decyear by year+(month-1)/12, which converts the months into decimal values. This means we can then use plot() and points() to draw simple times series charts into the two graphic windows generated by mfcol=c(2,1).

8.3.6   Monthly Values with Monthly Labels

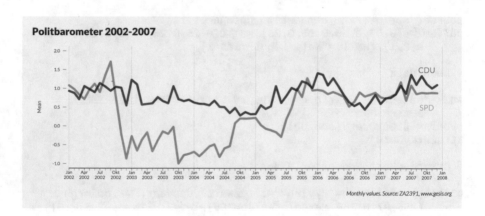

The figure once more shows data from the Politbarometer, in this case for the Social Democratic Party of Germany (SPD) and the Christian Democratic Union (CDU), two major German parties. Fewer points of time are given in this case, so that labelling below year level is possible. With the chosen font size, we can accommodate four months per year. In this example, the x-axis labelling matches the example "Daily values with monthly labels" (Sect. 8.3.3). Vertical auxiliary lines are added at the turn of the year to aid orientation; we have also added one at the start so that we can omit the line for the y-axis.

Data: see appendix A, ZA2391: Politbarometer 1977–2011 (partial accumulation).

```
pdf_file<-"pdf/timeseries_monthly_monthly_labels.pdf"
cairo_pdf(bg="grey98", pdf_file,width=11,height=4.8)

source("scripts/inc_datadesign_dbconnect.r")
par(cex=0.75,bg=rgb(240,240,240,maxColorValue=255))
par(omi=c(0.75,0.25,0.5,0.25),mai=c(0.25,0.55,0.25,0),mgp=c↘
    (2,1,0),family="Lato Light",las=1)
myColour1<-rgb(255,0,0,150,maxColorValue=255)
myColour2<-rgb(0,0,0,150,maxColorValue=255)

# Read data

sql<-"select concat_ws('-',jahr,monat,'01') monat, v8, v9 from t↘
    _za2391_zeitreihen where jahr >= 2002"
myData<-dbGetQuery(con,sql)
attach(myData)

# Create chart and other elements

plot(type="n",axes=F,xlab="",ylab="Mean",as.Date(monat),v8,col=↘
    rgb(255,97,0,150,maxColorValue=255),lwd=2,ylim=c(-1,2))

mbegin<-seq(as.Date("2002-01-01"),as.Date("2008-01-01"),by="1 ↘
    months")
ybegin<-seq(as.Date("2002-01-01"),as.Date("2008-01-01"),by="1 ↘
    years")
abline(v=ybegin,col="lightgrey")

points(as.Date(monat),v8,col=myColour1,lwd=5,type="l")
points(as.Date(monat),v9,col=myColour2,lwd=5,type="l")

text(as.Date("2007-10-01"),0.5,"SPD",col=myColour1,cex=1.5)
text(as.Date("2007-10-01"),1.5,"CDU",col=myColour2,cex=1.5)

axis(1,col=rgb(60,60,60,maxColorValue=255),at=mbegin,labels=↘
    format(mbegin,"%b\n%Y"),cex.axis=0.85,lwd.ticks=0.1,tck=-↘
    0.02)
axis(2,col=rgb(240,240,240,maxColorValue=255),col.ticks=rgb↘
    (60,60,60,maxColorValue=255),lwd.ticks=0.5,cex.axis=0.85,↘
    tck=-0.025,pos=as.Date("2001-12-15"))
```

```
# Titling

mtext("Politbarometer 2002-2007",3,line=0,adj=0,cex=1.5,family="↘
    Lato Black",outer=T)
mtext("Monthly values. Source: ZA2391, www.gesis.org",1,line=2,↘
    adj=1.0,cex=0.85,font=3,outer=T)
dev.off()
```

The script begins by reading the data from a database. From the columns year and month and the fictional day "01", we create a new column month, which can be easily converted to a month variable in R. We again start by generating an "empty" plot, then define a monthly and yearly sequence from 2002 to 2008, and then draw vertical orientation lines at start-of-the-year positions. Then come the time series v8 and v9 with points(), and the labels. The x-axis is created as before in Sect. 8.3.3. For the y-axis, we use pos to specify the exact position of the axis for the January 1988 value, so that the scale lines reach the first vertical orientation line. There is an alternative for the x-axis labels: here, years are spelt out with two digits, and the names of the months, abbreviated to the first letter, come in between (with the exception of January). To do this, we can slightly modify the code used as an explanation in the zoo package.

```
library(zoo)
tt<-as.Date(month)
ym <- as.yearmon(tt)
mon <- as.numeric(format(ym, "%m"))
yy <- format(ym, "%y")
mm <- substring(month.abb[mon], 1, 1)
if (any(mon == 1)) axis(side = 1, at = tt[mon == 1], labels = yy↘
    [mon == 1], cex.axis = 0.9) axis(1, at = tt[mon > 1], ↘
    labels = mm[mon > 1], cex.axis = 0.5, tcl = -0.3)
```

The result looks like this:

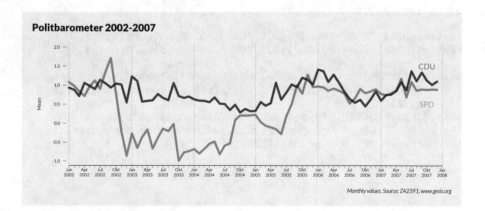

8.3.7   Monthly Values with Monthly Labels (Layout)

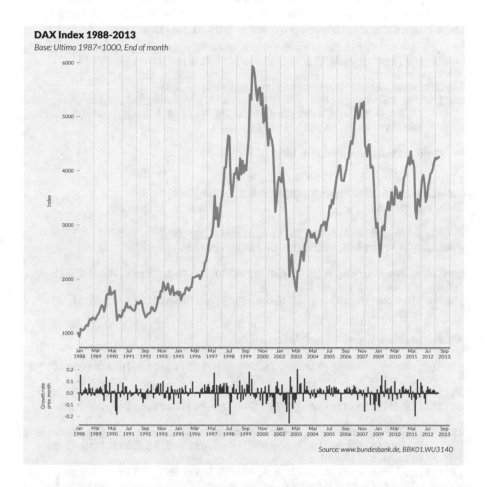

About the figure: the development of the DAX from 1988 to 2013 is the last variant of a monthly label: on the one hand, the absolute development, on the other the growth rate compared with the previous month. The aspect ratio of the absolute development is again guided by William S. Cleveland's 45° principle. Growth rates are represented as columns, again with a suitable aspect ratio. Overall, this results vertically in an asymmetrical layout. Since the upper chart is very tall, vertical auxillary lines are added for every x-axis label. x-Axis labelling matches Sect. 8.3.5, but given the larger amount of data in this example, we only generate one date output every 14 months. Data derive from the website of the German Central Bank http:// www.bundesbank.de. They are provided as BBK01.WU3140.xlsx.

```
pdf_file<-"pdf/timeseries_monthly_monthly_labels_2x1_layout.pdf"
cairo_pdf(bg="grey98", pdf_file,width=11,height=11)

layout(matrix(c(1,2),ncol=1),heights=c(80,20))
par(cex=0.75,bg=rgb(240,240,240,maxColorValue=255),omi=c\
    (0.75,0.25,0.5,0.25),mai=c(0.25,0.75,0.25,0),mgp=c(2,1,0),\
    family="Lato Light",las=1)

#   Read data and prepare chart

library(gdata)
myData<-read.xls("mydata/BBK01.WU3140.xlsx")
attach(myData)

myColour1<-rgb(255,0,0,150,maxColorValue=255)
myColour2<-rgb(0,0,0,150,maxColorValue=255)

monthbegin<-seq(as.Date("1988-01-01"),as.Date("2014-01-01"),by="\
    1 months")
yearbegin<-seq(as.Date("1988-01-01"),as.Date("2014-01-01"),by="1\
    years")

# Create chart and other elements

plot(type="n",axes=F,xlab="",ylab="Index",as.Date(paste(Monat,"\
    01",sep="-")),Wert)
abline(v=yearbegin,col="lightgrey")
points(as.Date(paste(Monat,"01",sep="-")),Wert,col=myColour1,lwd\
    =5,type="l")
axis(1,col=rgb(60,60,60,maxColorValue=255),at=monthbegin,labels=\
    format(monthbegin,"%b\n%Y"),cex.axis=0.95,lwd.ticks=0.1,\
    tck=-0.005)
axis(2,col=rgb(240,240,240,maxColorValue=255),col.ticks=rgb\
    (60,60,60,maxColorValue=255),lwd.ticks=0.5,cex.axis=0.95,\
    tck=-0.01,pos=as.Date("1988-01-01"))
myRate<-rep(0,nrow(myData))
```

```
for (i in 2:nrow(myData)) myRate[i]<-(Wert[i]-Wert[i-1])/Wert[i-↘
    1]
plot(type="h",axes=F,xlab="",ylab="Growth rate\nprev. month",as.↘
    Date(paste(Monat,"01",sep="-")),myRate,col=myColour2,lwd↘
    =3)
axis(1,col=rgb(60,60,60,maxColorValue=255),at=monthbegin,labels=↘
    format(monthbegin,"%b\n%Y"),cex.axis=0.95,lwd.ticks=0.1,↘
    tck=-0.02)
axis(2,col=rgb(240,240,240,maxColorValue=255),col.ticks=rgb↘
    (60,60,60,maxColorValue=255),lwd.ticks=0.5,cex.axis=0.95,↘
    tck=-0.025,pos=as.Date("1988-01-01"))

# Titling

mtext("DAX Index 1988-2013",3,line=1,adj=0,cex=1.5,family="Lato ↘
    Black",outer=T)
mtext("Base: Ultimo 1987=1000, End of month",3,line=-1,adj=0,cex↘
    =1.25,font=3,outer=T)
mtext("Source: www.bundesbank.de, BBK01.WU3140",1,line=2,adj↘
    =1.0,cex=1.05,font=3,outer=T)
dev.off()
```

In the script, we use layout() to split the figure into two windows, with the upper window making up 80% of the figure and the lower one 20%. Using mgp, the distance of axis titles is slightly reduced. After reading the XLS file, two sequences are specified: one monthly and one yearly series, each of which goes from January 1988 to January 2014. We again draw a type "n" plot with plot(); in this case, the x-variable is defined as a date variable. The variable is read in "1987–12" form from the data frame; as in the previous example, a "–01" is appended so the expression can easily be converted into a date. In a next step, we first draw vertical orientation lines at the respective turn of years. Then, we add the data with points() and the axes. For the y-axis, the exact position of the axis at the January 1988 value is specified with pos, so that the scale lines reach to the first vertical orientation line. Next, the growth rate grate is calculated. To this end, we first specify a vector of the same length as the data frame and fill it with the following loop. The growth rate can immediately be plotted with plot() since no orientation lines have to be added under the series. We draw the series, as is common practice for growth rates, as thin columns. The plot() function has set aside the type "h" for that. Axes are generated as in the chart above.

8.4   Exceptions and Special Cases

8.4.1   Time Series as Scatter Plot (Panel)

Inequality in 40 countries 1950-2010

Gini-coefficient of income distribution (yearly values)

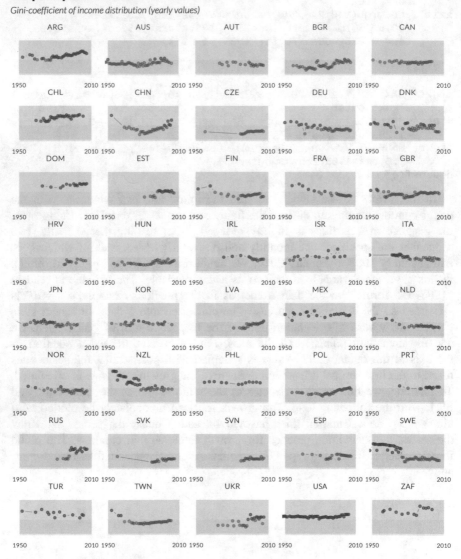

Blue: up to 0.36. Red: more than 0.36   *Source: World Income Inequality Database V2.0c May 2008*

The figure shows the disparity in income distribution in 40 countries from 1950 to 2010 in the form of the Gini coefficient. The Gini coefficient is a measure of the power of the disparity, which varies between 0 and 1. It is defined as the area between a Lorenz curve and the diagonal, divided by half the rectangular chart area (see Sect. 7.3). Since 40 countries are represented, the individual charts have to be greatly reduced. That is why only country codes are used as titles, and x-axis labels are limited to the first and last year. Since not all values for all years and countries are available, a scatter plot in the form of a point representation with connecting lines best reflects the actual data. To achieve a good overall impression, the colours of points and background were configured dependent on the power of the Gini coefficient: blue for a value up to 0.36, red for a greater value. There is no theoretical basis for this limit; it was chosen so approximately half the points lie below and the other half above it. Data derive from the World Income Inequality Database V2.0c May 2008, which is available from http://www.wider.unu.edu/research/database/. The file WIID2C.xls is provided for download and was read into a SQL database for this example.

```
pdf_file<-"pdf/timeseries_scatterplots_8x5.pdf"
cairo_pdf(bg="grey98", pdf_file,width=8.27,height=11.7)
par(omi=c(0.25,0.1,0.7,0.1),mai=c(0.1,0.1,0.45,0.1),mfrow=c(9,5)\
    ,family="Lato Light",las=1)

source("scripts/inc_datadesign_dbconnect.r")

# Read data and prepare chart

myCountry<-c("ARG","AUS","AUT","BGR", "CAN","CHL", "CHN", "CZE",\
        "DEU", "DNK",
    "DOM", "EST", "FIN", "FRA", "GBR", "HRV", "HUN", "IRL", "ISR"\
        , "ITA",
    "JPN", "KOR", "LVA", "MEX", "NLD", "NOR", "NZL", "PHL", "POL"\
        , "PRT",
    "RUS", "SVK", "SVN", "ESP", "SWE", "TUR", "TWN", "UKR", "USA"\
        , "ZAF")
myPart1<-"select Year, Gini from v_wiid2c_Gini where Country3='"
myPart2<-"'"
for (i in 1:length(myCountry))
{
sql1<-paste(myPart1,myCountry[i],myPart2,sep="")
sql2<-paste(myPart1,myCountry[i],myPart2," and Gini > 36",sep=""\
    )
myData<-dbGetQuery(con,sql1)
myData2<-dbGetQuery(con,sql2)

# Create chart and other elements

plot(myData,xlim=c(1950,2010),ylim=c(10,75),main=myCountry[i], \
    type="n",axes=F)
rect(1950,36,2010,75,col=rgb(255,0,0,50,maxColorValue=255),lwd\
    =0)
```

```
rect(1950,0,2010,36,col=rgb(68,90,111,50,maxColorValue=255),lwd
    =0)
points(myData,col=rgb(68,90,111,100,maxColorValue=255),pch=19,
    type="b")
axis(1,at=c(1950,2010),col=par("bg"),cex.axis=0.95)
if (length(myData2) > 0)
{
points(myData2,pch=19,col=rgb(255,0,0,100,maxColorValue=255))
}
}

# Titling

mtext("Inequality in 40 countries 1950-2010",3,outer=T,line=2,
    xpd=T,adj=0,family="Lato Black",cex=1.3)
mtext("Gini coefficient of income distribution (yearly values)"
    ,3,outer=T,line=0,xpd=T,adj=0,font=3)
mtext("Source: World Income Inequality Database V2.0c May 2008"
    ,1,outer=T,line=-3,xpd=T,adj=0.9,font=3)
mtext("Blue: up to 0.36. Red: more than 0.36",1,outer=T,line=-3,
    xpd=T,adj=0,font=3)
dev.off()
```

In the script, the figure is first split into 40 (8 times 5) parts. Then, we specify country codes for the plots we want to include. Using those codes, we put together two SQL commands for each country. The first one grabs all Gini coefficients for one country from the database, the second one only those that are greater than 36. Data are first drawn completely using plot() with the same y-range. The heading is added with the main parameter within plot(), and the background colour is then set with rect(), so that the lower area up to 36 becomes blue and the area above red. Afterwards, the type is set with points() using a "b" for "both". That means that both lines and points are plotted, but the lines do not touch the points; between the points and lines, a small space is left. In a second step, the Gini coefficients greater than 36 are overlaid in red over the first data. At the end come the usual titles and sources.

8.4.2   Time Series with Missing Values

About the figure: especially with historical time series, odd missing values should be taken into account. When representing time series as line charts, two cases have to be distinguished: if individual or sequential values within a series are missing, then these missing values are perceived as a gap in the line. This is not a big problem for the representation. However, it becomes more difficult if individual values are surrounded by gaps. In these cases, a combination of points and lines has to be chosen. The point size should be greater than the line weight to ensure points do not get lost in the graphic. The figure below shows such a gappy series on the example of the development of weekly working hours in different European countries from 1850 to 2000.

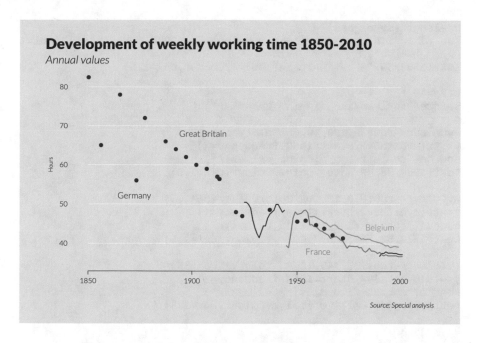

Development of weekly working time 1850-2010

Annual values

Data derive from a special evaluation and can be downloaded as an XLS table at https://link.springer.com/chapter/10.1007/978-3-030-28444-2_1.

```
pdf_file<-"pdf/timeseries_missing_values.pdf"
cairo_pdf(bg="grey98", pdf_file,width=13,height=9)

par(omi=c(0.65,0.75,0.95,0.75),mai=c(0.9,0.85,0.25,0.02),bg="↘
      antiquewhite2",family="Lato Light",las=1)

# Import data and prepare chart

library(gdata)
myData<-read.xls("myData/Work_hours_data.xls", cncoding="latin1"↘
      )
myColour<-rgb(139,35,35,maxColorValue=255)
y<-ts(myData$v1,start=1850,frequency=1)

# Define chart

plot(y,typ="n",axes=F,xlim=c(1850,2010),ylim=c(35,85),xlab="",↘
      ylab="Hours")

# Other elements

axis(1,cex.axis=1.25)
axis(2,cex.axis=1.25,col=par("bg"),col.ticks="grey81",lwd.ticks↘
      =0.5,tck=-0.025)

myHeights<-c(40,50,60,70,80)
```

```
n<-length(myHeights)
for (i in 1:n) segments(1850,myHeights[i],2000,myHeights[i],col=\
     "white")
text(1905,68,"Great Britain",col=myColour,cex=1.5)

ptyp=19
source("scripts/inc_missing_values.r")

myColour<-rgb(39,139,16,maxColorValue=255)
y<-ts(myData$v2,start=1850,frequency=1)
source("scripts/inc_missing_values.r")
text(1960,38,"France",col=myColour,cex=1.5)

myColour<-rgb(0,0,139,maxColorValue=255)
y<-ts(myData$v3,start=1850,frequency=1)
source("scripts/inc_missing_values.r")
text(1872,52,"Germany",col=myColour,cex=1.5)

myColour<-rgb(205,149,12,maxColorValue=255)
y<-ts(myData$v4,start=1850,frequency=1)
source("scripts/inc_missing_values.r")
text(1990,44,"Belgium",col=myColour,cex=1.5)

# Titling

mtext("Development of weekly working time 1850-2010",3,line=0.2,\
     adj=0,cex=2.6,family="Lato Black",outer=T)
mtext("Annual values ",3,line=-2,adj=0,cex=2,font=3,outer=T)
mtext("Source: Special analysis",1,line=0,adj=1,cex=1.25,font=3,\
     outer=T)
dev.off()
```

and

```
# inc_missing_values.r
lines(y,lwd=3,col=myColour)
von<-2; bis<-length(y)-1
for (i in von:bis)
{
if (is.na(y[i-1]) & !is.na(y[i]) & is.na(y[i+1]))
points(time(y)[i],y[i],pch=ptyp,cex=1.5,col=myColour)
}
# ... and the margins
if (!is.na(y[1]) & is.na(y[2])) points(time(y)[1],y[1],pch=ptyp,\
     cex=1.5,col= myColour)
if (!is.na(y[length(y)]) & is.na(y[length(y)-1])) points(time(y)\
     [length(y)],y[length(y)],pch=ptyp,cex=1.5,col= myColour)
```

The script starts by converting the time series v1, and later also v2, v3, and v4, to a time series object using ts(). The horizontal orientation lines are supposed to only reach from the first to the last value shown on the x-axis. That is why we use segments() rather than abline() to draw lines, which allows us to set the start and end points. For the four time series, the file inc_missing_values.r is sequentially embedded and checks the respective series for missing values: the time series is drawn first, using lines(). Next, data are checked for any values that are neighboured by missing values on both the left and right side. Such values would not be drawn with lines() or points() and the type "l". On the other hand, we cannot draw all points before or after the line using points(), as these individual points—in the case of missing values—are meant to be thicker than the line. To achieve this, we use a loop to plot a point with points() from the second to the second to last value, whenever there is a missing value to the right and left of the current value. While a vector of the time series type can easily be drawn with lines(), we do require x-values for points(). These are provided by the time() function. Finally we have to have a look at our borders. It could be that just the first value is available but the second one is missing, or that the second to last one is missing but the last one is available. Under these circumstances, the line function would not draw a line either. This means we require one condition each for the second and second to last value.

8.4.3 Seasonal Ranges (Panel)

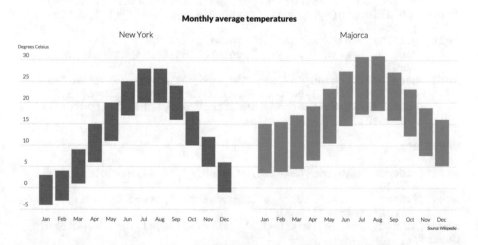

About the figure: vertical range charts are widely used, particularly for representation of temperature trends. Here, too, a panel presentation is a good choice for comparison of two developments. The width of the columns should be selected to make the development easily recognisable. A single y-scale for both presentations is sufficient if auxillary lines are drawn over the width of both charts. The auxillary lines can be drawn behind the columns. In this case, y-axis labelling is attached above the auxillary lines rather than before them. This is a matter of taste. The axis title is written horizontally across the axis to utilise the width of the figure. As can be seen in direct comparison, the temperature range over the course of 1 year is greater in New York than it is on Majorca. On the other hand though, the range over the course of a month is greater on Majorca. Data were taken from the German Wikipedia articles on "New York" and "Mallorca".

```
pdf_file<-"pdf/timeseries_seasonal_minmax_2x1.pdf"
cairo_pdf(bg="grey98", pdf_file,width=14,height=7)

par(omi=c(0.25,0.25,0.5,0.25),mai=c(0.45,0.35,0.5,0),mfcol=c
        (1,2),family="Lato Light",las=1)
library(gplots)
library(gdata)

# Import data and prepare chart

myData<-read.xls("myData/Climate.xlsx", encoding="latin1")
attach(myData)
myLines<-c(-5,0,5,10,15,20,25,30)

# Create chart and other elements

myT1<-barplot2(t(cbind(NY_min,NY_max-NY_min)),col=c(NA,"coral3")
        ,border=NA,names.arg=Month,ylim=c(-5,35),panel.first=
        abline(h=myLines,col="grey",lwd=1,lty="dotted"),axes=F)
for (i in 1:length(myLines)) {text(-0.8,myLines[i]+1.1,myLines[i
        ],xpd=T)}
text(-0.25,33,"Degrees Celsius",xpd=T,cex=0.8)
mtext(side=3,"New York",cex=1.5,col=rgb(64,64,64,maxColorValue
        =255))
myT2<-barplot2(t(cbind(MAJ_min,MAJ_max-MAJ_min)),col=c(NA,"
        cornflowerblue"),border=NA,names.arg=Month,ylim=c(-5,35),
        panel.first=abline(h=myLines,col="grey",lwd=1,lty="dotted"
        ),axes=F)

# Titling

mtext(side=3,"Majorca",cex=1.5,col=rgb(64,64,64,maxColorValue
        =255))
mtext(side=3,"Monthly average temperatures",cex=1.5,family="Lato
        Black",outer=T)
mtext(side=1,"Source: Wikipedia",cex=0.75,adj=1,font=3,outer=T)
dev.off()
```

In the script, we select a layout of two charts next to each other (mfol=c(1,2)). To have the bars suspended in air, e.g., to show their range, we have to put the difference of NY behind the bars. In the framework of the classic graphic model, R draws everything on top of each other; that means, lines have to be drawn first. However, this is not possible with the intended function, abline(), as it is based on an existing graphic. The solution is the panel.first parameter, which can be incorporated into the call of the bar chart function and causes its contents to be drawn as the lowermost layer. Here, a function or a list of functions can be handed over. The parameter can be used within the plot() function, but unfortunately not within the barplot() function. However, the gplots library contains the barplot2() function, which allows the use of panel.first. We do not plot axes; instead, we use the loop for (i in 1:length(myLines)) to draw the y-values slightly above the dotted line in the left chart. We do not have to repeat that in the right chart, since the dotted lines from the left chart are continued there.

8.4.4 Seasonal Ranges Stacked

Monthly average temperatures

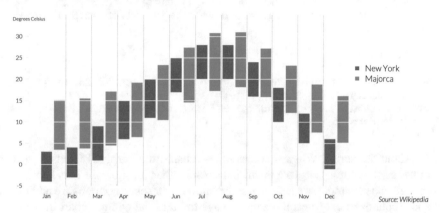

Source: Wikipedia

About the figure: With this variant, we can more immediately compare the individual months than with the previous representation form. We leave the x- and y-axes as they were, but this time we draw vertical auxillary lines between the months, and put horizontal auxillary lines in the background colour over the columns. With regard to temperature (warmer=better), it pays for New Yorkers to fly to Majorca from October to March, but it can even be colder in Majorca from April to September. Data were taken from the German Wikipedia articles on "New York" and "Mallorca" and, unlike in the previous example, complemented with empty lines.

```
pdf_file<-"pdf/timeseries_seasonal_minmax_overlay.pdf"
cairo_pdf(bg="grey98", pdf_file,width=14,height=7)

par(omi=c(0.25,0,0.75,0.25),mai=c(0.5,2,0.5,2),family="Lato ⤸
     Light",las=1)

# Import data and prepare chart

library(gdata)
myData<-read.xls("myData/Climate2.xlsx", encoding="latin1")
myLines<-c(-5,0,5,10,15,20,25,30)
attach(myData)
```

```
# Create chart and other elements

myT1<-barplot(t(cbind(NY_min,NY_max-NY_min)),col=c("white","↘
      coral3"),border=NA,ylim=c(-5,35),axes=F,axisnames=F)
myT2<-barplot(t(cbind(MAJ_min,MAJ_max-MAJ_min)),col=c("white","↘
      cornflowerblue"),border=NA,add=T,axes=F,names.arg=Month)
axis(2,at=myLines,col=par("bg"),col.ticks="grey81",lwd.ticks↘
      =0.5,tck=-0.025)
abline(h=myLines,col="white",lwd=2)
abline(v=seq(2.5,28.8,by=2.4),col="grey")
text(-0.95,34,"Degrees Celsius",xpd=T,cex=0.8)
legend(34,25,c("New York","Majorca"),col=c("coral3","↘
      cornflowerblue"),pch=15,bty="n",xjust=1,cex=1.5,pt.cex↘
      =1.5,xpd=T)

# Titling

mtext(side=3,"Monthly average temperatures",cex=2.25,adj=0.1,↘
      family="Lato Black",outer=T)
mtext(side=1,line=-1,"Source: Wikipedia",cex=1.25,adj=1,font=3,↘
      outer=T)
dev.off()
```

About the script: When it comes to bar charts, the barplot() function in R can either stack columns or place them next to each other. To show the bars floating as in the previous example, we have to use the option for stacked columns. To additionally show columns next to each other, we have to build the column chart in two steps. To that end, the XLSX file has to be constructed slightly differently. The minimal and maximal values of the second variable have to be offset by one row, respectively (Fig. 8.4).

A23	⏶ × ✓ ƒx					
	A	**B**	**C**	**D**	**E**	**F**
1	Month	NY_min	NY_max	MAJ_min	MAJ_max	
2	Jan	-4	3			
3				3.5	15.1	
4	Feb	-3	4			
5				3.8	15.5	
6	Mar	1	9			
7				4.5	17.1	
8	Apr	6	15			
9				6.5	19.2	

Fig. 8.4 Data structure for the figure

With these data, we first draw the minimum and maximum values for New York; during this procedure, we leave gaps next to each column. In a second step, we fill these gaps with the values for Majorca by using a second barplot() call and the add=TRUE parameter.

8.4.5 Season Figure (Seasonal Subseries Plot) with Data Table

Google trend for 'chicken soup'
Jan 2004 to Feb 2012

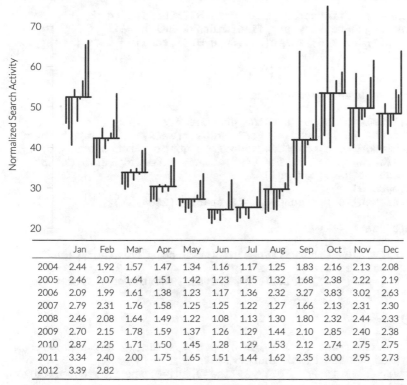

	Jan	Feb	Mar	Apr	May	Jun	Jul	Aug	Sep	Oct	Nov	Dec
2004	2.44	1.92	1.57	1.47	1.34	1.16	1.17	1.25	1.83	2.16	2.13	2.08
2005	2.46	2.07	1.64	1.51	1.42	1.23	1.15	1.32	1.68	2.38	2.22	2.19
2006	2.09	1.99	1.61	1.38	1.23	1.17	1.36	2.32	3.27	3.83	3.02	2.63
2007	2.79	2.31	1.76	1.58	1.25	1.25	1.22	1.27	1.66	2.13	2.31	2.30
2008	2.46	2.08	1.64	1.49	1.22	1.08	1.13	1.30	1.80	2.32	2.44	2.33
2009	2.70	2.15	1.78	1.59	1.37	1.26	1.29	1.44	2.10	2.85	2.40	2.38
2010	2.87	2.25	1.71	1.50	1.45	1.28	1.29	1.53	2.12	2.74	2.75	2.75
2011	3.34	2.40	2.00	1.75	1.65	1.51	1.44	1.62	2.35	3.00	2.95	2.73
2012	3.39	2.82										

Source: www.google.com/trends

About the figure: A "seasonal subseries plot" is a figure that may take some getting used to, but that can, in individual cases, be useful for multi-year monthly time series. Generally speaking, there are not many examples that would make this representation form appear useful: it really only is if (1) one is interested in the "season shape" of a time series, (2) this shape is much more pronounced than the trend, and (3) one also wants to visualise the individual yearly values to see if the season shape changes over time. In this case, data are first sorted by month, and only within the month by years and plotted. A horizontal line shows the monthly average, any deviant columns the individual yearly values for the respective month. If we go back to the figure in Sect. 8.3.4, we notice a strong seasonal pattern with maximal values around the turn of the year and minimal values during summer. This seasonal pattern is even more obvious in the present figure: here, the January and October values are greater than the values for November and December. The peaks in August, September, and October of 2006 also stand out. A seasonal subseries plot as shown here can easily be complemented with a contingency table that lists the yearly values for the respective months. Data can be downloaded as a text file from http://www.google.com/trends. For the current example, they were entered into an SQL table google_trends and aggregated at monthly level. The table is available as an SQL dump at https://link.springer.com/chapter/10.1007/978-3-030-28444-2_1.

```
pdf_file<-"pdf/timeseries_seasonalsubseries_inc.pdf"
cairo_pdf(bg="grey98", pdf_file,width=8,height=8)
par(omi=c(1,0,1,0.5),mai=c(2,0.80,0,0.5),family="Lato Light",las↘
    =1)


# Import data and prepare chart

source("scripts/inc_datadesign_dbconnect.r")
sql<-"select left(week, 4) year, substr(week, 6, 2) month, avg(↘
    chicken_soup) chicken_soup from google_trends group by ↘
    left(week, 4), substr(week, 6, 2) order by year, month"
myData<-dbGetQuery(con,sql)
attach(myData)
y<-ts(chicken_soup,frequency =12,start=c(2004,1))

# Create chart

monthplot(y,axes=F,box=F,type="h",lwd=3,col="darkred",ylab="↘
    Normalized Search Activity")
axis(2,col=par("bg"),col.ticks="grey81",lwd.ticks=0.5,tck=-↘
    0.025)

# Titling

mtext("Google trend for 'chicken soup'",3,line=2,adj=0,cex=2.0,↘
    family="Lato Black",outer=T)
mtext("Jan 2004 to Feb 2012",3,line=0,adj=0,cex=1.5,font=3,outer↘
    =T)
mtext("Source: www.google.com/trends",1,line=3,adj=1.0,cex=0.95,↘
    font=3,outer=T)
dev.off()
```

Here is the LaTeX code:

```
\documentclass{article}
\usepackage[paperheight=18.0cm, paperwidth=19.27cm, top=2cm, left=0.25cm, ↘
    right=0.25cm, bottom=2cm]{geometry}
\usepackage{color}
\usepackage{booktabs}
\renewcommand{\baselinestretch}{1.2}
\usepackage{fontspec}
\setmainfont[Mapping=text-tex]{Lato Light}
\definecolor{myBackground}{rgb}{0.99,0.99,0.99}
\pagecolor{myBackground}
\begin{document}
\thispagestyle{empty}
\begin{picture}(0,0)
\put(0,-445)
{\includegraphics[scale=0.895]{../pdf/timeseries_seasonalsubseries_inc.pdf}}
```

```
\end{picture}
\begin{picture}(0,0)
\put(25,-340)
{
\begin{tabular}{rrrrrrrrrrrrr}
  \toprule
 & Jan & Feb & Mar & Apr & May & Jun & Jul & Aug & Sep & Oct & Nov & Dec \\
  \midrule
2004 & 2.44 & 1.92 & 1.57 & 1.47 & 1.34 & 1.16 & 1.17 & 1.25 & 1.83 & 2.16 & ↘
     2.13 & 2.08 \\
  2005 & 2.46 & 2.07 & 1.64 & 1.51 & 1.42 & 1.23 & 1.15 & 1.32 & 1.68 & 2.38 ↘
       & 2.22 & 2.19 \\
  2006 & 2.09 & 1.99 & 1.61 & 1.38 & 1.23 & 1.17 & 1.36 & 2.32 & 3.27 & 3.83 ↘
       & 3.02 & 2.63 \\
  2007 & 2.79 & 2.31 & 1.76 & 1.58 & 1.25 & 1.25 & 1.22 & 1.27 & 1.66 & 2.13 ↘
       & 2.31 & 2.30 \\
  2008 & 2.46 & 2.08 & 1.64 & 1.49 & 1.22 & 1.08 & 1.13 & 1.30 & 1.80 & 2.32 ↘
       & 2.44 & 2.33 \\
  2009 & 2.70 & 2.15 & 1.78 & 1.59 & 1.37 & 1.26 & 1.29 & 1.44 & 2.10 & 2.85 ↘
       & 2.40 & 2.38 \\
  2010 & 2.87 & 2.25 & 1.71 & 1.50 & 1.45 & 1.28 & 1.29 & 1.53 & 2.12 & 2.74 ↘
       & 2.75 & 2.75 \\
  2011 & 3.34 & 2.40 & 2.00 & 1.75 & 1.65 & 1.51 & 1.44 & 1.62 & 2.35 & 3.00 ↘
       & 2.95 & 2.73 \\
  2012 & 3.39 & 2.82 &  &  &  &  &  &  &  &  &  & \\
   \bottomrule
\end{tabular}
}
\end{picture}
\end{document}
```

About the script: Once we have read the SQL data aggregated on a monthly level, we only need one function, which is provided by R: monthplot() can immediately generate the seasonal subseries plot. We still need to align the LaTeX table to make the columns match the distances of the x-axis. To do this, we can first draw an x-axis,

```
axis(1,at=c(1:12),
        labels=c("Jan","Feb","Mar","Apr","May","Jun","Jul","Aug"↘
             ,"Sep","Oct","Nov","Dec"))
```

which we can then use to align the table using the put() command and the scaling factor of \includegraphics. Once the correct specifications have been found, the axis can be removed and the figure can be generated. The LaTeX code needed for generation of the table can largely be created in R. The xtable() function from the package of the same name is a handy tool for this purpose, as it already attaches the necessary table markers to the monthly time series (abridged version below):

```
> xtable(y)
% latex table generated in R 3.0.1 by xtable 1.7-1 package
% Wed Oct  2 11:44:25 2013
\begin{table}[ht]
\centering
\begin{tabular}{rrrrrrrrrrrrr}
  \hline
 & Jan & Feb & Mar & Apr & May & Jun & Jul & Aug & Sep & Oct & ↘
       Nov & Dec \\
  \hline
2004 & 2.44 & 1.92 & 1.57 & 1.47 & $\ldots\ldots$  \\
   \hline
```

```
\end{tabular}
\end{table}
```

We only need the part enclosed by tabular. In that part, we replace the \hline commands with \toprule, \midrule and \bottomrule from the booktabs package, which generates more aesthetically pleasing line spacing.

8.4.6 Temporal Ranges

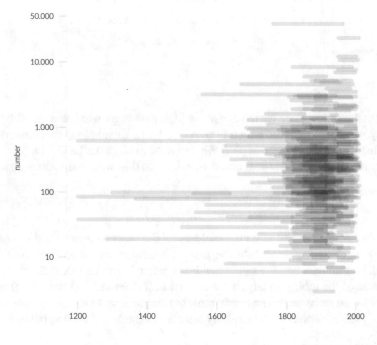

histat time series

Start, end, and number of time series per study, annual firgures

Source: gesis.org/histat

The figure shows the length and number of time series from different studies contained in the "histat" database. There are around 400 studies. For each study, the start and end points and the number of time series included in the study are described. The number is displayed logarithmically. Since there are a lot of overlaps, line colours are transparent. This means the distribution is clearly visible: most of the studies span the nineteenth century and contain between 100 and 1000 time series. The longest studies start as early as the thirteenth century. Data derive from an XLS table that was created in the context of a special evaluation. It lists the time series contained in the online database http://www.gesis.org/histat, where they are organised by study. For each study, the number of time series and the first and last date were captured.

```
pdf_file<-"pdf/timeseries_spans.pdf"
cairo_pdf(bg="grey98", pdf_file,width=9,height=9)

par(omi=c(0.75,0.5,1,0.5),mai=c(0.5,1.25,0.5,0.1),mgp=c(4.5,1,0)\
    ,family="Lato Light",las=1)

# Import data and prepare chart

library(gdata)
myData<-read.xls("myData/histat_studies.xlsx", encoding="latin1"\
    )
attach(myData)
n<-nrow(myData)
myColour<-rgb(240,24,24,30,maxColorValue=255)

# Define chart and other elements

plot(1:1,type="n",axes=F,xlab="Study start and end",ylab="number\
    ",xlim=c(min(from),max(to)),ylim=c(log10(min(number_\
    timeseries)),log10(max(number_timeseries))))
axis(1,col=par("bg"),col.ticks="grey81",lwd.ticks=0.5,tck=-\
    0.025)
axis(2,at=c(log10(10),log10(100),log10(1000),log10(10000),log10\
    (50000)),labels=c("10","100","1.000","10.000","50.000"),\
    col=par("bg"),col.ticks="grey81",lwd.ticks=0.5,tck=-0.025)
for (i in 1:n) segments(from[i],log10(number_timeseries)[i],to[i\
    ],log10(number_timeseries)[i],col=myColour,lwd=8)

# Titling

mtext("histat time series",3,line=2,adj=0,family="Lato Black",\
    outer=T,cex=2)
mtext("Start, end, and number of time series per study, annual \
    firgures",3,line=0,adj=0,cex=1.35,font=3,outer=T)
mtext("Source: gesis.org/histat",1,line=2,adj=1.0,cex=1.1,font\
    =3,outer=T)
dev.off()
```

In the script, data are read from the XLSX table, then a plot is specified and the x- and y-axes are drawn using axis. Data are plotted logarithmically on the y-axis; the labels "10", "100", etc., are individually given. Data are plotted as segments with the segments() function. To this end, the individual data are passed row by row, and lines are drawn from the starting position to the end position at height number_timeseries. Colour is set to very transparent with transparency of 30 so that, even with strong overlap of almost 400 plotted lines, the data distribution is easily visible.

Chapter 9
Scatter Plots

Up to four variables can be plotted in a scatter plot: two numerical variables on the x- and y-axis, a numerical or ordinal variable for definition of point size, and a nominal variable for colour definition. Additional elements can be:

- Any type of smoothing, e.g., a regression line
- Labels of individual data points
- A cross identifying the average
- An area or line (ellipse) that identifies a bivariate distribution
- A line that connects the individual points

The chapter first introduces five variants of scatter plots, each of which illustrates a different aspect and its implementation. This is again followed by "special cases".

General Aspects

Irrespective of the position of axis lines (be it at the borders or in the middle), axis labels should always be attached at the borders (right chart) and never in the middle (left chart) to avoid compromising the view of the data:

© Springer Nature Switzerland AG 2019
T. Rahlf, *Data Visualisation with R*, https://doi.org/10.1007/978-3-030-28444-2_9

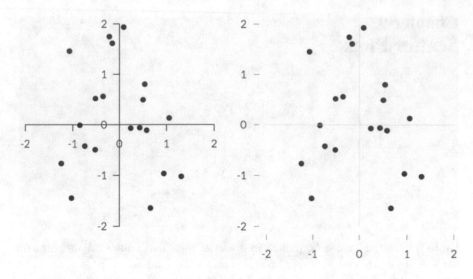

Point Size

It is not always easy to find the best size, since the range of the parameters should not be too far apart, but not too close either. Generally speaking, scaling should ensure that the areas are proportional to the attributes, and doubling of the value therefore results in a doubling of the area. This is not a rule set in stone. Any variation in parameters should definitely be clearly explained in a legend. It should generally be considered whether the additional information given by parameter variations really does add benefit or only contributes to confusion.

Point Labels

A big problem with the labelling of individual points in a scatter plot is overlap of labelling elements. Nowadays, there are various packages that provide functions for automatic positioning and use different calculation methods to avoid such overlaps. In reality, this never works perfectly, but it is always a good start. As a rule, the positions have to be manually adjusted—or alternatively, two columns that define the positions for the labels have to be included in the input data.

9.1 Variants

9.1.1 Scatter Plot Variant 1: Four Quadrants Differentiated by Colour

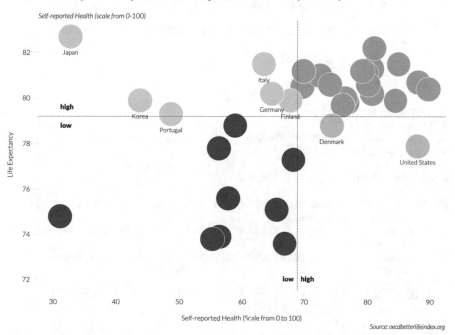

Life Expectancy and Self-reported Health (OECD)

Source: oecdbetterlifeindex.org

About the figure: The first example of a scatter plot shows the relationship between self-perception of health and life expectancy in OECD countries. In this case, we divide the range of values of both presented variables into "high" and "low", and thereby the two-dimensional range of values into four quadrants. Borders are defined by the respective average. Colouring depends on the content: high life expectancy and high self-perception of health make up the green area, low life expectancy and low self-perception values the red area. An above-average life expectancy with simultaneous below-average self-perception, or a below-average life expectancy with simultaneous above-average self-perception are the definitions of the light-red or light-green areas. Both case groups contain "unusual" cases: these points are therefore identified by labels below each point.

Data: see annex A, BetterLifeIndex_Data_2011V6.xls.

```
pdf_file<-"pdf/scatterplots_quadrants.pdf"
cairo_pdf(bg="grey98", pdf_file,width=11.69,height=9)

par(mar=c(4,4,0.5,2),omi=c(0.5,0.5,1,0),family="Lato Light",las↘
      =1)
library(RColorBrewer)

# Import data and prepare chart

mydata<-read.csv(file="myData/BetterLifeIndex_Data_2011V6.csv",↘
      head=F,
  sep=";",dec=",",skip=6)
mydata<-mydata[2:36,]
attach(mydata)

myX<-as.numeric(V16)
myY<-as.numeric(V15)
myX_des<-"Self-reported Health (Scale from 0 to 100)"
myY_des<-"Life Expectancy"

# Define chart and other elements

plot(type="n",xlab=myX_des,ylab=myY_des,myX, myY,xlim=c(30,90),↘
      ylim=c(72,83),axes=F)
axis(1,col=par("bg"),col.ticks="grey81",lwd.ticks=0.5,tck=-↘
      0.025)
axis(2,col=par("bg"),col.ticks="grey81",lwd.ticks=0.5,tck=-↘
      0.025)

myC1<-brewer.pal(5,"PiYG")[5]
myC2<-brewer.pal(5,"PiYG")[4]
myC3<-brewer.pal(5,"PiYG")[1]
myC4<-brewer.pal(5,"PiYG")[2]

myP1<-subset(mydata[c("V2","V16","V15")],myX > mean(myX) & myY >↘
      mean(myY))
myP2<-subset(mydata[c("V2","V16","V15")],myX < mean(myX) & myY >↘
      mean(myY))
myP3<-subset(mydata[c("V2","V16","V15")],myX < mean(myX) & myY <↘
      mean(myY))
myP4<-subset(mydata[c("V2","V16","V15")],myX > mean(myX) & myY <↘
      mean(myY))

myN1<-nrow(myP1)
myN2<-nrow(myP2)
myN3<-nrow(myP3)
myN4<-nrow(myP4)

symbols(myP1[,2:3],bg=myC1,circles=rep(1,myN1),inches=0.3,add=T,↘
      xpd=T,fg="white")
symbols(myP2[,2:3],bg=myC2,circles=rep(1,myN2),inches=0.3,add=T,↘
      xpd=T,fg="white")
```

```
symbols(myP3[,2:3],bg=myC3,circles=rep(1,myN3),inches=0.3,add=T, ⬎
    xpd=T,fg="white")
symbols(myP4[,2:3],bg=myC4,circles=rep(1,myN4),inches=0.3,add=T, ⬎
    xpd=T,fg="white")

text(myP2[,2:3],as.matrix(myP2$V2),cex=0.9,pos=1,offset=1.75)
text(myP4[,2:3],as.matrix(myP4$V2),cex=0.9,pos=1,offset=1.75)

abline(v=mean(myX,na.rm=T),col="black",lty=3)
abline(h=mean(myY,na.rm=T),col="black",lty=3)

text(min(V16),mean(V15)+0.005*mean(V15),"high",family="Lato ⬎
    Black",adj=0)
text(min(V16),mean(V15)-0.005*mean(V15),"low",family="Lato Black⬎
    ",adj=0)
text(mean(V16)-0.001*mean(V16),72,"high",family="Lato Black",pos⬎
    =4)
text(mean(V16)+0.001*mean(V16),72,"low",family="Lato Black",pos⬎
    =2)

# Titling

mtext("Life Expectancy and Self-reported Health (OECD)",3,adj=0,⬎
    line=2.5,cex=2.0,family="Lato Black")
mtext("Self-reported Health (scale from 0-100)",3,adj=0,line=0,⬎
    cex=1.0,font=3)
mtext("Source: oecdbetterlifeindex.org",1,line=4,adj=1,cex=0.95,⬎
    font=3)
dev.off()
```

The script begins by reading the data and limiting them to rows 2–36; x and y variables are specified and so are the labels. These are:

- V2: country name
- V15: life expectancy
- V16: self-reported health

This is followed by definition of the plot and drawing of axes. After that, four colours myC1 to myC4 are specified, which are individually taken from the Brewer palette "PiYG"; we also create four partial data frames myP1 to myP4 containing data of the individual quadrants. Then, the function symbols() is called for each partial data frame. We could have drawn the points using points() or text(), but symbols() immediately provides a symbol with a white border. Point size is not specified using cex, but with circles, a vector whose length has to correspond to the number of symbols to be plotted. We do not use a variable, as we do not want variations in point size. Instead, we specify 1, which is repeated using rep() until its length matches the number of data. Additionally, the parameter inches is required to convert the data values to the correct scaling. A value of 0.3 yields the desired size. In contrast to points() or text(), add = T has to be used with symbols(), since the latter is a high-level function. Now, we can use text() for the output of point labels for the second and fourth quadrant. Since 2 was read as a factor, the variable has to be converted before being displayed; this is done with as.matrix(). At the end, median lines are plotted using abline() and labelled on the left and right, or above and below, and then title and sources are printed.

9.1.2   Scatter Plot Variant 2: Outliers Highlighted

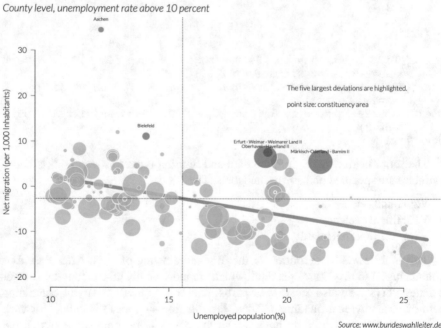

Unemployed population, migration in Germany 2005
County level, unemployment rate above 10 percent

Source: www.bundeswahlleiter.de

The figure shows the unemployed population, and immigration and emigration in Germany in 2005. There is a negative correlation between these two parameters. In this instance, we use a second colour for outliers and the same colour for a regression line. This may sound contradictory, but follows the logic that outliers are defined by the regression line, or rather the largest distance from it. Point size is defined by a third variable. Since the population is by definition approximately the same across constituencies, it would not be a good idea to use them for orientation when selecting point size. We choose the area instead; given the approximately equal population numbers, the area is an indicator for population density, which may also play a role. For orientation purposes, medians are plotted as a dotted line.

Data can be downloaded in CSV file format as struktbtwkr2005.csv from http://github.com/gka/nr17-r-workshop.

Headers were removed for the example.

```
pdf_file<-"pdf/scatterplots_outliers.pdf"
cairo_pdf(bg="grey98", pdf_file,width=11.69,height=9)


par(mar=c(4,4,0.5,2),omi=c(0.5,0.5,1,0),family="Lato Light",las↘
    =1)


# Import data and prepare chart

library(gdata)
myStructuralData<-read.csv(file="myData/struktbtwkr2005.csv",↘
    head=F,sep=";",dec=".")
myData<-subset(myStructuralData,V2 > 0 & V34 > 10)
attach(myData)

myXDes<-"Unemployed population(%)"
myYDes<-"Net migration (per 1,000 Inhabitants)"

# Define chart and other elements

plot(type="n",xlab=myXDes,ylab=myYDes,V34,V21,xlim=c(10,26),ylim↘
    =c(-20,35),axes=F,cex.lab=1.2)
axis(1,lwd.ticks=0.5,cex.axis=1.15,tck=-0.015)
axis(2,lwd.ticks=0.5,cex.axis=1.15,tck=-0.015)

myC1<-rgb(0,208,226,200,maxColorValue=255)
myC2<-rgb(255,0,210,150,maxColorValue=255)

fit<-lm(V21 ~ V34)
myData$fit<-fitted(fit)
points(V34,myData$fit,col=myC2,type="l",lwd=8)

myData$resid<-residuals(fit)
myData.sort<-myData[order(-abs(myData$resid)) ,]
myData.sort_begin<-myData.sort[1:5,]

myP1<-myData.sort[5+1:length(myData$fit),c("V34","V21")]
myP2<-myData.sort_begin[c("V34","V21")]

myR1<-sqrt(myData.sort$V6)/10
myR2<-sqrt(myData.sort_begin$V6)/10

symbols(myP1,circles=myR1,inches=0.3,bg=myC1,fg="white",add=T)
symbols(myP2,circles=myR2,inches=0.3,bg=myC2,fg="white",add=T)

text(myP2,iconv(as.matrix(myData.sort_begin["V3"]),"LATIN1","UTF↘
    -8"),cex=0.65,pos=3,offset=1.1)
```

```
abline(v=mean(V34,na.rm=T),col="black",lty=3)
abline(h=mean(V21,na.rm=T),col="black",lty=3)


text(20,20, "The five largest deviations are highlighted. \n\
      npoint size: constituency area", adj=0)
# Titling
mtext("Unemployed population, migration in Germany 2005",3,adj
      =0,line=2,cex=2.5,outer=T,family="Lato Black")
mtext("County level, unemployment rate above 10 percent",3,adj
      =0,line=0,cex=1.5,outer=T,font=3)
mtext("Source: www.bundeswahlleiter.de",1,line=4,adj=1,cex=1.15,
      font=3)
dev.off()
```

In the script, data are read from a CSV file. The following variables are used:

V2:   the constituency code; rows coded 0 represent the state total.
V3:   name of the constituency
V6:   area as of 31 December 2004
V21:   net migration
V34:   unemployed population in percent

Only constituencies with an unemployment rate exceeding 10% were selected. Since the data also include aggregated statistics for all administrative levels, they have to be filtered using V2 > 2. After definition of a type "n" plot(), axes with slightly shorter increment markings are plotted, and two colours myC1 and myC2 are specified. The regression line is calculated using the lm() function, saved as fit, and plotted using points(). Now we also obtain the residuals using residuals() and sort the data frame in descending order by the size of the residuals. The first five of these are saved as data.sort_start. We want to plot two sets of points: on the one hand these residues (myP1) and on the other hand the "rest", that is the entire data minus the residues (myP2). Lastly, we need myR1 and myR2 for the point size. Since doubling the radius means quadrupling the area, we have to use a square root. With all of these data, the two point sets can now be plotted using symbols(). The "outliers" still have to be labelled, and the medians are added as a dotted line.

9.1.3 Scatter Plot Variant 3: Areas Highlighted

Relationship between height and weight

Selected celebrities

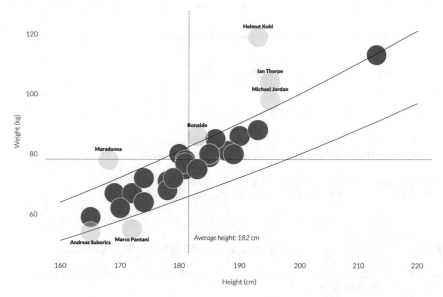

Source: celebrityheights.com, howmuchdotheyweigh.com

About the figure: In the third variant, we again use different colours to highlight extraordinary cases. This instance deals with the relationship between height and weight of selected celebrities. Height and weight averages are added as lines for orientation, in addition to body mass index (BMI) curves for a BMI of 20 and 25. Values below 20 and above 25 are colour-coded. If the main focus is on "normal" values, then these should be represented in a bolder colour. Again, extraordinary cases are labelled.

Data: see annex A, persons.xlsx.

```
pdf_file<-"pdf/scatterplots_areas.pdf"
cairo_pdf(bg="grey98", pdf_file,width=11.69,height=9)

par(mai=c(0.85,1,0.25,0.25),omi=c(1,0.5,1,0.5),family="Lato ⌐
    Light",las=1)

# Import data and prepare chart

library(gdata)
myPersons<-read.xls("myData/persons.xlsx", encoding="latin1")
attach(myPersons)
myData<-subset(myPersons,W>0 & s=="m" & name!="Max Schmeling")
```

```
attach(myData)

# Define chart and other elements

plot(type="n",xlab="Height (cm)",ylab="Weight (kg)",h,w,xlim=c↘
     (160,220),ylim=c(50,125),axes=F)
axis(1,col=par("bg"),col.ticks="grey81",lwd.ticks=0.5,tck=-↘
     0.025)
axis(2,col=par("bg"),col.ticks="grey81",lwd.ticks=0.5,tck=-↘
     0.025)

myC1<-rgb(255,0,210,maxColorValue=255)
myC2<-rgb(0,208,226,100,maxColorValue=255)

myP1<-subset(myData[c("h","w")],w>20*(h/100*h/100) & w<25*(h/100↘
     *h/100))
myP2<-subset(myData[c("h","w")],w<20*(h/100*h/100))
myP3<-subset(myData[c("h","w")],w>25*(h/100*h/100))

myDes2<-as.matrix(subset(name,w<20*(h/100*h/100)))
myDes3<-as.matrix(subset(name,w>25*(h/100*h/100)))

symbols(myP1,bg=myC1,fg="white",circles=rep(1,nrow(myP1)),inches↘
     =0.25,add=T)
symbols(myP2,bg=myC2,fg="white",circles=rep(1,nrow(myP2)),inches↘
     =0.25,add=T)
symbols(myP3,bg=myC2,fg="white",circles=rep(1,nrow(myP3)),inches↘
     =0.25,add=T)

text(myP2,myDes2,cex=0.75,pos=1,offset=1.1,family="Lato Black")
text(myP3,myDes3,cex=0.75,pos=3,offset=1.1,family="Lato Black")

curve(20*(x/100*x/100),xlim=c(160,220),add=T)
curve(25*(x/100*x/100),xlim=c(160,220),add=T)

abline(v=mean(h,na.rm=T),lty=3)
abline(h=mean(w,na.rm=T),lty=3)
text(182.5,52,"Average height: 182 cm",adj=0,font=3)

# Titling

mtext("Relationship between height and weight",3,adj=0,line=2,↘
     cex=2.1,outer=T,family="Lato Black")
mtext("Selected celebrities",3,adj=0,line=0,cex=1.4,outer=T,font↘
     =3)
mtext("Source: celebrityheights.com, howmuchdotheyweigh.com",1,↘
     line=1,adj=1,cex=0.95,outer=T,font=3)
dev.off()
```

In the script, the XLSX table is read and data are filtered: we choose only men (s ="m") and only those for whom a weight is specified (w>0). Then the plot is defined, axes are plotted, and three point sets myP1, myP2, and myP3 are created: the first one contains individuals with a BMI between 20 and 25, the second those with BMI below 20, and the third set those with BMI above 25. Labels for point sets two and three are saved in myDes2 and myDes3. Now we can use symbols() to plot all three point sets. The parameter circles is defined as rep(1,nrow(myP1)) and therefore constant. Any points outside of the BMI 20–25 corridor are labelled; similarly, average lines are plotted and labelled. At the end come the titles and sources.

9.1.4 Scatter Plot Variant 4: Superimposed Ellipse

Number of men and women in Germany

Municipalities, between 12,000 and 16,000 inhabitants

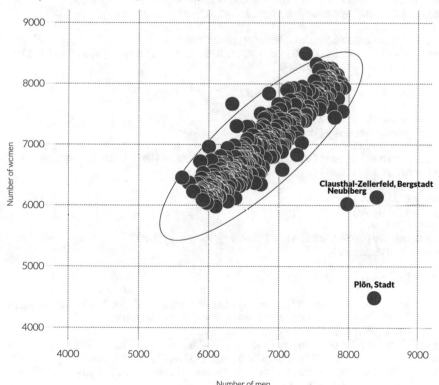

Source: GV-ISys

The figure shows the number of men and women in Germany for municipalities with 12–16,000 inhabitants. It is not surprising that there is a strong correlation. However, there are three extraordinary cases that are again labelled. Since a close match between the numbers of men and women can be assumed, axes should have the same range and the data area should be square. If we further assume a Gaussian distribution for the numbers of both men and women, then a bivariate Gaussian distribution can be plotted, whose parameters are specified by the averages and data correlation. Given the data distribution, it makes sense to plot a point-grid intersection.

Data: the table municipalities derives from the "Gemeindeverzeichnis Informationssystem (GV-ISys)" of destatis.de [website in German]. In the section "Administrative Gebietsgliederungen" (administrative area structure) there is an item "Gemeinden mit 5000 und mehr Einwohnern nach Fläche und Bevölkerung" (municipalities of 5000 or more inhabitants by area and population), which hides an XLS table 07Gemeinden.xls. The first nine rows and the last three rows of the table were deleted and the data saved in CSV format before being imported into the SQL database.

```
pdf_file<-"pdf/scatterplots_path_ellipse.pdf"
cairo_pdf(bg="grey98", pdf_file,width=10,height=10)

source("scripts/inc_datadesign_dbconnect.r")
par(mai=c(1,1,0.25,0.25),omi=c(1,0.5,1,0.5), mgp=c(4,1,0),family⟍
    ="Lato Light",las=1)

# Import data and prepare chart

sql<-"select * from gemeinden"
myData<-dbGetQuery(con,sql)
attach(myData)
mySelection<-subset(myData,bevinsg >= 12000 & bevinsg <= 16000)
attach(mySelection)
library(ellipse)
mx<-mean(bevm)
my<-mean(bevw)
mxy<-c(mx,my)
sxy<-sapply(mySelection[,c("bevm","bevw")],sd)
r<-cor(bevm,bevw)

# Define chart and other elements

plot(bevm,bevw,type="n",axes=F,xlab="Number of men",ylab="Number⟍
    of women",xlim=c(4000,9000),ylim=c(4000,9000))
lines(ellipse(r,scale=sxy,centre=mxy))
abline(v=seq(4000,9000,by=1000),col="black",lty=3,lwd=1)
abline(h=seq(4000,9000,by=1000),col="black",lty=3,lwd=1)
axis(1,cex.axis=1.2,col=par("bg"),col.ticks="grey81",lwd.ticks⟍
    =0.5,tck=-0.025)
axis(2,cex.axis=1.2,col=par("bg"),col.ticks="grey81",lwd.ticks⟍
    =0.5,tck=-0.025)
```

```
symbols(bevm,bevw,bg="red",circles=rep(2,nrow(mySelection))),↘
    inches=0.15,add=T,xpd=T,fg="white")
mySelection2<-subset(mySelection,bevm >= 7500 & bevw <= 6500)
attach(mySelection2)
text(bevm,bevw,iconv(gemeinde,"LATIN1","UTF-8"),family="Lato ↘
    Black",pos=3,offset=1.1)

# Titling

mtext("Number of men and women in Germany",3,adj=0,line=1.5,cex↘
    =2.5,family="Lato Black",outer=T)
mtext("Municipalities, between 12,000 and 16,000 inhabitants",3,↘
    adj=0,line=-0.25,cex=1.5,font=3,outer=T)
mtext("Source: GV-ISys",1,line=2,adj=1,cex=1.25,font=3,outer=T)
dev.off()
```

In the script, data are read from a MySQL database and limited to municipalities with a population 12–16,000. The ellipse package is loaded, and both averages are calculated for the ellipse plot. The standard deviation for the variables bevm and bevw is calculated using sapply(), and the correlation with cor(). The ellipse is drawn after specification of plot(), and abline() then plots a dotted grid on each thousand mark from 4000 to 9000. After the axes, the actual points are plotted using the symbols() function; as above, point circumferences are again constant. Scaling is adjusted with the inches parameter. Points of municipalities are labelled if their male population exceeds 7500 and their female population is below 6500. As usual, the title and sources make up the end.

9.1.5   Scatter Plot Variant 5: Connected Points

GDP and life expectancy in Greece

Correlation of growth rates, 1986-2010

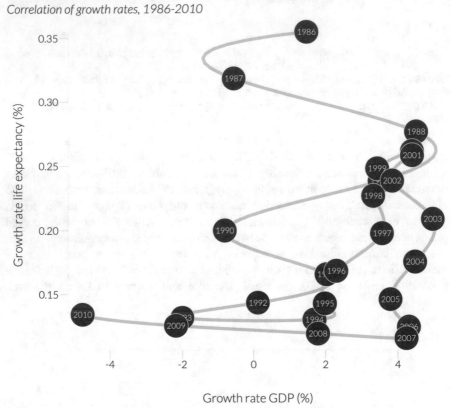

Source: gapminder.org

About the figure: If the data are time series, then the chronological development can also be shown in a scatter plot if consecutive points are connected. This is best done using a spline function. This leads to a "path effect" that describes the chronological development better than a simple (linear) point-to-point connection would. The example shows the correlation between the development of the national product and life expectancy in Greece between 1986 and 2010. Even though both parameters have the same dimension in this case, axis lengths do not necessarily have to be proportional.

Data: The XLS files for the scatter plot can be downloaded from http://www.
gapminder.org/data/. The XLS file Greece.xlsx was generated from those data and
can be downloaded at https://link.springer.com/chapter/10.1007/978-3-030-28444-
2_1.

```
pdf_file<-"pdf/scatterplots_connected.pdf"
cairo_pdf(bg="grey98", pdf_file,width=10,height=10)

par(mai=c(1.1,1.25,0.15,0),omi=c(1,0.5,1,0.5), mgp=c(4.5,1,0),\
      family="Lato Light",las=1)

# Import data and prepare chart

library(gdata)
myData<-read.xls("myData/gapminder/Greece.xlsx", encoding="\
      latin1")
myData<-myData[myData$Year>=1985, ]

attach(myData)
n<-nrow(myData)
grGDP<-vector()
grLEXP<-vector()
for (i in 2:n)
{
grGDP[i]<-(GDP[i]-GDP[i-1])/GDP[i-1]
grLEXP[i]<-(LEXP[i]-LEXP[i-1])/LEXP[i-1]
}
myData$grGDP<-grGDP*100
myData$grLEXP<-grLEXP*100
myData<-myData[2:n, ]

n<-nrow(myData)

t <- 1:n
ts <- seq(1, n, by = 1/10)
xs <- splinefun(t, myData$grGDP)(ts)
ys <- splinefun(t, myData$grLEXP)(ts)

# Define chart and other elements

plot(myData$grGDP, myData$grLEXP, type="n", xlab="Growth rate \
      GDP (%)", ylab="Growth rate life expectancy (%)", cex.lab\
      =1.5, axes=F)
axis(1,col=par("bg"),col.ticks="grey81",lwd.ticks=0.5,tck=-\
      0.025,cex.axis=1.25)
axis(2,col=par("bg"),col.ticks="grey81",lwd.ticks=0.5,tck=-\
      0.025,cex.axis=1.25)

lines(xs, ys,lwd=7,col="grey")
for (i in 1:n)
{
symbols(myData$grGDP[i],myData$grLEXP[i],bg="brown",fg="white",\
      circles=1,inches=0.25,add=T)
```

```
text(myData$grGDP[i],myData$grLEXP[i], myData$Year[i],col="white⤸
    ")
}

# Titling

mtext("GDP and life expectancy in Greece",3,adj=0,line=1.5,cex⤸
    =2.5,family="Lato Black",outer=T)
mtext("Correlation of growth rates, 1986-2010",3,adj=0,line=-⤸
    0.25,cex=1.5,font=3,outer=T)
mtext("Source: gapminder.org",1,line=2,adj=1,cex=1.25,font=3,⤸
    outer=T)
dev.off()
```

After reading the data, the script continues by calculating growth rates for the variables GDP and LEXP in a loop; the results are saved as grGDP and grLEXP. The procedure for calculation of the bivariate spline derives from Kihske Takahashi's response to a forum question. The splinefun() function from the stats package provides the necessary connection between the data. First, we calculate the individual function values for both variables; for "smooth" resolution, the values are calculated in steps of tenths.

The chart is defined using plot(), and the spline is plotted first; the points and labels are then placed across. It is important that we call the symbols() and text() functions consecutively for each point. If we plotted all symbol points first and then all labels, then there would be overlaps in the labels. With this stepwise plotting on the other hand, closely adjacent labels are covered by the symbols. Overall, this gives a much more readable impression. As usual, the title and sources come last.

9.2 Exceptions and Special Cases

9.2.1 Scatter Plot with Few Points

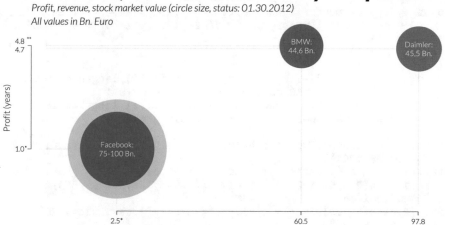

Facebook, BMW and Daimler by comparison

Profit, revenue, stock market value (circle size, status: 01.30.2012)
All values in Bn. Euro

Revenue (years)

* Estimated
** Result before tax   *Source www.spiegel.de*

This figure can only be called a scatter plot in the formal sense; in fact, it is more of a schematic overview. It shows the "correlation" between revenue and profit for Daimler, BMW, and Facebook. Point size is a third variable that shows the actual or estimated trading volume. For Facebook, January 2012 estimates ranged from 75 to 100 billion Euro. The exceptional position of Facebook when compared with the other two companies is very obvious in this representation form: even though revenue and profit are markedly lower, the expected trading volume is considerable higher. Data were taken from an article published in the German newspaper SPIEGEL.

```
pdf_file<-"pdf/scatterplots_three_points.pdf"
cairo_pdf(bg="grey98", pdf_file,width=14,height=9)

par(mai=c(2,1,1,1),omi=c(0,0,0,0),xpd=T,family="Lato Light",las↘
    =1)

# Define data and prepare chart

names<-c("BMW:\n44,6 Bn.","Daimler:\n45,5 Bn.","","Facebook:\n75↘
    -100 Bn.")
myValue<-c(44.6,45.5,100,75)
myRevenue<-c(60.5,07.0,2.5,2.5)
myProfit<-c(4.8,4.7,1,1)
```

```
myC1<-rgb(80,80,80,maxColorValue=255)
myC2<-rgb(255,97,0,maxColorValue=255)
myC3<-"grey"
myC4<-rgb(58,87,151,maxColorValue=255)

# Define chart and other elements

plot(myRevenue,myProfit,axes=F,type="n",xlab="Revenue (years)",\
     ylab="Profit (years)",xlim=c(-20,100),ylim=c(-1,6),cex.lab\
     =1.5)
for (i in 1:3)
{
arrows(myRevenue[i],-1,myRevenue[i],myProfit[i],length=0.10,lty=\
     "dotted",angle=10,code=0,lwd=1,col="grey70")
arrows(-20,myProfit[i],myRevenue[i],myProfit[i],length=0.10,lty=\
     "dotted",angle=10,code=0,lwd=1,col="grey70")
}
points(myRevenue,myProfit,pch=19,cex=myValue/2.6,col=c(myC1,myC2\
     ,myC3,myC4))
text(myRevenue,myProfit,names,col="white",cex=1.3)
axis(1,at=c(2.5,60.5,97.8),labels=c("2.5*","60.5","97.8"),cex.\
     axis=1.25)
axis(2,at=c(1,4.8),labels=c("1.0","4.8\n4.7"),cex.axis=1.25)
text(-25.5,5.08,"**")
text(-26.5,1.08,"*")

# Titling

mtext(line=1,"Facebook, BMW and Daimler by comparison",cex=3.5,\
     adj=0,family="Lato Black")
mtext(line=-1,"Profit, revenue, stock market value (circle size,\
      status: 01.30.2012)",cex=1.75,adj=0,font=3)
mtext(line=-3,"All values in Bn. Euro",cex=1.75,adj=0,font=3)
mtext(side=1,line=6.5,"Source www.spiegel.de",cex=1.75,adj=1,\
     font=3)
mtext(side=1,line=4.5,"* Estimated",cex=1.75,adj=0)
mtext(side=1,line=6.5,"** Result before tax",cex=1.75,adj=0)
dev.off()
```

The script starts by direct specification of names, plottable values, and the colours myC1 to myC4. After chart definition with plot(), dotted lines in x- and y-direction are drawn with the arrows() function from within a loop. The actual points are plotted in size cex=myValue/2.6 and labelled using text(). Axis labels are manually added at suitable positions. Annotation asterisks also have to be added manually to the axis labels with the text() function. In addition to the usual titles and sources, the individual border labels "Estimated" and "Result before tax" are added in this instance.

9.2.2   Scatter Plot with User-Defined Symbols

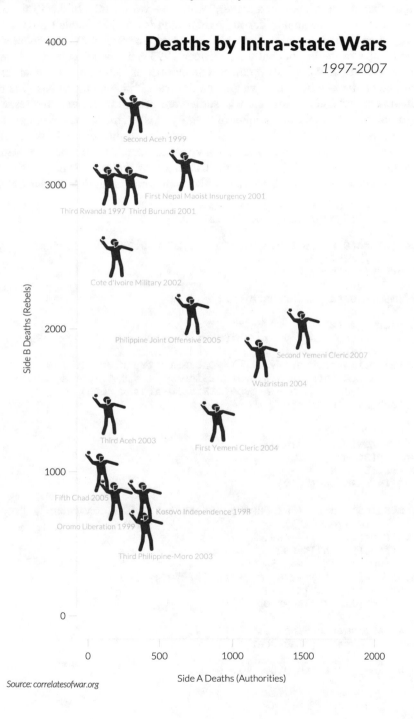

Deaths by Intra-state Wars

1997-2007

Side B Deaths (Rebels)

4000
3000
2000
1000
0

Second Aceh 1999

First Nepal Maoist Insurgency 2001

Third Rwanda 1997 Third Burundi 2001

Cote d'Ivoire Military 2002

Philippine Joint Offensive 2005

Second Yemeni Cleric 2007

Waziristan 2004

Third Aceh 2003

First Yemeni Cleric 2004

Fifth Chad 2005

Kosovo Independence 1998

Oromo Liberation 1999

Third Philippine-Moro 2003

0   500   1000   1500   2000

Side A Deaths (Authorities)

Source: correlatesofwar.org

About the figure: If there are comparatively few data and if circumstances are right, then using symbols in a scatterplot may be worthwhile. In this example, we use the "Protest" symbol by Jakob Vogel, which is freely available from the Noun Project and which has been embedded in a font (see Sect. 4.2). The figure shows the number of victims of civil wars between 1995 and 1999; data were collated by the Correlates of War (COW) project. The number of victims on the "state" side is plotted on the x-axis, and the victims on the "rebel" side on the y-axis. The name of the conflict and the year the war started are printed underneath the respective symbol. In this case, equal scaling of x and y axis is crucial; with a double range on the y-axis, this results in an aspect ratio of 1:2. In this example, we align the title and subtitle to the right to achieve a visual counterbalance to the data. Data can be downloaded as a CSV file from the Correlates of War (COW) project's website (Intra-StateWarData_v4.1.csv). The symbol font is created as described in Sect. 4.2.

```
pdf_file<-"pdf/scatterplots_symbols.pdf"
cairo_pdf(bg="grey98", pdf_file,width=7.5,height=12)

par(omi=c(0.5,0.5,0,0),mai=c(0.5,1.25,0,0.25),family="Lato Light↘
      ",las=1)
library(maptools)

# Import data and prepare chart

library(gdata)
myData<-read.xls("myData/Intra-StateWarData_v4.1.xlsx", encoding↘
      ="latin1")
mySelection<-subset(myData, myData$StartYear1>=1995 & myData$↘
      SideADeaths > 0 & myData$SideADeaths < 2000 & myData$↘
      SideBDeaths > 0 & myData$SideBDeaths < 4000)
attach(mySelection)

myColour<-"darkred"
myN<-nrow(mySelection)
h<-rep(0, myN)
v<-rep(0, myN)
myOffset<-cbind(h, v)

# mySelection[, c("WarName", "StartYear1", "SideADeaths", "↘
      SideBDeaths")]
myOffset[1, "h"]<--400
myOffset[5, "h"]<-232
myOffset[4, "h"]<--275
myOffset[2, "h"]<-270; myOffset[2, "v"]<-100;
myOffset[13, "h"]<--275
myOffset[12, "h"]<--300

myX<-as.numeric(SideADeaths)
myY<-as.numeric(SideBDeaths)

# Define chart and other elements
```

```
plot(myX, myY, typ="n", xlab="", ylab="", axes=F, xlim=c(0, ↘
     2000), ylim=c(0, 4000))
axis(1,col=par("bg"),col.ticks="grey81",lwd.ticks=0.5,tck=-↘
     0.025)
axis(2,col=par("bg"),col.ticks="grey81",lwd.ticks=0.5,tck=-↘
     0.025)
text(myX+130+myOffset[, "h"], myY-180+myOffset[, "v"], paste(↘
     WarName, StartYear1, sep=" "), cex=0.8, xpd=T, col="grey")

mtext(side=1, "Side A Deaths (Authorities)", adj=0.5, line=3)
mtext(side=2, "Side B Deaths (Rebels)", las=0, adj=0.5, line=4)

# Titling

mtext("Deaths by Intra-state Wars",3,adj=1,line=-3,cex=2.1,↘
     family="Lato Black")
mtext("1997-2007",3,adj=1,line=-5,cex=1.4,font=3)
mtext("Source: correlatesofwar.org",1,line=1,adj=0,cex=0.95,↘
     outer=T,font=3)

# Other elements of chart

par(family="Datendesign")
text(myX, myY, "b", col=myColour, cex=5, xpd=T)
dev.off()
```

In the script, data are first read and filtered to the required values (conflicts that started in or after 1995 and their stated number of resulting casualties); then, figure specifications are applied using plot(). No axis lines are plotted. The name of the conflict and the year it began are then added slightly staggered. In individual cases, the offset has to be entered manually. To do this, we first define a vector offset that is individually filled. After selection of our symbol font "Datadesign", the symbols are also drawn with the text() function.

9.2.3   Map of Germany as Scatter Plot

Relation of men and women in Germany 2005

for 100 men there are
- fewer than 90 women
- 90 to 110 women
- more than 110 women

Point size: Population size
Max.: 3.4 Mill. (Berlin, Stadt)   •   Median: 0.011 Mio.

Source: www.destatis.de, opengeodb.giswiki.org, www.lichtblau-it.de

The last example leads over to the next chapter. About the figure: If length and width parameters are available for the points, then a scatter plot is sufficient for drawing a map. This concept is illustrated by an example that shows the almost 3000 German municipalities with more than 5000 inhabitants. Point size is not proportional to the area size of the municipalities or cities in this case, but reflects the population. The size of the points is explained in the legend on the basis of the largest point and the point size median. Point colour represents the men-to-women ratio: grey points show a balanced ratio, light blue ones a surplus in men, and red points a surplus in women. This is explained in a second legend. The large number of points makes the outline of Germany visible without an added line.

Data: see annex A, v_women_men.

```
pdf_file<-"pdf/maps_germany_scatterplot.pdf"
cairo_pdf(bg="grey98", pdf_file,width=8,height=11)

par(mai=c(1.1,0,0,0),omi=c(0.25,0.5,0.75,0.5),family="Lato Light↘
    ",las=1)

# Import data and prepare chart

source("scripts/inc_datadesign_dbconnect.r")
sql<-"select * from v_women_men"
myDataset<-dbGetQuery(con,sql)
attach(myDataset)

legmaxsize<-max(transbev); legmaxvalue<-max(bevinsg)
legmaxtext<-myDataset[which(myDataset$transbev==legmaxsize),"↘
    gemeinde"]
if (length(legmaxtext) > 1)
{
n<-length(legmaxtext)
for (i in 2:n) legmaxtext<-c(legmaxtext,paste(",",legmaxtext[i↘
    ,]))
}
legmidsize<-quantile(transbev,0.5); legmidvalue<-quantile(↘
    bevinsg,0.5)

# Define chart and other elements

plot(lng,lat,type="n",axes=F,xlab="",ylab="")
low<-subset(myDataset,wm < 0.90)
medium<-subset(myDataset,wm >= 0.90 & wm <= 1.10)
high<-subset(myDataset,wm > 1.10)

c1<-rgb(0,191,255,200,maxColorValue=255)
c2<-rgb(150,150,150,80,maxColorValue=255)
c3<-rgb(128,0,0,200,maxColorValue=255)

attach(low)
points(lng,lat,pch=19,col=c1,cex=transbev)
attach(medium)
points(lng,lat,pch=19,col=c2,cex=transbev)
```

```
attach(high)
points(lng,lat,pch=19,col=c3,cex=transbev)

l1<-paste("Max.: ",format(legmaxvalue,digits=2)," Mill. (",\
    legmaxtext,")",sep="")
l2<-paste("Median: ",format(legmidvalue,digits=2)," Mio.",sep=""\
    )
legend(6.2,47.1,c(l1,l2),text.col="azure4",title="Point size: \
    Population size",title.adj=0.3,border=F,pch=19,col=rgb\
    (150,150,150,80,maxColorValue=255),bty="n",cex=1.1,pt.cex=\
    c(legmaxsize,legmidsize),xpd=T,ncol=2)
legend(13,50.25,text.col="azure4",c("fewer than 90 women","90 to\
     110 women","more than 110 women"),title="for 100 men \
    there are",title.adj=0,pt.cex=1,xpd=T,border=F,pch=19,col=\
    c(c1,c2,c3),bty="n",cex=0.8)

# Titling

mtext("Relation of men and women in Germany 2005",line=0,adj=0,\
    cex=1.8,family="Lato Black",outer=T)
mtext("Source: www.destatis.de, opengeodb.giswiki.org, www.\
    lichtblau-it.de",side=1,line=-1,adj=1,cex=0.9,font=3,outer\
    =T)
dev.off()
```

The script begins with determination of the median and maximal sizes, and their respective values, for the legend; the total data frame is used for this. Legend labels are added with the useful which() function, which can determine the municipality with the maximal population. The function returns the position in the data frame and can therefore be used as a row number within the square brackets of dataset. The slightly heavy-handed loop is required for legmaxname, since the maximum is a value that can theoretically occur more than once, unlike the median. Once this is done, we use plot() to define the figure, and create the three partial data frames low, medium, and high, which present a low, medium, or high men/women ratio. Each one is plotted in a different colour using points(). The legend function is called twice for creation of the legends. The first call generates a legend underneath the map, the second one the label on the right.

Chapter 10
Maps

R is not only particularly useful for generation of graphical representations, but also very specifically for maps. The collection site http://cran.r-project.org/web/views/spatial.html lists more than 100 packages for processing of geo data. After introductory examples, we first distinguish between those maps that visualise points, icons or entire diagrams in maps, and finally so-called choropleth maps, in which the areas within the maps illustrate the information. Finally, we consider two relevant examples that cannot be assigned to the above-mentioned categories.

10.1 Introductory Examples

10.1.1 Maps of Germany: Local Telephone Areas and Postcode Districts

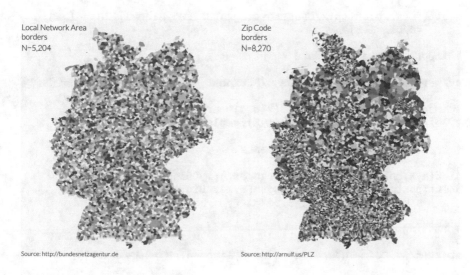

Local Network Area borders N=5,204

Zip Code borders N=8,270

Source: http://bundesnetzagentur.de

Source: http://arnulf.us/PLZ

© Springer Nature Switzerland AG 2019
T. Rahlf, *Data Visualisation with R*, https://doi.org/10.1007/978-3-030-28444-2_10

The figure shows the German local telephone areas on the left and postcode (PCD) districts on the right. In Germany, there are more than 5000 local telephone areas and more than 8000 postcode districts. As is clear from the figure, local telephone areas are approximately the same size, while postcode districts consider population density. As we only want to show the structure of the geographical units, we filled the shapes using randomly selected colour gradations from Brewer palettes. Both maps are shown in the Mercator projection.

Map files for the postcodes can be downloaded as ESRI shapefiles from https://arnulf.us/PLZ. At the time this book was written, map files for the local telephone areas were available for download as ESRI shapefiles from the German Bundesnetzagentur's website.

```
pdf_file<-"pdf/maps_germany_shp_onb_zipcode.pdf"
cairo_pdf(bg="grey98", pdf_file,width=16,height=9)

par(mar=c(0,0,0,0),oma=c(1,1,1,0), mfcol=c(1,2),family="Lato ⬊
    Light",las=1)
library(maptools)
library(rgdal)
library(RColorBrewer)

# Import data and prepare chart

myX<-readShapeSpatial("myData/ONB_Grenzen/ONB_Grenzen.shp")
myColour<-sample(1:7,length(myX),replace=T)

# Create chart and other elements

plot(myX,col=brewer.pal(7,"Greens")[myColour],border=F)
mtext(paste("N=", format(length(myX), big.mark=","), sep=""),⬊
    side=3,line=-6,adj=0,cex=1.7)
mtext("Local Network Area \nborders",side=3,line=-4,adj=0,cex⬊
    =1.7)
mtext("Source: http://bundesnetzagentur.de",side=1,line=-1,adj⬊
    =0,cex=1.3)

# Import data and prepare chart

myY<-readShapeSpatial("myData/PLZ/post_pl.shp",proj4string=CRS("⬊
    +proj=longlat"))
myX=spTransform(myY,CRS=CRS("+proj=merc"))
myColour<-sample(1:7,length(myX),replace=T)

# Create chart and other elements

plot(myX,col=brewer.pal(7,"Oranges")[myColour],border=F)
mtext(paste("N=", format(length(myX), big.mark=","), sep=""),⬊
    side=3,line=-6,adj=0,cex=1.7)

# Titling

mtext("Zip Code \nborders",side=3,line=-4,adj=0,cex=1.7)
```

```
mtext("Source: http://arnulf.us/PLZ",side=1,line=-1,adj=0,cex↘
    =1.3)
dev.off()
```

In the script, we require the following libraries: maptools for readShapeSpatial(), rgdal for the spTransform() function, and RColorBrewer for the use of colour palettes with brewer.pal(). Once the shapefile for the local network districts has been read, a vector colour is generated and filled with uniformly distributed random numbers between 1 and 7. That way, the map myX is plotted with a sevenary Brewer palette. The number of cases is printed left of the map, and then the title and subtitle are added. The same is repeated for the postcodes.

10.1.2 Filtered Postcode Map

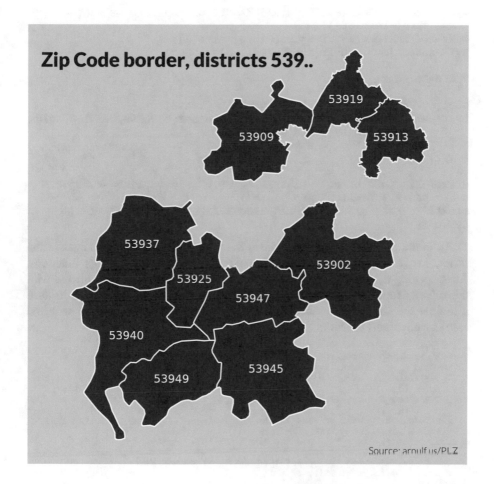

About the figure: For this example, all areas whose postcodes begin with 539 are filtered from the postcode map. These areas are plotted in dark blue on a light-green background. Such colour combinations are often found on survey maps of linked transport systems. Postcodes are written at the geographical centre of the districts.

About the data: map files for the postcodes can be downloaded as ESRI shapefiles from http://arnulf.us/PLZ.

```
pdf_file<-"pdf/maps_germany_zipcodes_sample.pdf"
cairo_pdf(bg="grey98", pdf_file,width=10,height=10)

par(bg="darkolivegreen1",mai=c(0,0,0,0),oma=c(1,1.5,1,1),family=\
      "Lato Light",las=1)
library(maptools)
library(rgdal)

# Import data and prepare chart

myY<-readShapeSpatial("myData/PLZ/post_pl.shp",proj4string=CRS("\
      +proj=longlat"))
myX<-spTransform(myY,CRS=CRS("+proj=merc"))
myY<-subset(myX,substr(myX$PLZ99,1,3)=="539")

# Create chart

plot(myY,col="darkblue",border="white",lwd=3)
text(getSpPPolygonsLabptSlots(myY),labels=myY$PLZ99,cex=1.5,col=\
      "white",family="Lato Bold")

# Titling

mtext("Zip Code border, districts 539..",side=3,line=-4,adj=0,\
      cex=2.7,family="Lato Black")
mtext("Source: arnulf.us/PLZ",side=1,line=-1,adj=1,cex=1.3)
dev.off()
```

As in the previous example, we need the maptools and rgdal packages in the script, so we can execute the readShapeSpatial() and spTransform() functions. After opening the shapefile and the Mercator transformation, data are filtered with the subset() function. For ease of comprehension, we first use str(myX) to have a look at the structure of the object that R created from the shapefile with the readShapeSpatial() function:

```
> str(myX)
Formal class 'SpatialPolygonsDataFrame' [package "sp"] with 5 slots
  ..@ data       :'data.frame': 8270 obs. of  3 variables:
  .. ..$ PLZ99   : Factor w/ 8270 levels "01067","01069",..: 1 2 3 4 5 6 7 8 \
        9 10 ...
  .. ..$ PLZ99_N : int [1:8270] 1067 1069 1097 1099 1109 1127 1129 1139 1157 \
        1159 ...
  .. ..$ PLZORT99: Factor w/ 6359 levels "\xd6hningen",..: 1229 1229 1229 1229 \
        1229 1229 1229 ↩1229 1229 1229 ...
  .. ..- attr(*, "data_types")= chr [1:3] "C" "N" "C"
  ..@ polygons   :List of 8270
  .. ..$ :Formal class 'Polygons' [package "sp"] with 5 slots
  .. .. .. ..@ Polygons :List of 1
  . . .
```

The object contains five slots. The first one is a data.frame that can be addressed with @data and contains the variables PLZ99, PLZ99_N, PLZORT99. We can display the first five rows:

```
> myX@data[1:4,]
  PLZ99 PLZ99_N PLZORT99
```

We can also directly address the variables of the data slot, i.e., using myX$PLZ99 intead of myX@data$PLZ99, and use subset(myX,substr(myX$PLZ99,1,3)== "539") to filter for data whose postcode begins with 539. This automatically filters for the respective polygons. The getSpPPolygonsLabptSlots() function delivers the coordinates for the centre of the polygons, which is where the postcodes are written.

10.1.3 Map of Europe NUTS 2006 (Cut-out)

As in Sect. 10.1.1, the figure illustrates the size and location of European regions with a random colour distribution. The map only displays the NUTS3 layer in the Mercator projection, but it shows the strong size variation of these regions within the individual EU countries. For, e.g., Germany, it is evident that the approximately 400 districts (administrative districts and district-free cities) shown as the lowest level represent a comparatively fine regional structure. By the by, this is only a section of the map, since the original map also includes overseas territories of EU countries.

Data: see annex A, NUTS.

```
pdf_file<-"pdf/maps_nuts2006.pdf"
cairo_pdf(bg="grey98", pdf_file,width=10,height=7)

par(omi=c(0,0,0,0),mai=c(0,0,0,0),family="Lato Light",las=1)
library(maptools)
library(rgdal)
library(RColorBrewer)

# Import data and prepare chart

myX<-readShapeSpatial("myData/NUTS-2006/NUTS_RG_03M_2006.shp",↘
      proj4string=CRS("+proj=longlat"))
m=spTransform(myX,CRS=CRS("+proj=merc"))
colour<-sample(1:7,length(m),replace=T)
m$colour<-colour
palette<-brewer.pal(7,"Purples")

# Create chart

plot(m,xlim=c(-1000000,3000000),ylim=c(4000000,10000000),border=↘
      F,col=palette[m$colour])
dev.off()
```

In the script, the NUTS shapefile is read with specification of the projection +proj=longlat; this enables transformation of the created shape object into the Mercator projection. We then generate an evenly distributed random variable between 1 and 7 for the colour. In this instance, we select the Brewer palette "purples". After that, the map of Europe is plotted; however, we limit the map selection using xlim and ylim, as territories lying far from the European mainland would otherwise be shown as well.

10.2 Points, Diagrams, and Symbols in Maps

10.2.1 Map of Germany with Selected Locations and Outline (Panel)

About the figure: For maps, too, panel representations are a good choice for temporal comparisons. In this example, the number of UFO sightings in Germany in 2006 and 2012 is illustrated in two maps. Sightings are differentiated into identified and non-identified sightings. Identified sightings are those that ultimately turned out to be models, hot-air balloons, smudges on camera lenses, butterflies, rubbish bags or others. Sightings are plotted as symbols on a map of Germany with federal state borders.

Data were taken from the website http://www.ufo-datenbank.de maintained by the German language society for UFO research (Deutschsprachigen Gesellschaft für UFO-Forschung e.V.: DEGUFO). For our purpose, information for the years 2006 and 2012 was transferred into a database table ufos and linked with the zipcode and city tables of Daneil Lichtblau's postcode database:

```
CREATE VIEW v_ufos AS
(
select
right(c.sighting_date, 5) AS year,
a.name AS name,
a.lat AS lat,
a.lng AS lng,
b.zipcode AS zipcode,
c.Sichtungsdatum AS sighting_date,
c.classification AS classification,
c.assessment AS assessment,
c.identification AS identification
from
```

```
(
(city a join zipcode b) join ufos c)
where ((a.id = b.city_id) and (b.zipcode = c.PLZ)));
```

For map data, see annex A, gadm.org.

```
pdf_file<-"pdf/maps_germany_cities_points_outline_1x2.pdf"
cairo_pdf(bg="lavender", pdf_file,width=12,height=6)

par(mai=c(0,0,0,0), omi=c(0.15,0.25,0.55,0.25),mfcol=c(1,2),↘
     family="Lato Light",las=1)
library(sp); library(RColorBrewer); library(geoR)
source("scripts/inc_datadesign_dbconnect.r")

# Import data and prepare chart

sql<-"select substr(sighting,7,4) year, lng, lat, identification↘
      from v_ufos where substr(sighting,7,4) in (2006, 2012)"
myDataset<-dbGetQuery(con,sql)
myDataset$myYear<-as.numeric(myDataset$year)
attach(myDataset)

myYear1<-2006; myYear2<-2012

myYear1id<-subset(myDataset,myYear == myYear1 & identification !↘
     = "_")
myYear1ui<-subset(myDataset,myYear == myYear1 & identification ↘
     == "_")

myYear2id<-subset(myDataset,myYear == myYear2 & identification !↘
     = "_")
myYear2ui<-subset(myDataset,myYear == myYear2 & identification ↘
     == "_")

myData<-c(myYear1ui,myYear2ui,myYear1id,myYear2id)
load("myData/gadm2/DEU_adm1.RData")
myColour<-"linen"
myColour_uid<-rgb(128,0,0,200,maxColorValue=255)
myColour_id<-rgb(0,153,0,120,maxColorValue=255)

# Create chart and other elements

plot(gadm,border="white",col=myColour)
par(family="Quivira")
text(jitterDupCoords(cbind(myYear1ui$lng,myYear1ui$lat),max↘
     =0.01),"",col=myColour_uid,cex=2.3)
points(jitterDupCoords(cbind(myYear1id$lng,myYear1id$lat),max↘
     =0.01),pch=15,col=myColour_id,cex=1.6)

mtext("■", side=1, line=-3, col=myColour_uid, cex=1.5, adj=0)
mtext("■", side=1, line=-1.9, col=myColour_id, cex=1.15, adj=0)
par(family="Lato Light")
mtext("not identified", side=1, line=-3, cex=1.1, adj=0.05)
mtext("identified", side=1, line=-2, cex=1.1, adj=0.05)
```

```
text(5,53,myYear1,cex=2.7,col="azure4")

plot(gadm,border="white",col=myColour)
par(family="Quivira")
text(jitterDupCoords(cbind(myYear2ui$lng,myYear2ui$lat),max↘
    =0.01),"■",col=myColour_uid,cex=2.3)
par(family="Lato Light")
points(jitterDupCoords(cbind(myYear2id$lng,myYear2id$lat),max↘
    =0.01),pch=15,col=myColour_id,cex=1.6)
text(5,53,myYear2,cex=2.7,col="azure4")

# Titling

mtext("UFO Sightings in Germany",line=0,adj=0,cex=2.2,family="↘
    Lato Black",outer=T)
mtext("Source: www.ufo-datenbank.de.",side=1,line=-1,adj=1,cex↘
    =0.9,font=3,outer=T)
dev.off()
```

The script starts by reading data from an SQL table; after that, two variables for the years are defined. If required, data can easily be displayed for other years. Two partial data frames are generated for both years: those with identified and those with non-identified flying objects. The map is read from a file DEU_adm1.RData; on gadm.org, data are already provided in this binary R format. After specification of the colours for the map and the symbols, the map is plotted. We use "Quiviera" by Alexander Lange as font for the alien. The jitterDupCoords() function uses a random term to cause a slight shift of any overlapping points, thereby ensuring that all points are visible. The unidentified flying objects are plotted as a text symbol with the text() function, the others as a point with the points() function. Since we need two different character sets in this instance, we cannot use the legend() function for the legend, but rather use mtext(). The green rectangle in the legend is drawn using the Unicode symbol "BLACK MEDIUM SQUARE" from the "Geometric Shapes" block, available as a glyph in the Quivira font. The second map is plotted analoguously to the first one.

10.2.2   Map of Germany with Selected Locations (Pie Charts) and Outline

Shares of agricultural, forest and settlement area

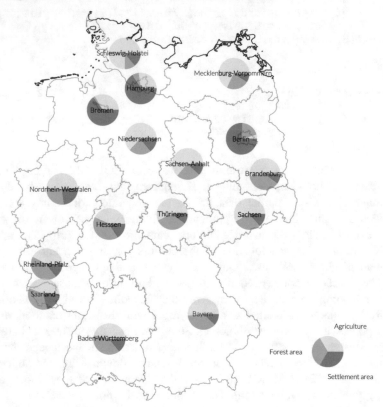

Source: Statistisches Bundesamt Fachserie 3 Reihe 5.1

About the figure: If the figure is not about the distribution of points but rather about structural characteristics for larger regions, then another diagram type can be placed at the centre of the geographical units. In the present case, the proportions of agriculture, woodland, and settlement areas in Germany are plotted as pie charts within the federal states. The legend is designed to describe the names for the three types of land use and a circle with equally sized segments. The rotation of the circle is selected to ensure that all three labels are easy to read. However, with these types of representation, one should always question whether they provide any benefit over a regular illustration form that does not include the geographical information. This means it would be feasible to only list the federal states and visualise the proportions in the form of a dot chart for three variables (Sect. 6.1.11).

Data derive from a publication by the Federal Office of Statistics (Statistisches Bundesamt; Fachserie 3, Reihe 5.1); these and the coordinates for the centres of the federal states were written into an XLS table.

Map data: see annex A, gadm.org.

```
pdf_file<-"pdf/maps_germany_cities_piecharts_outline.pdf"
cairo_pdf(bg="grey98", pdf_file,width=13,height=11)
myColour_h<-rgb(240,240,240,maxColorValue=255)

par(omi=c(0.5,0.5,0.5,0),mai=c(0,0,0,0),lend=1,family="Lato \
     Light",las=1)
library(sp)
library(plotrix)
library(gdata)

# Import data and create chart

myData<-"myData/data_maps_germany_cities_piecharts.xlsx"
myCountries<-read.xls(myData,head=T, encoding="latin1")
attach(myCountries)

load("myData/gadm2/DEU_adm1.RData")
plot(gadm,border=rgb(151,151,151,maxColorValue=255),lwd=0.5)

load("myData/gadm2/DEU_adm0.RData")
plot(gadm,border="black",lwd=0.95,add=T)

# Other elements

n<-nrow(myCountries)
for (i in 1:n)
{
myCircle<-c(agriculture[i],forest[i],cultivation[i])
floating.pie(long[i],lat[i],myCircle,radius=0.5,col=c(rgb\
     (215,215,0,150,maxColorValue=255),rgb(34,139,34,150,\
     maxColorValue=255),rgb(178,34,34,150,maxColorValue=255)),\
     border=F)
text(long[i],lat[i],name[i])
}
floating.pie(16,48.25,c(1,1,1),radius=0.5,col=c(rgb\
     (215,215,0,150,maxColorValue=255),rgb(34,139,34,150,\
     maxColorValue=255),rgb(178,34,34,150,maxColorValue=255)),\
     border=F)
text(16.75,47.75,"Settlement area")
text(14.7,48.25,"Forest area")
text(16.75,48.75,"Agriculture")

# Titling

mtext("Shares of agricultural, forest and settlement area",3,\
     line=-0.25,adj=0,cex=2.25,outer=T,family="Lato Black")
mtext("Source: Statistisches Bundesamt Fachserie 3 Reihe 5.1",1,\
     line=0,adj=0.9,cex=1.25,font=3)
dev.off()
```

The sp package has to be loaded in the script, as it extends the plot() function by a method for drawing of objects like gadm of SpatialPolygonsDataFrame type. The piecharts can be drawn using the pieGlyph() function from the Rgpraphviz package by Kasper Hansen and others. Unfortunately, that package is no longer available on the CRAN server; you will have to load it from http://www.bioconductor.org. We use the plotrix package by Jim Lemon as an alternative, which provides the floating.pie() function. We can now use a loop to position a pie chart for each federal state. Coordinates are in the XLSX table. After the loop, we draw another circle for the legend; this circle contains three segments of equal size and is labelled with the three area types.

10.2.3   Map of Germany with Selected Locations (Columns) and Outline

Prices for condominiums in selected cities 2011

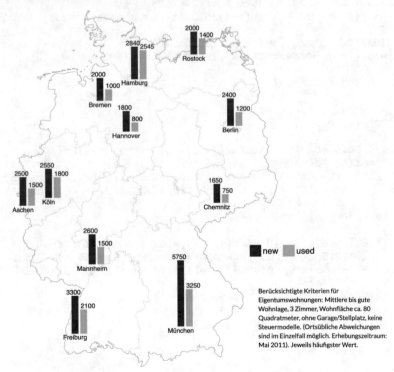

Source: real estate price index from the LBS brochure 'Markt für Wohnimmobilien 2011'

The figure shows the square metre prices for preowned and new condominiums in selected German cities in 2011. Values are depicted as columns with their respective values written above and the name of the city below. In addition to the national border, borders of federal states are drawn in a lighter colour. Data were taken from a website that references the real estate price level (Immobilienpreisspiegel) from the LBS brochure "Markt für Wohnimmobilien 2011" and entered into an XLS file. Longitude and latitude can be found in, e.g., the data file world.cities from the mapdata package. See annex A, gadm.org for map data.

```
pdf_file<-"pdf/maps_germany_cities_columns_outline_inc.pdf"
cairo_pdf(bg="grey98", pdf_file,width=13,height=11)

par(omi=c(0.5,0.5,0.5,0),mai=c(0,0,0,2),lend=1)
library(sp)
library(gdata)

# Import data and create chart

myLocations<-read.xls("myData/data_maps_germany_cities_columns.\
    xlsx",head=T, encoding="latin1")
attach(myLocations)

load("myData/gadm2/DEU_adm1.RData")
plot(gadm,border=rgb(151,151,151,50,maxColorValue=255),lwd=0.5)

load("myData/gadm2/DEU_adm0.RData")
plot(gadm,border="darkgrey",lwd=0.95,add=T)

# Other elements

n<-nrow(myLocations)
for (i in 1:n)
{
myBar1<-data.frame(c(long[i],long[i]),c(lat[i],lat[i]+new[i]/\
    4000))
lines(myBar1,lwd=17,col="darkred")
text(long[i]+0.2,lat[i]+0.08+new[i]/4000,new[i],adj=1)

myBar2<-data.frame(c(long[i]+0.3,long[i]+0.3),c(lat[i],lat[i]+\
    used[i]/4000))
lines(myBar2,lwd=17,col="darkgrey")
text(long[i]+0.2,lat[i]+0.08+used[i]/4000,used[i],adj=0)
text(long[i],lat[i]-0.09,name[i],adj=0.5)
}
legend(14,50,c("new","used"),border=F,pch=15,col=c("darkred","\
    darkgrey"),bty="n",cex=1.3,pt.cex=4,xpd=NA,ncol=2)

# Titling

mtext("Prices for condominiums in selected cities 2011",3,line=-\
    0.5,adj=0,cex=2.25,family="Lato Black",outer=T)
```

```
mtext("Source: real estate price index from the LBS brochure '\
      Markt für Wohnimmobilien 2011'",1,line=0,adj=0.9,cex=1.25,\
      font=3)
dev.off()
```

Here is the integration into LaTeX:

```
\documentclass{article} \usepackage[paperheight=27.94cm,\
      paperwidth=33.02cm,top=0cm,left=0cm,right=0cm,bottom=0cm]{\
      geometry}
\usepackage{multicol}
\usepackage{graphicx,color}
\usepackage{fontspec}
\usepackage[abs]{overpic}
\setmainfont[Mapping=text-tex]{Lato Regular}
\definecolor{background}{gray}{.99}
\pagecolor{background}
\linespread{1.2}
\begin{document}
\pagestyle{empty}
\begin{center}
\fontsize{12pt}{14pt}\selectfont
\begin{overpic}[scale=1.00,unit=1mm]{maps_germany_cities_columns\
      _outline_inc.pdf}
\put(220,70){\begin{minipage}[t]{8.5cm}
\raggedright Criteria taken into account for condominiums: ...
\end{minipage}}
\end{overpic}
\end{center}
\end{document}
```

In the script, we add lend=2 at the start of the parameter settings; this causes angular endings of the lines that we draw later on using line and makes them look like columns. Data are read from an XLS file. Both maps are again accessed from the gadm website; no download is necessary. To show the gadm object with plot, we have to load the sp library, as gadm is an object of SpatialPolygonsDataFrame type. Using Deu_adm1.RData, the borders of the federal states are plotted first, with the national border plotted with Deu_adm0.RData over the top. We use a light grey for the state borders and a dark grey for the national border. We then call a loop that first defines a data frame for each location. This data frame is really nothing more than a line, a connection of two points. The first point is made up of the coordinates of the location, while the second point is the coordinates of the point extended by the attributes of the variable in y-direction; in this case, the attribute is the value for new condominiums. However, this value still has to be adjusted to the figure's scaling and is therefore divided by 4000, a number that we arrived at through trial and error. The second bar is generated in the same way, only the value for preowned condominiums is displayed, and the "bar" is shifted to the right by 0.3 points to set it immediately next to the first one. We add the respective value at 0.8 points above each bar using the text function, shifted by 0.2 points; values are right-aligned for the left bar and left-aligned for the right bar. At 0.09 points underneath the bars and centred, the location is printed as text. At the end come the legend, caption, and title. The explanatory text is written in LaTeX, and the PDF file created by R is integrated into a short LaTeX file; in that file, the displayed text is placed in an 8.5-cm-wide minipage using put(220,70). The procedure is described in Sect. 4.1.

10.2.4   Map of Germany as Three-Dimensional Scatter Plot

Where we live...

Städte sind aus kulturwissenschaftlicher Perspektive der Idealfall einer Kulturraumverdichtung und aus Sicht der Soziologie vergleichsweise dicht und mit vielen Menschen besiedelte, fest umgrenzte Siedlungen mit vereinheitlichenden staatsrechtlichen oder kommunalrechtlichen Zügen wie einer sozial stark differenzierter Einwohnerschaft. Eine grundlegende

Theorie zur Verteilung zentraler Nutzungen im Raum stammt von Walter Christaller. „Zentrale Orte" sind Standort von Angeboten, die nicht nur von den eigenen Bewohnern sondern regelmäßig auch von Einwohnern der Nachbargemeinden genutzt werden.

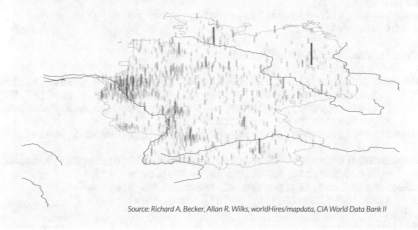

Source: Richard A. Becker, Allan R. Wilks, worldHires/mapdata, CIA World Data Bank II

About the figure: In three-dimensional figures, a series of issues have to be considered. The effort for creation is much higher than for two-dimensional figures, and their use should be well thought through. This example is meant to demonstrate a way of creating 3D illustrations with comparatively minimal effort and is inspired by an American example published in Time Magazine. The example shows a map of Germany which large rivers have been added for orientation. The map is presented in top view from the front. German cities and municipialities of more than approximately 15,000 inhabitants are depicted as column charts. This means the local distribution and larger cities can be taken in at a glance. Aside from the location of the major cities of Berlin, Hamburg, and Munich, the strong concentration around the Rhine and in the Ruhr district is also very obvious. The map is complemented with an explanatory text above the map. Data are provided by R in the world.cities data frame. The data frame comprises six variables: name, country.etc, pop, lat, long, and capital as of January 2006. Map data and river courses are also provided by R.

```
pdf_file<-"pdf/maps_germany_cities_3d_90_inc.pdf"
cairo_pdf(bg="grey98", pdf_file,width=13,height=11)

par(omi=c(0.5,0,0.25,0.25),mai=c(0,0,0,0),lend=2,family="Lato ⬎
      Light",las=1)
library(scatterplot3d)
library(mapdata)

# Import data

dt.map<-map("worldHires","Germany",plot=F)
dt.map2<-map("rivers",plot=F,add=T)
data(world.cities)
Germany<-subset(world.cities,country.etc=="Germany")
attach(Germany)

# Create chart and other elements

s3d<-scatterplot3d(long,lat,pop**0.42,box=F,axis=F,grid=F,scale.⬎
      y=2.2,mar=c(0,1.5,2,0),type="n",xlim=c(5,15),ylim=c(47,55)⬎
      ,zlim=c(0,2000),angle=90,color="grey",pch=20,cex.symbols⬎
      =2,col.axis="grey",col.grid="grey")
s3d$points3d(dt.map$x,dt.map$y,rep(0,length(dt.map$x)),col="grey⬎
      ",type="l")
s3d$points3d(dt.map2$x,dt.map2$y,rep(0,length(dt.map2$x)),col=⬎
      rgb(0,0,255,170,maxColorValue=255),type="l")
s3d$points3d(long,lat,pop**0.42,col=rgb(255,0,0,pop**0.36,⬎
      maxColorValue=255),type="h",lwd=5,pch=" ")

# Titling

mtext("Where we live...",adj=0.0,cex=3.5,line=-5,family="Lato ⬎
      Black",outer=T)
mtext("Source: Richard A. Becker, Allan R. Wilks, worldHires/⬎
      mapdata, CIA World Data Bank II",1,adj=0.9,cex=1.5,line=0,⬎
      font=3,outer=T)
dev.off()
```

Here is the integration into LaTeX:

```
\documentclass{article}
\usepackage[paperheight=27cm, paperwidth=35cm,
        top=0cm, left=0cm, right=0cm, bottom=0cm]{geometry}
\usepackage{multicol}
\usepackage[german]{babel}
\usepackage{graphicx,color}
\usepackage{fontspec}
\usepackage[abs]{overpic}
\setmainfont[Mapping=text-tex]{Lato Regular}
\setlength{\parindent}{0in}
\setlength{\columnsep}{1.5pc}
\definecolor{background}{rgb}{0.99,0.99,0.99}
\pagecolor{background}
\begin{document}
\pagestyle{empty}
\begin{center}
\fontsize{16pt}{24pt}\selectfont
\begin{overpic}[scale=0.95,unit=1mm]{maps_Germany_locations_3d_\
    90_inc.pdf}
\put(0,180){\begin{minipage}[t]{30.5cm}
\begin{multicols}{2}
From a cultural science point of view, cities are the ideal case\
    of...
\end{multicols}
\end{minipage}}
\end{overpic}
\end{center}
\end{document}
```

Before turning to the script, we should have a look at how to create three-dimensional figures in R. A suitable package is scatterplot3d by Uwe Ligges. We use the coordinates from the map of Germany taken from R's geo database worldHires. The columns x and y are longitudes and latitudes, respectively. If we specify 0 as the third dimension of the three-dimensional scatter plot throughout, then we are virtually putting the map on the floor of the three-dimensional figure. For illustration purposes, we take the same two steps as before: we first specify the (empty) chart using scatterplot3d() and then plot our points inside using s3d$points3d(). The angle parameter of the scatterplot3d() function is used to specify a different angle with each pass: first 65, then 90, and lastly 110.

```
par(mfcol=c(1,3))
library(scatterplot3d)
library(mapdata)
dt.map <- map("worldHires", "Germany", plot=F)
angle<-c(65,90,110)
for (i in 1:3)
{
s3d <- scatterplot3d(dt.map$x,dt.map$y,rep(0,length(dt.map$x)),
        xlab="", ylab="", zlab="", zlim=c(0,5), scale.y=3.5, ⬎
             type="n",
  mar=c(0,0.35,0,0.35), box=T,axis=T,grid=FALSE,angle=winkel[i])
s3d$points3d(dt.map$x,dt.map$y,rep(1,length(dt.map$x)),type="l")
mtext(angle[i])
}
```

The result can be seen in Fig. 10.1. Now we can have a look at the actual figure. We begin by setting angular line endings using lend=2. Then the packages scatterplot3d and maptools are loaded, the latter of which provides the map of Germany from the worldHires database and the rivers. City data are loaded from the world.cities data frame that is provided by R; from this we extract all cities in Germany. Both the map of Germany and the rivers are called using map() and assigned to an object, but not plotted yet.

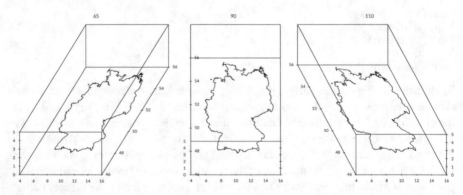

Fig. 10.1 Three different angles

Then comes the call of the scatterplot3d() function. The third dimension is now pop, the population of the cities; for representation, this parameter is scaled with **0.42. That means long and lat are the x and y coordinates, pop the z coordinates. We specify 90° as the angle so that one has a top view of the map from the front (see Fig. 10.1). The scaling factor scale.y determines the "depth", the extent to which the map reaches into space. It is important that we use type="n" again to prevent plotting of the chart. The call is only used to specify the dimensions; in a second step, we plot the actual data into these dimensions with three calls of the s3d$points3d() function. The first call plots the map of Germany dt.map at height of 0 (which is on the floor). The second call plots the rivers dt.map2, also at height 0. The third call plots the population data using type="h" as columns. The pch="" ensures that no symbol is plotted, but only the column. The rest is business as usual. The LaTeX integration was described in Sect. 4.1.

10.2.5   Map of North Rhine-Westphalia with Selected Locations (Symbols) and Outline

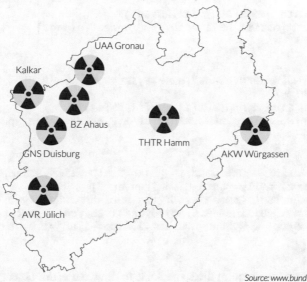

Nuclear Country North Rhine-Westphalia
important companies of the nuclear industry are located in NRW

The figure shows important sites of the nuclear economy in the German federal state of North Rhine-Westphalia. Locations of respective industries are added as radioactivity symbols to the outline map of the federal state. As with this example, it is often difficult to determine an exact position for the labels. Rule-based positioning usually leads to overlaps.

Data: The locations of nuclear power plants in North Rhine-Westphalia were obtained from http://www.bund-nrw.de. See annex A, gadm.org, for map data.

```
pdf_file<-"pdf/maps_nrw_symbols.pdf"
cairo_pdf(bg="grey98", pdf_file,width=8,height=6.5)
```

```
par(mai=c(0.5,0.0,0.0,0.5),omi=c(0,0.5,0.75,0),family="Lato ⬎
      Light")
library(sp); library(plotrix); library(gdata)

# Import data and prepare chart

akwnrw<-read.xls("myData/akwnrw.xlsx",head=T, encoding="latin1")
myX<-akwnrw$long; myY<-akwnrw$lat
dm<-rep(0.2,length(myX))
myNpp<-function(myX,myY,dm,... ){
floating.pie(myX,myY,c(1,1,1,1,1,1),radius=dm,startpos=45,border⬎
      =F,col=c("yellow","black","yellow","black","yellow","black⬎
      "))
points(myX,myY,pch=19,cex=2,col="yellow")
points(myX,myY,pch=19,cex=1,col="black")
}
load("myData/gadm2/DEU_adm1.RData")
myNrw<-gadm[gadm$NAME_1=="Nordrhein-Westfalen",]

# Create chart

plot(myNrw,border="black",axes=F,lwd=0.5)
n<-length(myX)
for (i in 1:n) myNpp(myX[i],myY[i],dm[i])
text(akwnrw$namlong,akwnrw$namlat-0.2,akwnrw$name,xpd=T)

# Titling

mtext("Nuclear Country North Rhine-Westphalia",3,line=1,adj=0,⬎
      cex=2,family="Lato Black",outer=T)
mtext("important companies of the nuclear industry are located ⬎
      in NRW",3,line=-0.5,adj=0,cex=1.25,font=3,outer=T)
mtext("Source: www.bund-nrw.de",1,line=-1,adj=1.0,cex=0.95,font⬎
      =3)
dev.off()
```

In the script, we require the sp package to draw the gadm object, and the plotrix package for the floating.pie() function. Data are read from the akwnrw.xlsx file. myX and myY save the dataframe's variables long and lat, and dm the diameter, which is set to 0.2. We require the latter as a vector for each value that repeats locations. We then specify a function myNpp for generation of the NPP symbol. The symbol is a pie chart with six slices, alternating yellow and black, which is overlaid with first a yellow and then a smaller black point. Outlines are again read from the adm1.RData file. From that file, gadm[gadm$NAME_1=="Nordrhein-Westfalen",] extracts the polygon that defines the borders for this federal state. Be aware that, by using square brackets and the comma before the closing square bracket, all rows of the data frame for which the condition before the comma is true are selected. After plotting of the state border with plot, the NPP symbols with all values from the data frame are drawn for each location by calling the myNpp() function. The locations are written underneath with text(). However, we do not use the original coordinates—even shifted by constants—but individual positions that are obtained from the data frame. To this end, we created an individual column in the data frame, since the specific constellation of the locations meant that an automated solution was out of the question. Alternatively, we could have corrected the automatically created positions with Inkscape or Adobe Illustrator. The usual labels come at the end.

10.2.6 Map of Tunisia with Self-Defined Symbols

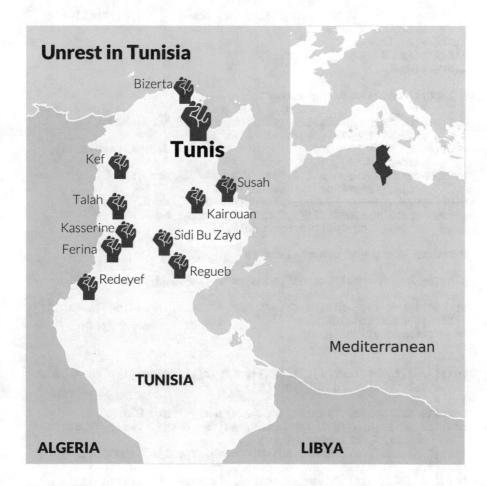

About the figure: The map was adapted from a figure published in the German magazine ZEIT. Displayed is a detailed map of northern Africa that shows the unrest in Tunisia in 2011. Locations where unrest was reported are marked with a "speaking" symbol. The symbol is a fist that was created during a Noun Project workshop. In such detailed maps, it is generally common to use two colours: one for the country concerned and another for the remaining countries added for orientation purposes. It is also common to present the observer with a second schematic map akin to a legend, showing the site of the presented country in a larger, known context. Data for the locations were taken from the map available at http://www.zeit.de/ politik/ausland/2011-01/tuniesen-tunis-unterstuetzer-gegenprotest; the coordinates were found in Wikipedia articles for these locations. Map data are shapefiles for Libya, Tunisia, and Algeria, respectively, available on http://gadm.org.

```
pdf_file<-"pdf/maps_tunisia_symbols.pdf"
cairo_pdf(bg="grey98", pdf_file,width=10,height=10)


par(bg="lightskyblue1",mai=c(0,0,0,0),oma=c(0,0,0,0),family="\
      Lato Light",las=1)
library(maptools)
library(rgdal)
library(gdata)


# Import data and prepare chart

myTun<-readShapeSpatial("myData/TUN_adm/TUN_adm0.shp",\
      proj4string=CRS("+proj=longlat"))
plot(myTun,col="mintcream",border="white",lwd=3, xlim=c(8,14), \
      ylim=c(32,38))
myDza<-readShapeSpatial("myData/DZA_adm/DZA_adm0.shp",\
      proj4string=CRS("+proj=longlat"))
plot(myDza,col="burlywood1",border="white",lwd=3, add=T)
myLby<-readShapeSpatial("myData/LBY_adm/LBY_adm0.shp",\
      proj4string=CRS("+proj=longlat"))


# Create chart and other elements

plot(myLby,col="burlywood1",border="white",lwd=3, add=T)


myLocations<-read.xls("myData/tunisia.xlsx", encoding="latin1")
attach(myLocations)
n<-nrow(myLocations)
for (i in 1:n)
{
text(long[i]+hoffset[i], lat[i]+voffset[i], place[i], cex=1.75,\
      col="black", adj=adjust[i])
}
text(10.12, 36.43, "Tunis", cex=3, family="Lato Black")
text(12.5, 33.5, "Mediterranean", adj=0, cex=2, family="Lato \
      Regular", col="darkblue")
text(7.25, 32, "ALGERIA", adj=0, cex=2, family="Lato Black")
text(9,33, "TUNISIA", adj=0, cex=2, family="Lato Black")
text(12, 32, "LIBYA", adj=0, cex=2, family="Lato Black")


par(family="Datendesign")
```

```
text(long, lat, "a", cex=size, col="red")

# Titling

mtext("Unrest in Tunisia",side=3,line=-4,adj=0.05,cex=2.7,family↖
    ="Lato Black",col="black")

# Separate figure

par(mai=c(6,6,0,0), bg="white",new=T)
data(wrld_simpl)
w=wrld_simpl[wrld_simpl@data[,"NAME"] != "Antarctica",]
m=spTransform(w,CRS=CRS("+proj=merc"))
plot(m,xlim=c(-900000,2800000),ylim=c(3300000,7000000),col=rgb↖
    (160,160,160,100,maxColorValue=255),border=F)
w=wrld_simpl[wrld_simpl@data[,"NAME"] == "Tunisia",]
m=spTransform(w,CRS=CRS("+proj=merc"))
plot(m, add=T,col="red",border=F)
dev.off()
```

In the script, we set our background colour as lightskyblue1, which is already available in R, and then set all borders to 0. We need the maptools and rgdal packages to read the data for the projection transformations. With these, we first plot Tunisia, then Algeria and Libya into the same figure. The map section that we have to draw is obtained by first specifying the borders and drawing the axes. Once usable values have been read and entered into xlim or ylim, we set the borders to 0 and remove the axis commands; the latter is not really necessary as they are now outside our plot area—we only do this to keep things tidy. The locations are read from an XLSX table, and each location's name is immediately written into the figure. Since positioning for each location is guided by the longitude and latitude and would therefore result in overlaps, there are two columns in the XLSX table—hoffset and voffset—which allow for individual adjustments. Another column, adjust, contains information on text alignment (left or right). Now come the individual labels for Tunis, Algeria, Tunisia, and Libya. The actual points are drawn as symbols by first embedding the self-made symbol font "Datadesign" as a font (see Sect. 4.2); remember: we assigned the fist symbol to the letter "a". Using text(), the symbols are drawn into the map and the title is added. The last step is the display of another map at the top right. To this end, we use the new=T parameter of the par() function to define a new graphic window that we put on top of the first one. The lower-resolution wrld_simple map that comes with R can be used for the map; this resolution is sufficient for the survey map. First, we have to delete Antarctica. Even though it would not be drawn, the spTransform() function would otherwise throw out an error and abort the script. Borders are once more identified by trial and error. Finally, Tunisia is extracted from the map in a second step, and added to the existing map using add=T.

10.3   Choropleth Maps

Up to now, we have only visualised information derived from individual points in maps; now, we turn to illustrations in which entire areas within a map reveal the statistical information. These illustration forms, called choropleth maps, are often encountered after political elections to demonstrate the regional distribution of votes, but there are also other structural characteristics that are described in this way.

10.3.1   Choropleth Map of Germany at District-Level

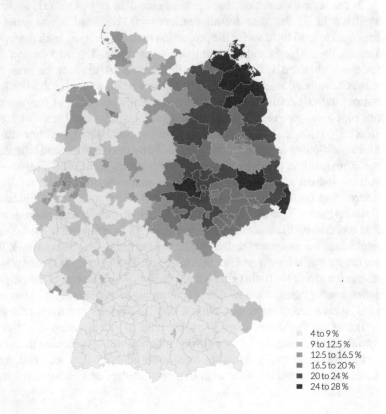

Unemployed at electoral disctrict level 2005

- 4 to 9 %
- 9 to 12.5 %
- 12.5 to 16.5 %
- 16.5 to 20 %
- 20 to 24 %
- 24 to 28 %

Source (map geometry and data): www.bundeswahlleiter.de

About the figure: In this example, the map shows the percentages of unemployed population of Germany at district level. To do this, the constituencies of 2005 are used. Constituencies are different from administrative districts, as they legally have to be adjusted to population size. Every constituency has to comprise approximately the same number of inhabitants. The unemployed population was divided into six classes. Generally, classifications for choropleth maps are best chosen so all classes are either of the same size or class limits are chosen so they result in similar frequencies among the classes. However, both are only guidelines that can be deviated from in individual cases. For the present example, we have chosen a classification that considers frequencies. For colours, we chose a single colour palette with appropriate colour gradations. In this case, a Brewer colour palette with six gradations in orange is a good choice. In choropleth maps, the limits of the territorial units should generally be drawn in a light or even in the background colour; the black value in small-scale territorial units would be too high otherwise. The legend with the respective colours is drawn at the bottom right.

Data: the structural data can be downloaded as the CSV file struktbtwkr2005.csv from

http://www.bundeswahlleiter.de/

Headers were removed for this example, and the file was saved anew. The shapefiles containing the constituencies of 2005 are no longer provided by the Bundeswahlleiter (Federal Returning Officer), but are freely available from

http://www.datendieter.de/item/shapefiles_der_Wahlkreise_zur_Bundestagswahl

There, users are asked to add the following copyright notice: "©Bundeswahlleiter, Statistisches Bundesamt, Wiesbaden, 2005, Wahlkreiskarte für die Wahl zum 16. Deutschen Bundestag. District boundaries are from: VG 1000, Bundesamt für Kartographie und Geodäsie."

We need the following variables:

- x$CTCY_NO: constituency number in the shapefile
- y$V2: constituency number in the structure file
- y$V34: unemployed population in the district

```
pdf_file<-"pdf/maps_germany_choropleth_counties.pdf"
cairo_pdf(bg="grey98", pdf_file,width=8,height=9)

par(mai=c(0,0,0,0),omi=c(1,0.25,1,0.25),family="Lato Light",las↘
     =1)
library(maptools)
library(RColorBrewer)
library(gdata)

# Import data and prepare chart

x<-readShapeSpatial("myData/Geometrie_Wahlkreise_16DBT_VG1000.↘
     shp")
y<-read.csv(file="myData/struktbtwkr2005.csv",head=F,sep=";",dec↘
     =".")
myDistrict_structure_data<-subset(y,V2 > 0)
n<-length(x)
```

```
position<-vector()
for (i in 1:n){
  position[i]<-match(x$WKR_NR[i], myDistrict_structure_data$V2)
}
myColour_no<-cut(myDistrict_structure_data$V34[position],c↘
    (4,9,12.5,16.5,20,24,28))
levels(myColour_no)<-c("4 to 9 %","9 to 12.5 %","12.5 to 16.5 %"↘
    ,"16.5 to 20 %","20 to 24 %","24 to 28 %")
myColours<-brewer.pal(6,"Oranges")

# Create chart

plot(x,col=myColours[myColour_no],border=grey(.8),lwd=.5)
legend("bottomright",levels(myColour_no),cex=0.95,border=F,bty="↘
    n",fill=myColours)

# Titling

mtext("Unemployed at electoral disctrict level 2005",3,line=1,↘
    adj=0,family="Lato Black",outer=T,cex=2)
mtext("Source (map geometry and data): www.bundeswahlleiter.de"↘
    ,1,line=2.6,adj=0,cex=0.95,font=3,outer=T)
dev.off()
```

About the script: The basic idea for the following procedure stems from a blog entry by Mark Heckmann. After reading the shapefile with the constituencies and the CSV file with the structural data, the aggregated data of the federal states are filtered with subset(y,V2 > 0) so that only the constituencies remain. Then, the number of data frames n is determined and an empty vector position is generated/initialised. Within a loop, the position of the data frame within the structure file is determined for each constituency in the shapefile using the match() function. The result position is the row positions of the shapefile constituencies within the structure file. These can be used to define a colour vector myColour_no, which contains the numbers of the unemployed population of the structure file (V34) in order of the constituencies' appearance in the shapefile. The cut() function sorts the unemployed population into one of six classes. Since cut() returns factor values, the labels for the individual classes can then be immediately specified with levels(). The Brewer palette with orange values is selected for coloration. Finally, the map and legend can be plotted with plot() and legend(), respectively.

10.3.2 Choropleth Map of Germany at District-Level (Panel)

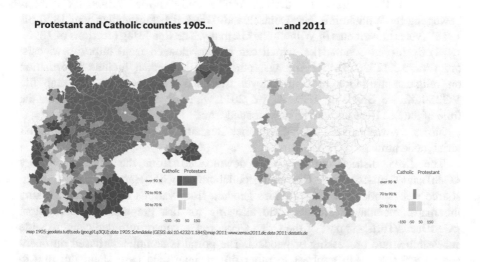

Protestant and Catholic counties 1905...   **... and 2011**

map 1905: geodata.tuffts.edu (goo.gl/Lq3QU); data 1905: Schmädeke (GESIS: doi:10.4232/1.1845);map 2011: www.zensus2011.de; data 2011: destatis.de

About the figure: The second choropleth example compares two points in time. In this instance, it is about the distribution of religious affiliation in Germany in 1905 and in 2011. In both years, religious affiliation was established as part of the census. The first thing that should be considered is that Germany has undergone considerable changes concerning its borders, meaning we require two different maps. Data are not empirically split according to their distribution or by selection of identical distances, but according to substantial considerations. The intent of the figure is to illustrate two questions: (1) How were religious affiliations distributed back in that day? and (2) What is the distribution like today? On the basis of these two intents, two construction attributes for the figure arise: on the one hand, we do not select a single continuous colour palette, but two. In 1905, at our starting date, German districts were either at least 50% Catholic, or at least 50% Protestant. This means we can create the classes "(over) 50–70% Catholic", "(over) 70–90% Catholic" and "(over) 90% Catholic", and the same classes for "Protestant districts". This is based on the assumption that there are no districts whose Catholic or Protestant population makes up less than 50%. In 1905, that was not the case. For these three respective classes, we select three blue and two orange grades. The legend identifies the frequencies in the form of bar charts. The second map uses the same classifications. The drastic regression in religious affiliation is immediately visible: the number of districts with Protestant or Catholic religious affiliation of over 50% has immediately halved. Remember that we are consciously omitting a large amount of information in this second map. This is not an isolated illustration of the distribution of religious affiliation in 2011; with that goal, class limits would have had to be lowered considerably. In this case, our aim is a direct comparison with 1905.

Data: the information contained in religious_affiliation_vz_2011.xlsx was provided on the information pages for the 2011 census. The information contained in the ZA8145_1912.xlsx file derives from a study entitled "ZA8145: Wähler-bewegung im Wilhelminischen Deutschland. Die Reichstagswahlen von 1890 bis 1912" (voter movements in Wilhelmine Germany. The Reichstag elections of 1890–1912) by Jürgen Schmädeke, which can be downloaded from http://www.gesis. org (doi: 10.4232/1.8145). Amongst others, available data include information on religious affiliation at district level from the 1905 census. The map file VG250_Kreise_Shapefile_UTM32_VZ_2011, which contains district borders at the time of the 2011 census, can be downloaded from

https://www.zensus2011.de/DE/Infothek/Begleitmaterial_Ergebnisse/Begleitma terial_node.html

The 1895 district borders were downloaded from the meta repository <GeoData@Tufts>. Tufts University collaborates with Harvard on an open-source web application that provides geodata from various repositories, making the maps' metadata searchable and allowing for fast graphical representation. <GeoData@Tufts> is part of the Open Geoportal, a consortium dedicated to wide accessibility and processing of geodata. The portal is completely based on open-source technology. In contrast to other offers, map data here stand out due to their expansive complementation with metadata. The map available from that site, SDE2_GERMAN1895ELECTORALDISTRICTS.shp, was originally published by Jürgen Schmädeke and digitalised by Daniel Ziblatt.

```
pdf_file<-"pdf/maps_germany_choropleth_counties_1x2.pdf"
cairo_pdf(bg="grey98", pdf_file,width=15,height=8)

par(omi=c(0.5,0,0,0), mai=c(0,6,1,1),family="Lato Light",las=1)
library(maptools)
library(RColorBrewer)
library(gdata)

myC0<-rgb(0,0,0,0,maxColorValue=255)
myC1a<-rgb(68,90,111,50,maxColorValue=255)
myC1b<-rgb(68,90,111,150,maxColorValue=255)
myC1c<-rgb(68,90,111,250,maxColorValue=255)
myC2a<-rgb(255,97,0,50,maxColorValue=255)
myC2b<-rgb(255,97,0,150,maxColorValue=255)
myC2c<-rgb(255,97,0,250,maxColorValue=255)

# first the right map

x<-readShapeSpatial("myData/VG250_Kreise_Shapefile_UTM32_VZ_↘
    2011.shp")
myDistrict_structure_data<-read.xls("myData/konfession_vz_2011.↘
    xls",sheet=1, encoding="latin1")
n<-length(x)
myNo<-as.numeric(as.character(x$RS))
position<-vector()
for (i in 1:n){
  position[i]<-match(myNo[i], myDistrict_structure_data$AGS)
```

```
}
myColour_no<-cut(100*myDistrict_structure_data$RKAT[position]/ ⟍
     myDistrict_structure_data$BEV[position],c(0,50,70,90,100))
myColourPalette<-c(myC0,myC1a,myC1b,myC1c)
plot(x,col=myColourPalette[myColour_no],border=grey(.8),lwd=.3)

myColour_no<-cut(100*myDistrict_structure_data$EVANG[position]/ ⟍
     myDistrict_structure_data$BEV[position],c(0,50,70,90,100))
myColourPalette<-c(myC0,myC2a,myC2b,myC2c)
plot(x,col=myColourPalette[myColour_no],border=grey(.8),lwd=.3,⟍
     add=T)

# then the left map

par(mai=c(0,0,0,7),plt=c(0,0.5,0,0.95),new=T)
x<-readShapeSpatial("myData/OGP86/SDE2_⟍
     GERMAN1895ELECTORALDISTRICTS.shp")
myDistrict_structure_data<-read.xls("myData/ZA8145_1912.xlsx",⟍
     sheet=1, encoding="latin1")
n<-length(x)
myNo<-x$DISTRICT_N
position<-vector()
for (i in 1:n){
  position[i]<-match(myNo[i], myDistrict_structure_data$wkr_nr)
}
myColour_no<-cut(myDistrict_structure_data$kat05p[position],c⟍
     (0,50,70,90,100))
myColourPalette<-c(myC0,myC1a,myC1b,myC1c)
plot(x,col=myColourPalette[myColour_no],border=grey(.8),lwd=.3,⟍
     xpd=T)

myColour_no<-cut(myDistrict_structure_data$ev05p[position], c⟍
     (0,50,70,90,100))
myColourPalette<-c(myC0,myC2a,myC2b,myC2c)
plot(x,col=myColourPalette[myColour_no],border=grey(.8),lwd=.3,⟍
     add=T,xpd=T)

# Titling

mtext("Protestant and Catholic counties 1905... ",3,line=-2.75,⟍
     adj=0.05,family="Lato Black",outer=T,cex=2,xpd=T)
mtext("... and 2011",3,line=-2.75,adj=0.65,family="Lato Black",⟍
     outer=T,cex=2,xpd=T)
mtext("map 1905: geodata.tutfts.edu (goo.gl/Lq3QU); data 1905: ⟍
     Schmädeke (GESIS: doi:10.4232/1.1845);map 2011: www.⟍
     zensus2011.de; data 2011: destatis.de",1,line=-0.7,adj⟍
     =0.04,cex=0.95,font=3,outer=T,xpd=T)

# Lastly, the bar charts as legend

par(mai=c(0.75,5,5.5,9),fg="white",new=T)
myDistrict_structure_data<-read.xls("myData/ZA8145_1912.xlsx",⟍
     sheet=1, encoding="latin1")
```

```
protestant<-myDistrict_structure_data[myDistrict_structure_data$⬎
     ev05p>50,]
attach(protestant)
myClasses<-c(50,70,90,100)
levels(myClasses)<-c("50 to 70 %","70 to 90 %","over 90  %")
barplot(table(cut(ev05p,myClasses)),col=myColourPalette<-c(myC2a⬎
     ,myC2b,myC2c),xlim=c(-150,150),names.arg=levels(myClasses)⬎
     ,main="",cex.axis=0.75,cex.names=0.75,cex.main=0.75,col.⬎
     axis="black",col.main="red",horiz=T)
text(130,4.5,"Protestant",pos=1,col="black",xpd=T,cex=1)

catholic<-myDistrict_structure_data[myDistrict_structure_data$⬎
     kat05p>50,]
attach(catholic)
myClasses<-c(50,70,90,100)
levels(myClasses)<-c("50 to 70 %","70 to 90 %","over 90  %")
barplot(-table(cut(kat05p,myClasses)),col=myColourPalette<-c(⬎
     myC1a,myC1b,myC1c),names.arg=levels(myClasses),main="",cex⬎
     .axis=0.75,cex.names=0.75,cex.main=0.75,col.axis="black",⬎
     col.main="red",horiz=T,add=T)
text(-120,4.5,"Catholic",pos=1,col="black",xpd=T,cex=1)

par(mai=c(0.75,13,5.5,1),fg="white",new=T)
myDistrict_structure_data<-read.xls("myData/konfession_vz_2011.⬎
     xlsx",sheet=1)

protestant<-myDistrict_structure_data[myDistrict_structure_data$⬎
     PEVANG>50,]
attach(protestant)
myClasses<-c(50,70,90,100)
levels(myClasses)<-c("50 to 70 %","70 to 90 %","over 90  %")
barplot(table(cut(PEVANG,myClasses)),col=myColourPalette<-c(⬎
     myC2a,myC2b,myC2c),xlim=c(-150,150),names.arg=levels(⬎
     myClasses),main="",cex.axis=0.75,cex.names=0.75,cex.main⬎
     =0.75,col.axis="black",col.main="red",horiz=T)
text(130,4.5,"Protestant",pos=1,col="black",xpd=T,cex=1)

catholic<-myDistrict_structure_data[myDistrict_structure_data$⬎
     PRKAT>50,]
attach(catholic)
myClasses<-c(50,70,90,100)
levels(myClasses)<-c("50 to 70 %","70 to 90 %","over 90  %")
barplot(-table(cut(PRKAT,myClasses)),col=myColourPalette<-c(⬎
     myC1a,myC1b,myC1c),names.arg=levels(myClasses),main="",cex⬎
     .axis=0.75,cex.names=0.75,cex.main=0.75,col.axis="black",⬎
     col.main="red",horiz=T,add=T)
text(-120,4.5,"Catholic",pos=1,col="black",xpd=T,cex=1)
dev.off()
```

The script is constructed as in the previous example. However, the structure is more extensive in this case, as we are drawing two maps and two legends in the form of bar charts. We plot the map from right to left, which means we start with the 2011 map and then add the 1905 map on the left, followed by the bar charts as legend. For the first map, we set the left margin to 6 inches; with a total width of 15 inches, this will take us into the right area of the figure. First, seven colours are specified from myC0 to myC2c. Then the shapefile and the XLSX table with the structure data are read and linked as in the previous example. However, in contrast to the last example, the maps are here drawn in two steps: first all Catholic, then all Protestant administrative districts. With cut(), classes ranging from greater than 0 to 50, greater than 50 to 70, greater than 70 to 90, and greater 90 to 100 are established. We do not have to define levels() at this point; we take care of that later with the bar charts. All administrative districts belonging to the first class (up to 50) are plotted in white. The administrative districts that remained white in the first pass, whose Catholic population made up a maximum of 50%, are "painted over" with the protestant electorate colours during the second pass: here, the second colour palette is c(myC0, myC2a, myC2b, myC2c) instead of c(myC0, myC1a, myC1b, myC1c). The left map is generated the same way. new=T creates the new figure within the old one. The right 7-inch margin causes positioning within the left half. This variant of splitting a figure is preferred over the one using mfcol=c(1,2), as the latter does not allow overlap of the charts within the figure, or in other words, the margins would be too wide. Finally, the bar charts that we need for the legend are generated. To this end, the margin settings are again set to make the bar charts appear at the bottom right, next to the maps. The bars depicting the frequencies of the Catholic and Protestant districts are presented back to back. To do this, the data frame myDistrict_structure_data is first limited to those districts whose proportion of Protestants exceeds 50. Classifications match those of the maps. The barplot() function then draws the frequency table for these classes; with xlim=c(-150, 150), the space for the catholic frequencies is already considered. These are generated in the same way in the next step: -table(cut(cath05p,myClasses)). The minus symbol reverses the direction so that the bars for Catholic districts point in the other direction. We already encountered this method with population pyramids (Sect. 7.2). Ultimately, the frequencies of the 2011 districts are plotted using the same pattern.

10.3.3   Choropleth Map of Europe at Country-Level

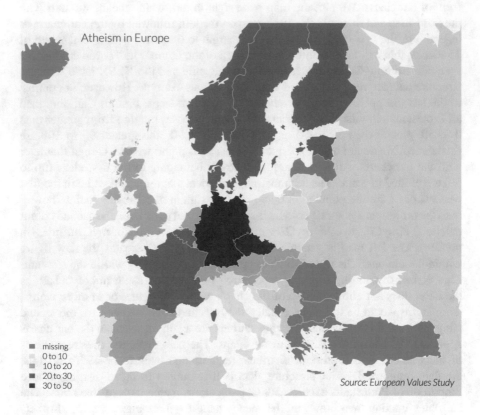

About the figure: Choropleth maps do not necessarily have to show small-scale data. Characterisation of individual countries with variable attributes in different colours can also be useful. In this example, we colour European countries according to a value concerning atheism. Again, we restrict ourselves to one colour and select appropriate gradations with a Brewer palette. In this instance, an additional second colour is used to indicate "missing values". This is useful if we want to point out that certain countries were included in the analysis, but no concrete data for the depicted variable are available. In the chosen layout, not only the legend, but also the title are witten within the map. That way, the figure itself can be bigger (see Sect. 9.2.2). Data derive from the European Values Study and were kindly provided by Mira Hassan. They are based on results by Siegers, Pascal (2012): Alternative Spiritualitäten: neue Formen des Glaubens in Europa: eine empirische Analyse. Akteure und Strukturen, Frankfurt/New York: Campus. See annex A, NUTS, for map data.

```
pdf_file<-"pdf/maps_europe_choropleth_countries.pdf"
cairo_pdf(bg="grey98", pdf_file,width=13,height=11)

par(omi=c(0,0,0,0),mai=c(0,0,0,0),family="Lato Light",las=1)
```

```
library(maptools)
library(RColorBrewer)
library(sp)
library(rgdal) # for spTransform
library(gdata)

# Import data and prepare chart

myData<-read.csv("myData/prop.table EVS cntr.csv",sep=";",dec=",\
     ")
myData[is.na(myData$ATHE),"ATHE"]<--9
data(wrld_simpl)
w=wrld_simpl[!(wrld_simpl@data[,"ISO2"] %in% myData$Country),]
w=w[w@data[,"NAME"] != "Antarctica",]
m=spTransform(w,CRS=CRS("+proj=merc"))

# Create chart

plot(m,xlim=c(-2000000,5000000),ylim=c(4000000,10000000),col=rgb\
     (160,160,160,100,maxColorValue=255),border=F)

x<-readShapeSpatial("myData/NUTS-2010/NUTS_RG_60M_2010.shp",\
     proj4string=CRS("+proj=longlat"))
y<-x[x$NUTS_ID %in% myData$Country,]
m=spTransform(y,CRS=CRS("+proj=merc"))

myClasses<-c(-10,0,10,20,30,50)
myColourPalette<-c("cornflowerblue",brewer.pal(4,"Reds"))

id<-m$NUTS_ID
n<-length(id)
position<-vector()
for (i in 1:n){
  position[i]<-match(m$NUTS_ID[i], myData$Country)
}
myColour_no<-cut(myData$ATHE[position],myClasses)
levels(myColour_no)<-c("missing","0 to 10","10 to 20","20 to 30"\
     ,"30 to 50")
plot(m,col=myColourPalette[myColour_no],border="white",add=T)

legend("bottomleft",levels(myColour_no),cex=1.45,border=F,bty="n\
     ",fill= myColourPalette,text.col="black")

# Titling

mtext("Atheism in Europe",at=-1300000,cex=2,adj=0,line=-3)
mtext("",at=-1300000,cex=2,adj=0,line=-4.8)
mtext("Source: European Values Study",1,at=3200000,cex=1.7,adj\
     =0,line=-2.3,font=3)
dev.off()
```

We start the script by defining a window that is 13 inches wide and 11 inches high. The inner and outer margins are set to 0, and the background to white. We require four libraries: maptools, RColorBrewer, sp, and rgdal for the spTransform()

function. Data are read, and missing values are replaced with −9. The used EU map only comprises EU countries. Especially in the east of the Adriatic coast, this is a problem as the coastlines are missing. It is therefore a good idea to begin by plotting the R map wrld_simpl, and then overlay it with the EU map. As the resolutions of these maps are not identical and the boundary lines therefore marginally differ, we start by filtering the data that do not appear in the data frame. Antarctica has to be excluded, because spTransform would otherwise generate an error. Then the Marcator projection is used, and the result is plotted as a first layer. The EU shapefile is read and filtered for the countries occurring in the data frame; then the Mercator projection is applied as well. Now data have to be classified. Class limits are guided by studies conducted by Pascal Siegers (see above). It would be wise to use a Brewer colour palette for the available data, and a different colour such as cornflowerblue for the missing values (−9). The names correspond to the classifications. The connection between data and map is established as in the previous examples using match(). With plot(), the choropleth map is drawn over the existing map (add=TRUE); at the end, the legend, title, and subtitles are added.

10.3.4 Choropleth Map of Europe at Country-Level (Panel)

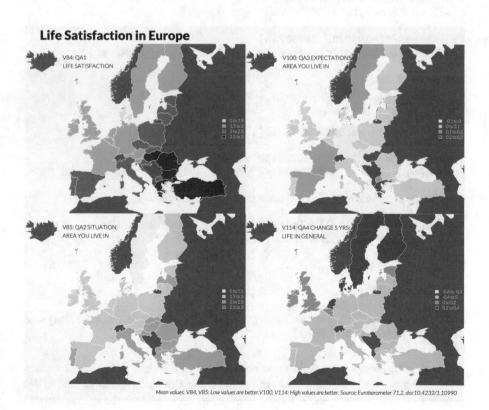

Mean values. V84, V85: Low values are better.V100, V114: High values are better. Source: Eurobarometer 71.2, doi:10.4232/1.10990

About the figure: Panel representations are also a good choice for maps of Europe. However, the level of detail here is restricted by the generally limited size for publication. The above example shows the attributes of four Eurobarometer variables. We are revisiting the example from Sect. 7.1.2, concerning the distribution of countries for four questions about life satisfaction. There, we calculated the frequency distributions of averages for the questions "On the whole, are you very satisfied, fairly satisfied, not very satisfied or not at all satisfied with the life you lead?" and "How would you judge the current situation in each of the following?", and for the two questions "What are your expectations for the next twelve months: will the next twelve months be better, worse or the same, when it comes to...?" and "Compared with 5 years ago, would you say things have improved, gotten worse or stayed about the same when it comes to...?"; those averages were then used to specify classes. These classes are here shown as a panel of 2x2 individual charts.

Data: see annex A, ZA4972: Eurobarometer 71.2 (May-Jun 2009). Map data: see annex A, NUTS.

```
pdf_file<-"pdf/maps_europe_choropleth_countries_2x2.pdf"
cairo_pdf(bg="grey98", pdf_file,width=13,height=11)

par(omi=c(0.5,0,0.5,0),mai=c(0,0,0,0))
par(mfcol=c(2,2),family="Lato Light")
library(maptools)
library(RColorBrewer)
library(sp)
library(rgdal)
library(gdata)

# Import data, prepare and create graphic

myplotFile<-"scripts/inc_plot_maps_europe_choropleth_countries.r\
    "

mySelection<-"v84"
myClasses<-c(0,1.5,2,2.5,3)
myClass_des<-c("0 to 1.5","1.5 to 2","2 to 2.5","2.5 to 3")
myColourPalette<-brewer.pal(4,"Reds")
source(myplotFile)
mtext("V84: QA1",at=-1300000,adj=0,line=-3)
mtext("LIFE SATISFACTION",at=-1300000,adj=0,line=-4.5)

mySelection<-"v85"
myClasses<-c(0,1.5,2,2.5,3)
myClass_des<-c("0 to 1.5","1.5 to 2","2 to 2.5","2.5 to 3")
myColourPalette<-brewer.pal(4,"Greens")
source(myplotFile)
```

```
mtext("V85: QA2 SITUATION:",at=-1300000,adj=0,line=-3)
mtext("AREA YOU LIVE IN",at=-1300000,adj=0,line=-4.5)

mySelection<-"v100"
myClasses<-c(-0.1,0,0.1,0.2,0.3)
myClass_des<-c("-0.1 to 0","0 to 0.1","0.1 to 0.2","0.2 to 0.3")
myColourPalette<-brewer.pal(4,"Blues")
source(myplotFile)
mtext("V100: QA3 EXPECTATIONS:",at=-1300000,adj=0,line=-3)
mtext("AREA YOU LIVE IN",at=-1300000,adj=0,line=-4.5)

mySelection<-"v114"
myClasses<-c(-0.6,-0.4,0,0.2,0.4)
myClass_des<-c("-0.6 to -0.4","-0.4 to 0","0 to 0.2","0.2 to 0.4↘
    ")
myColourPalette<-brewer.pal(4,"Purples")
source(myplotFile)
mtext("V114: QA4 CHANGE 5 YRS:",at=-1300000,adj=0,line=-3)
mtext("LIFE IN GENERAL",at=-1300000,adj=0,line=-4.5)

# Titling

mtext(side=3,"Life Satisfaction in Europe",adj=0.05,outer=T,line↘
    =0.5,cex=2.25,family="Lato Black")
mtext(side=1,"Mean values. V84, V85: Low values are better.V100,↘
    V114: High values are better. Source: Eurobarometer 71.2,↘
    doi:10.4232/1.10990",adj=0.95,outer=T,line=1,font=3)
dev.off()
```

and

```
# inc_plot_maps_europe_choropleth_countries.r
data(wrld_simpl)
w=wrld_simpl[wrld_simpl@data[,"NAME"] != "Antarctica",]
m=spTransform(w,CRS=CRS("+proj=merc"))

plot(m,xlim=c(-2000000,5000000),ylim=c(4000000,10000000),col=rgb↘
    (120,120,120,maxColorValue=255),border=F)

x<-readShapeSpatial("myData/NUTS-2006/NUTS_RG_03M_2006.shp",↘
    proj4string=CRS("+proj=longlat"))

xls<-read.xls("myData/eb_nuts.xlsx", encoding="latin1")

y<-x[x$NUTS_ID %in% xls$nuts_id,]

m=spTransform(y,CRS=CRS("+proj=merc"))

source("scripts/inc_datadesign_dbconnect.r")
sql<-paste("select ", mySelection," selection from v_za4972_↘
    countries",sep="")
myDataset<-dbGetQuery(con,sql)
attach(myDataset)

myNo<-m$NUTS_ID
```

```
myDataset$ref_NUTS_ID<-xls$nuts_id

position<-vector()
for (i in 1:30){
  position[i]<-match(m$NUTS_ID[i], myDataset$ref_NUTS_ID)
}

myColour_no<-cut(myDataset$selection[position],myClasses)
levels(myColour_no)<-myClass_des

plot(m,col=myColourPalette[myColour_no],border="white",add=T)
legend("right",levels(myColour_no),cex=0.95,border=F,bty="n",\
      fill= myColourPalette,text.col="white")
```

In this case, we begin the script by dividing the figure into four areas. For each area, we then define the variable that we want to plot, the classification and class names, and the colour palette. With those specifications, the four parts of the figure are then plotted. To this end, a file inc_plot_maps_europe_choropleth_countries.r is called to take over this task. In the embedded file, the wrld_simpl map is read first, Antarctica is filtered, and the Mercator projection is applied. A section of the map is then specified by xlim and ylim and drawn as the "base map". In a second step, the EU map x is read; from that map, we only require those countries that also appear in the Eurobarometer data frame. To this end, we read an XLS table that contains the NUTS-IDs for the country IDs in the Eurobarometer data frame. By using %in% to create a new object y that only contains the countries listed in the XLS table, the shapefile can be limited to only those countries. y still has to be adjusted with a Mercator transformation. However, we have to use the proj4string=CRS("+proj=longlat") parameter with readShapeSpatial() to make the Mercator projection work here. Now, the data we want to plot can be read from a MySQL table. The SQL command is put together with paste() as the selection variable is handed over, respectively. The rest matches the previous examples: The values are linked with the map data from the shapefile, classes are created, and finally the map and legend are plotted.

10.3.5 *World Choropleth Map: Regions*

Colour by region

About the figure: The last example choropleth map shows a world map in which the continents or countries are grouped into world regions. The map is based on data from the GAPMINDER project. We will need this map later as a legend for a scatter plot (Sect. 11.6). To this end, the world regions are coloured differently.

Data: The XLS file regions.xlsx can be downloaded at https://link.springer.com/chapter/10.1007/978-3-030-28444-2_1. The map derives from the wrld_simpl data frame that comes with R.

```
pdf_file<-"pdf/maps_world_gapminder_regions.pdf"
cairo_pdf(bg="grey98", pdf_file,width=8.4,height=6)

par(omi=c(0,0,0,0),mai=c(0,0.82,0.82,0),family="Lato Light",las↘
    =1)

library(maptools) # contains wrld_simpl
library(rgdal) # for spTransform
library(gdata)

# Import data and prepare chart

data(wrld_simpl)
w<-wrld_simpl[wrld_simpl@data[,"NAME"] != "Antarctica",]
m<-spTransform(w,CRS=CRS("+proj=merc"))
```

```
myCountries<-m@data$ISO2
n<-length(myCountries)
myMapColours<-numeric(n)

myR1<-"Sub-Saharan Africa"
myR2<-"South Asia"
myR3<-"Middle East & North Africa"
myR4<-"America"
myR5<-"Europe & Central Asia"
myR6<-"East Asia & Pacific"

myC1<-rgb(0,115,157,150,maxColorValue=255)
myC2<-rgb(158,202,229,150,maxColorValue=255)
myC3<-rgb(84,196,153,150,maxColorValue=255)
myC4<-rgb(255,255,0,150,maxColorValue=255)
myC5<-rgb(246,161,82,150,maxColorValue=255)
myC6<-rgb(255,0,0,150,maxColorValue=255)

myRegion<-c(myR1,myR2,myR3,myR4,myR5,myR6)
myColour<-c(myC1,myC2,myC3,myC4,myC5,myC6)

myRegions<-read.xls("myData/gapminder/regions.xlsx", encoding="\
    latin1")

for (i in 1:length(myRegion))
{
myRegionSelection<-subset(myRegions$ID,myRegions$Group==myRegion\
    [i])
myCountrySelection<-NULL
for (j in 1:length(myRegionSelection)) myCountrySelection<-c(\
    myCountrySelection, trim(as.character(myRegionSelection[j\
    ])))
for (j in 1:length(myCountrySelection))
{
myMapColours[grep(paste("^",myCountrySelection[j],"$",sep=""),\
    myCountries)]<-myColour[i]
}
}

# Create chart

plot(m,col=myMapColours,border=F)

# Titling

mtext("Colour by region",3,line=0,adj=0,cex=4,family="Lato Black\
    ")


dev.off()
```

About the script: We use two blog entries as guidance for this example. We again use the Mercator projection of the wrld_simpl map provided with R for our world map. In a first step, the ISO2 codes from that map are stored in a vector myCountries, and an empty vector of equal length mapcolours is initialised. Then, six regions and six colours are specified and assigned to the vectors region and colour. The XLSX table listing the regions and associated countries is read next. For each region, the data frame is then filtered for the countries of that region, and a vector with this country selection is created; this is done in a loop. In a second, inner, loop, the vector myMapcolours is filled for these recently generated country selections by using the grep() function; this function locates the country ID from the XLSX table in the country IDs of the wrld_simpl ISO2 codes. To find all codes, we extend the expression myCountry_selection[j] with ^ and $ to a regular expression. The generated vector myMapcolours can then be used to create the world map with the coloured world regions.

10.4   Exceptions and Special Cases

Two rather unusual examples wrap up this chapter: on the one hand the representation of so-called great circles, which have become widespread through Paul Butler's Facebook map, and on the other hand the use of data from the OpenStreetMap project.

10.4.1   World Map with Orthodromes

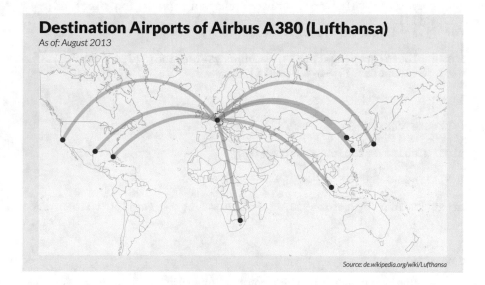

About the figure: The world map shows the eight flight routes Lufthansa serves
with their Airbus 380 model. Leaving Frankfurt, Germany, these are: Beijing,
Tokyo, Johannesburg, San Francisco, Miami, Singapore, Houston, and Shanghai.
The routes are displayed as great circles (aka "orthodromes"), which show the
shortest route between two points on a spherical surface. We do not want to delve
into the depths of spherical geometry or get caught up in the snares of different
projection transformations: it is simply descriptive. As not the entire world map is
shown (Arctic and Antarctic regions are problematic with all projections and have
been omitted), the shown section should be framed in colour. Data were taken from
the German Wikipedia article "Lufthansa". Mapdata are provided with the mapdata
package in R.

```
pdf_file<-"pdf/maps_world_great_circles.pdf"
cairo_pdf(bg="grey98", pdf_file,width=13.75,height=8)

par(omi=c(0.5,0.5,1.25,0.5),mai=c(0,0,0,0),lend=1,bg="\
     antiquewhite",family="Lato Light")
library(mapdata)
library(geosphere)
library(gdata)
# library(mapproj)

# Import data and prepare chart

myProj.type<-"mercator"
myProj.orient<-c(90,0,30)
x<-map(proj=myProj.type,orient=myProj.orient,wrap=T)

# Create chart and other elements

plot(x,xlim=c(-3,3),ylim=c(-1,2),type="n",axes=F,xaxs="i",yaxs="\
     i")
rect(-3,-1,3,3,col="aliceblue",border=NA)
map("worldHires","Germany",fill=T,add=T,col="antiquewhite",proj=\
     myProj.type,orient=myProj.orient)
lines(x,col="darkgrey")
data(world.cities)
myData<-read.xls("myData/orthodat.xlsx", encoding="latin1")
attach(myData)

myTColour<-rgb(128,128,128,100,maxColorValue=255)
for (i in 1:nrow(myData))
{
myStart<-world.cities[11769,] # Frankfurt
myDestination<-subset(world.cities,name==stadt[i] & country.etc\
     ==land[i])
myGC1<-gcIntermediate(c(myStart$long,myStart$lat),c(\
     myDestination$long,myDestination$lat),addStartEnd=T, n=50)
merc<-mapproject(myGC1[,1],myGC1[,2],projection=myProj.type,\
     orientation=myProj.orient)
lines(merc$x,merc$y,lwd=10,col=myTColour)
```

```
myDestP<-mapproject(myDestination$long,myDestination$lat,proj=\
    myProj.type,orient=myProj.orient)
points(myDestP,col="darkred",pch=19,cex=2)
}
myStartP<-mapproject(myStart$long,myStart$lat,proj=myProj.type,\
    orient=myProj.orient)
points(myStartP,col="darkblue",pch=19,cex=2)

# Titling

mtext("Destination Airports of Airbus A380 (Lufthansa)",3,line\
    =3,adj=0,cex=3,family="Lato Black",outer=T)
mtext("As of: August 2013",3,line=1,adj=0,cex=1.75,font=3,outer=\
    T)
mtext("Source: de.wikipedia.org/wiki/Lufthansa",1,line=0.8,adj\
    =1.0,cex=1.25,font=3)
dev.off()
```

In the script, we first specify the outer margin and set the inner margins to zero; the background becomes "antiquewhite". When it comes to packages, we need mapdata for the worldHires map (which also loads the maps package), but we also require geosphere, which provides the gcIntermediate() function. We then go on to specify the Mercator value within the variable myProj.type, and an orientation of the projection in myProj.orient. Using the map() function, the map is first saved in an object x, and is then plotted using plot() and with specification of xlim and ylim. xaxs and yaxs are set to i, so the margins do not go beyond the set limits. A rectangle filled with aliceblue is laid over the entire data area. The section of the worldHires map is added in antiquewhite with add=T, and lines() is used for the outlines of the world map.

Next, the data with the required coordinates are loaded from world.cities. Those cities that we actually want to plot are read from the XLSX table orthodat.xlsx; the cities in the table are passed in a loop. Frankfurt is defined as the start, and the destination is the filtered dataframe world.cities that contains the respective city from the read XLSX table. The great circles between the start and destination points are then calculated. However, they have to be converted before plotting; to this end, we use the mapproject() function and the parameters projection and orientation. After plotting the lines, we repeat the calculation for the start and destination points, which we then enlarge and plot over the line endings. Titles and captions come last.

10.4.2 City Maps with OpenStreetMap Data (Panel)

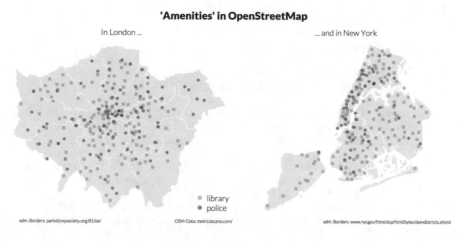

'Amenities' in OpenStreetMap

In London ...   ... and in New York

■ library
● police

adm. Borders: parlvid.mysociety.org:81/os/   OSM-Data: metro.teczno.com/   adm. Borders: www.nycgov/html/dcp/html/bytes/dwndistricts.shtml

About the figure: The free OpenStreetMap project offers huge potential for visualisation of geo data. By now, there are more than one million users that contribute
to the 2004-founded project, and they have captured more than three billion GPS
points. The figure is a true-to-scale comparative representation of the areas of
London and New York with their administrative borders. Public libraries and police
stations tagged as "amenities" in the OpenStreetMap database are added to the map.

Data: The shape data for London's administrative districts are provided at
http://parlvid.mysociety.org:81/os/

mysociety is an offering provided by the UK Citizens Online Democracy
movement, which collates data from the national survey offices, among others. The
administrative borders of New York are provided as shapefiles from the official site
https://www1.nyc.gov/site/planning/data-maps/open-data/districts-download-
metadata.page

different variants, such as election districts, municipal court districts, state senate
distrcits, etc., can be selected. The OpenStreetMap data are excerpts of the so-called
Planet Files. These files are available for numerous cities and can be be downloaded
from
http://download.geofabrik.de

Data are regularly updated "rectangular" sections from the Planet File; they
contain geographical or political area units and borders, but there are no selection
criteria. This means the available data are not for London, but for a rectangle around
London.

```
pdf_file<-"pdf/maps_cities_1x2.pdf"
cairo_pdf(bg="grey98", pdf_file,width=14,height=7)
```

```
par(omi=c(0,0,0.5,0),mai=c(0.5,0,0.5,0),mfcol=c(1,2),family="\
     Lato Light",
  las=1)
library(maptools) # also loads sp for over
library(rgdal) # for spTransform

# Prepare chart and import data

flib<-rgb(0,139,0,120,maxColorValue=255)
fpol<-rgb(139,0,0,120,maxColorValue=255)

to_myProj<-"+proj=merc +a=6378137 +b=6378137 +lat_ts=0.0 +lon_\
     0=0.0 +x_0=0.0 +y_0=0 +k=1.0 +units=m +nadgrids=@null +\
     wktext +over +no_defs"

from_myProj<-"+proj=tmerc +lat_0=49 +lon_0=-2 +k=0.999601272 +x_\
     0=400000 +y_0=-100000 +ellps=airy +towgs84=375,-\
     111,431,0,0,0,0 +units=m +no_defs"
myShapeFile<-"myData/london/greater_london_const_region.shp"
myLon_adminD<-readShapeSpatial(myShapeFile,proj4string=CRS(from_\
     myProj))
myLon_adminD<-spTransform(myLon_adminD,CRS=CRS(to_myProj))


# Create chart and other elements

plot(myLon_adminD,col=rgb(139,139,139,60,maxColorValue=255),\
     border="white")
lon_osm<-readShapeSpatial("myData/london/london.osm-amenities.\
     shp")
proj4string(lon_osm)<-proj4string(myLon_adminD)
inside1<-!is.na(over(lon_osm,myLon_adminD)) & lon_osm@data$type \
     == "library"
inside2<-!is.na(over(lon_osm,myLon_adminD)) & lon_osm@data$type \
     == "police"
points(lon_osm[inside1[ ,1],],col=flib,pch=15,cex=1.25,lwd=0)
points(lon_osm[inside2[ ,1],],col=fpol,pch=19,cex=1.25,lwd=0)

legend("bottomright",c("library","police"),col=c(flib,fpol),pch=\
     c(15,19),bty="n",pt.cex=1.5,cex=1.5)
mtext(side=3,"In London ...",cex=1.5,col=rgb(64,64,64,\
     maxColorValue=255))
mtext(side=1,"OSM-Data: metro.teczno.com/",adj=1,cex=0.85)
mtext(side=1,"adm. Borders: parlvid.mysociety.org:81/os/",adj\
     =0.1,cex=0.85)

# Import data and prepare chart

from_myProj<-"+proj=lcc +lat_1=40.66666666666666 +lat_\
     2=41.03333333333333 +lat_0=40.16666666666666 +lon_0=-74 +x\
     _0=300000 +y_0=0 +ellps=GRS80 +towgs84=0,0,0,0,0,0,0 +\
     units=us-ft +no_defs"
myShapeFile<-"myData/newyork/nybb.shp"
```

```
myNy_adminD<-readShapeSpatial(myShapeFile,proj4string=CRS(from_↘
    myProj))
myNy_adminD=spTransform(myNy_adminD,CRS=CRS(to_myProj))

# Create chart

plot(myNy_adminD,col=rgb(139,139,139,60,maxColorValue=255),↘
    border="white")

# Import data and other elements

ny_osm<-readShapeSpatial("myData/newyork/new-york.osm-amenities.↘
    shp")
proj4string(ny_osm)<-proj4string(myNy_adminD)
inside1<-!is.na(over(ny_osm,myNy_adminD)) & ny_osm@data$type == ↘
    "library"
inside2<-!is.na(over(ny_osm,myNy_adminD)) & ny_osm@data$type == ↘
    "police"
points(ny_osm[inside1[ ,1],],col=flib,pch=15,cex=1.25,lwd=0)
points(ny_osm[inside2[ ,1],],col=fpol,pch=19,cex=1.25,lwd=0)

# Titling

mtext(side=3,"... and in New York",cex=1.5,col=rgb(64,64,64,↘
    maxColorValue=255))
mtext(side=3,"'Amenities' in OpenStreetMap",outer=T,cex=2,family↘
    ="Lato Black")
mtext(side=1,"adm. Borders: www.nycgov/html/dcp/html/bytes/↘
    dwndistricts.shtml",adj=0.9,cex=0.85)
dev.off()
```

In this case, we need four shapefiles in the script: the administrative borders of London and New York, and also the OpenStreetMap data for both cities. Let us start with the data for the administrative borders of London and the OpenStreetMap data of the London section:

```
library(maptools)
par(mfcol=c(1,2))
x <- readShapeSpatial("greater_london_const_region.shp")
plot(x, axes=TRUE)
y <- readShapeSpatial("london.osm-amenities.shp")
plot(y, axes=TRUE, pch=1, col=rgb(100,100,100,60,maxColorValue↘
    =255))
```

The result is shown in Fig. 10.2. To be able to uniformly plot all data, two problems have to be solved: On the one hand, all data have to feature the same projection and the same coordinate system. On the other hand, the OpenStreetMap data have to be filtered so only data within the administrative border are plotted. The Quantum GIS application helps us with our first problem; R provides the package sp and the over() function for our second problem.

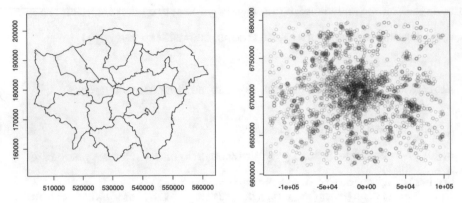

Fig. 10.2   Administrative borders of London (*left*) and OpenStreetMap data (*right*)

When calling the shapefile of London's administrative districts in Quantum GIS, we can display the properties by right-clicking the layer (Fig. 10.3).

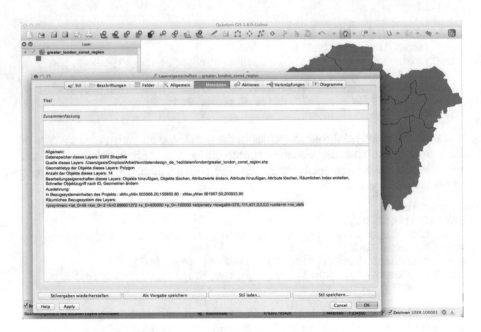

Fig. 10.3   Metadata of the greater_london_const_region.shp shapefile

In the last row, the reference frame for the layer is specified and can be copied and pasted into our R script. We can do the same for the administrative borders and OpenStreetMap data of New York. As the OpenStreetMap data for London and New York use identical projections as coordinate system, we use those as a starting point; this means we save them in the variable to_myProj. The respective information concerning the administrative districts is saved in from_myProj. Now we can read the administrative borders using the readShapeSpatial() function with the proj4string=CRS(from_myProj) parameter, and then bring it into the desired form with the spTransform() function and the CRS=CRS(to_myProj) parameter. The result can be plotted with plot() and generates London's administrative borders as grey areas. After that, we read the OpenStreetMap data. We first need the proj4string() function from the sp package to adjust the projections. Then the data are filtered: on the one hand, data of the type library or police are taken from the data slot of the shapefile, on the other hand, the over() function selects only those data that lie within the administrative borders. The determined points are plotted using points(). After the legend, and the titles and caption of the London part, we proceed analogously for the New York part.

10.4.3 Georeferenced Map in Grid Format

Today, there are a large number of historical maps which have been digitalized in high resolution and which are georeferenced and can thus be combined with vector maps. We demonstrate the usage options here with a map from the Harvard Geospatial Library which has been released for educational and non-commercial use (Fig. 10.4).

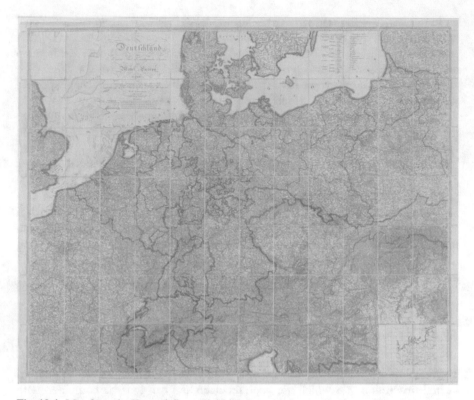

Fig. 10.4 Map from the Harvard Geospatial Library

The map bears the rather long-winded title:

„Deutschland und der gröste Theil der umliegenden Staaten : oder Mittel-Europa in 35 Blättern : nach astronomischen Ortsbestimmungen und den besten Special-Karten, mit Rücksicht auf die neuesten Grenz-Bestimmungen entworfen, zufolge der Wiener Congress-Akte, des Pariser Friedens vom 21ten Nov. 1815, und der neuesten Austauschungen 1816 von H.H. Gotthold ; geschrieben und gestochen von H. Kliewer ; sämtliche Gebürge im Atlas sind gezeichnet und gestochen von Paulus Schmidt, so wie auch die Sectionen 3,11,16,18,19,21,25,26,31 von demselben gestochen worden." [Germany and the majority of the surrounding states: or, Central Europe on 35 pages: drafted according to astronomical positioning and the best special maps, taking into account the latest border determinations as a consequence of the Act of the Congress of Vienna, of the Treaty of Paris dated November 21st, 1815, and the latest exchanges in 1816 by H.H. Gotthold; written and engraved by H. Kliewer; all mountains in the atlas are drawn and engraved by Paulus Schmidt, with Sections 3, 11, 16, 18, 19, 21, 25, 26, 31 also engraved by the same person.]

The map was published in 1816 by Simon Schropp & Co. to a scale of approx. 1:1,100,000. It can be downloaded without registration. A total of three files are downloaded: one file with the ending jp2, which contains the actual map as an image in the format JPG 2000, a "world file" with the ending jp2, which consists of six lines of text, each with one number, and which enables the geo-classification of the map, and finally, a file with comments about the map (origin, method of creation, etc.) in XML format. The map is 519.3 MB, with a width and height of 18,941 pixels × 15,702 pixels, and therefore has a resolution of 297 megapixels. As we can see from the image, it was folded intensely before the digitalisation.

10.4.3.1 Grid Overlay with QGIS

First we load the map to QGIS. There we can also consider the information from the world file (Fig. 10.5). As a first step, we define a new project (Project -> New). In the new project, we then add a grid layer (Layer -> Add Layer -> Add Grid Layer). When doing so, we have to select the coordinate reference system. As specified in the description of the map (both online and in the XML file), we select "Europe_Lambert_Conformal_Conic (EPSG: 102014)".

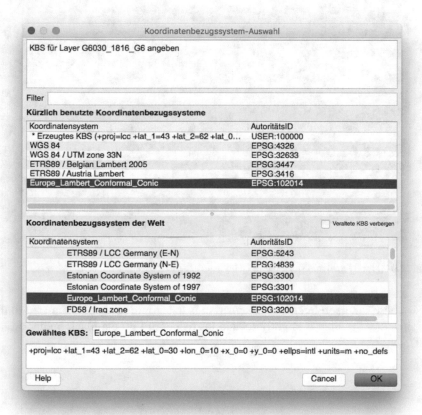

Fig. 10.5 Selection of the coordinate reference system

We use the following coordinate reference system string in R later to import the map there:

```
+proj=lcc +lat_1=43 +lat_2=62 +lat_0=30 +lon_0=10 +x_0=0 +y_0=0 ↘
    +ellps=intl +units=m +no_defs
```

In a second step, we add a current map of Germany as a vector layer (Layer -> Add Layer -> Add Vector Layer). We select "DEU_adm1.shp", which we previously downloaded from the Global Administrative Areas website (game.org). In the file "DEU_adm1.prj", the shapefile contains information about the coordinate reference system so that the map is positioned correctly on the historical map automatically. We then open the layer window (View -> Operating Fields -> Layer Window) and use the right mouse button on the entry "DEU_adm1" to select the menu item "Properties". Here we can adjust the colour and transparency of the current map. The result looks as in Fig. 10.6.

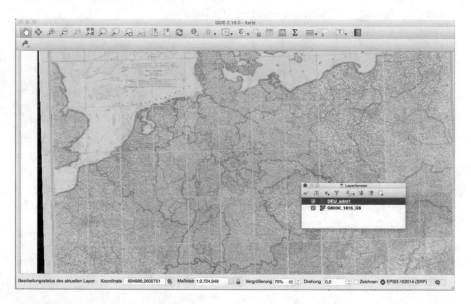

Fig. 10.6 Overlay grid and shapefile in QGIS

Because the georeferencing information is taken into account, compared to the original file, the map is slightly rotated (as evidenced by the black border on the left).

10.4.3.2   Grid Overlay with R

With R too, it is possible to overlay the grid map with other georeferenced data, for example vector maps. To do this, however, the JPG 2000 file imported into QGIS must be saved in GTiff format. To do this, in the layout window, with the right mouse button on the entry "G6030_1816_G6", we select "Save as...". Here, all of the details have been defaulted correctly, we just have to specify a file name. We enter "G6030_1816_G6_Export_QGIS.tif". The result is a TIFF file of 927.9 MB that we can import into R in the next step without any problems.

```
library(maptools)
library(raster)
library(Cairo)
png_datei<-"raster_overlay.png"
CairoPNG(png_datei,width=7000,height=6000)

x <- brick('G6030_1816_G6_Export_QGIS.tif')
projektion.raster<-"+proj=lcc +lat_1=43 +lat_2=62 +lat_0=30 +lon↘
    _0=10 +x_0=0 +y_0=0 +ellps=intl +units=m +no_defs"
projection(x)<-projektion.raster
plotRGB(x, maxpixels=42000000)

y<-readShapeSpatial("DEU_adm_shp/DEU_adm1.shp",proj4string=CRS("↘
    +proj=longlat +datum=WGS84 +no_defs"))
y=spTransform(y,CRS=CRS(projektion.raster))
plot(y, add=T, border=rgb(228,217,202,130,maxColorValue=255), ↘
    col=rgb(255,0,0,30,maxColorValue=255))
dev.off()
```

In R, we load the libraries maptools (for importing and transforming the shapefile), raster (for the import and display of the TIFF file) and Cairo (for saving the result as a PNG file). The TIFF file is imported using the function brick() and then the projection read out from QGIS is assigned to this TIFF file. The "grid" is displayed with the function plotRGB(). In the second step, the shapefile is imported and transformed such that the projection matches that of the grid. Finally, the transformed shapefile is drawn over the existing grid illustration (Fig. 10.7). The result is a high-resolution map (7000 × 6000 pixels, 91.7 MB, shown here only as a screenshot).

Fig. 10.7 Raster Overlay mit R

10.4.3.3 Map Extracts

Overlaying historical grid maps and vector data works not only for whole files, but also for extracts. The package raster provides the function crop() for this purpose. For illustration purposes, we use a map from eurostat, which provides the NUTS levels 1 and 2 for Germany. In Germany, the second level is made up of 38 regions (in particular, administrative regions).

Here we select the administrative region Arnsberg:

```
library(gdata)
nuts2013<-readShapeSpatial("NUTS_2013_03M_SH/data/NUTS_RG_03M_↘
    2013.shp", proj4string=CRS("+proj=longlat"))
m=spTransform(nuts2013,CRS=CRS("+proj=merc"))
DEU<-m[substr(m$NUTS_ID, 1, 3)=="DE",]
DEA5<-m[m$NUTS_ID=="DEA5",]
plot(DEU)
plot(DEA5, add=T)
```

The individual steps generally correspond to the previous example. The difference here is that the shapefile is filtered and a rectangle is defined with the outer limits of this administrative region and then "cut out" of the grid map. Finally, as in the previous example, both levels are laid over one another.

```
library(maptools)
library(raster)
library(Cairo)
library(gdata)
png_datei<-"DEA5_crop_raster_overlay.png"
CairoPNG(png_datei,width=1000,height=1000)

x <- brick('G6030_1816_G6_Export_QGIS.tif')
projektion.raster<-"+proj=lcc +lat_1=43 +lat_2=62 +lat_0=30 +lon↖
    _0=10 +x_0=0 +y_0=0 +ellps=intl +units=m +no_defs"
projection(x)<-projektion.raster

nuts2013<-readShapeSpatial("NUTS_2013_03M_SH/data/NUTS_RG_03M_↖
    2013.shp", proj4string=CRS("+proj=longlat"))
m=spTransform(nuts2013,CRS=CRS("+proj=merc"))
DEU<-m[substr(m$NUTS_ID, 1, 3)=="DE",]
DEA5<-m[m$NUTS_ID=="DEA5",]
DEA5p=spTransform(DEA5,CRS=CRS(projektion.raster))
ausschnitt<-crop(x, DEA5p)
```

```
plotRGB(ausschnitt, maxpixels=1000000)
plot(DEA5p, add=T, border=rgb(228,217,202,130,maxColorValue=255)
     , col=rgb(255,0,0,30,maxColorValue=255))
dev.off()
```

The result is the extract of the historical map that covers the area that is the
administrative region Arnsberg today. In the centre of the enlargement, we can
clearly see a folded edge of the map (Fig. 10.8).

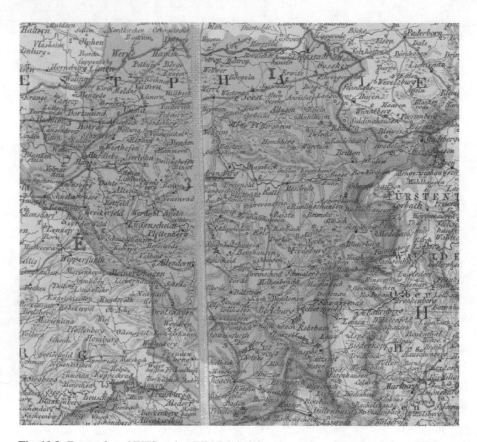

Fig. 10.8 Extract from NUTS region DEA5 (administrative region of Arnsberg) from 2006 on a
historical map from 1816

10.4.4   Cartogram (Panel)

In a cartogram, the geographical areas (continents, countries, or administrative
units) are distorted due to the definition of a statistical size. For example, if we
were to represent the population density of Germany in this way, the city states

would be significantly larger, and the federal states smaller in contrast. That is no easy undertaking, as the sizes of the areas cannot be varied independently of one another.

Gross domestic product of the federal states 2015

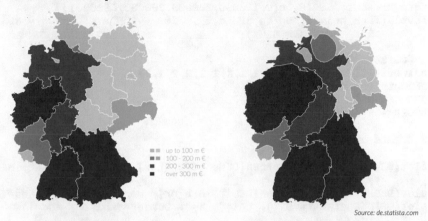

Source: de.statista.com

The illustration shows a comparison of the gross domestic product (GDP) of the federal states in Germany for 2015. For this purpose, we have shown the western and eastern federal states each with a Brewer colour palette such that in both cases (west and east), there is an underlying palette with four classes. However, for the eastern federal states, only the first two classes are assigned. For comparison purposes, on the left, we have selected a representation with the actual proportions. In contrast, on the right we see the distorted map. North Rhine-Westphalia and Bavaria, as well as Bremen, Hamburg and Berlin, become significantly larger, and in contrast, the five eastern federal states are smaller.

```
pdf_file<-"pdf/maps_germany_cartogram_1x2.pdf"
cairo_pdf(bg="grey98", pdf_file, width=12,height=6)
par(omi=c(0.15,0.25,0.55,0.15), mai=c(0,0,0,1), family="Lato ↘
    Light", mfcol=c(1,2))

library(maptools)
library(gdata)
library(cartogram)
library(RColorBrewer)

# Prepare chart and import data

nuts2013<-readShapeSpatial("myData/NUTS_2013_03M_SH/data/NUTS_RG↘
    _03M_2013.shp", proj4string=CRS("+proj=longlat"))
m=spTransform(nuts2013,CRS=CRS("+proj=merc"))
DE<-m[substr(m$NUTS_ID, 1, 2)=="DE" & m$STAT_LEVL_==1,]

bl<-read.xls("myData/bl.xlsx")
```

```
DE@data$BIP<-bl$BIP
DE@data$AUS<-bl$AUS

westfarben<-brewer.pal(5,"PuRd")[2:5]
ostfarben<-brewer.pal(5,"PuBu")[2:5]

farb_nr<-cut(DE$BIP, c(0,100000,200000,300000,650000))
levels(farb_nr)<-c("up to 100 m €", "100 - 200 m €", "200 - 300 ⬎
     m €", "over 300 m €")

DE@data$farb_nr<-farb_nr
ostwest<-c(2,1,1,1,1,0,2,1,1,1,1,1,2,2,1,2)
DE@data$ostwest<-ostwest
DEWest<-DE[DE$ostwest==1, ]
DEOst<-DE[DE$ostwest==2, ]

# Create chart

plot(DEWest, col=westfarben[DEWest$farb_nr], border="white", new⬎
     =T)
plot(DEOst, col=ostfarben[DEOst$farb_nr], border="white", add=T)
plot(DE[DE$ostwest==0, ], col="grey", border="white", add=T)

legend(1580000, 6400000, xpd=T, c("", "", "", ""), cex=0.95, ⬎
     border=FALSE, bty="n", fill=westfarben)
legend(1620000, 6400000, xpd=T, levels(farb_nr), cex=0.95, ⬎
     border=FALSE, bty="n", fill=c(ostfarben[1], ostfarben[2], ⬎
     "white", "white"), text.col="darkgrey")

DEC<-cartogram(DE, "BIP", 9)
DECWest<-DEC[DEC$ostwest==1, ]
DECOst<-DEC[DEC$ostwest==2, ]

plot(DECWest, col=westfarben[DECWest$farb_nr], border="white")
plot(DECOst, col=ostfarben[DECOst$farb_nr], border="white", add=⬎
     T)
plot(DEC[DEC$ostwest==0, ], col="grey", border="white", add=T)

# Titling

mtext("Gross domestic product of the federal states 2015", line⬎
     =0, adj=0, cex=2.2, family="Lato Black", outer=T)
mtext("Source: de.statista.com", side=1, line=-1, adj=1, cex⬎
     =0.9, font=3, outer=T)
dev.off()
```

The script uses the cartogram package from Sebastian Jeworutzki et al., apart from Rcartogram, one of two that are currently available for creating cartograms in R.[1] The two packages differ in the selection of the algorithm: the package Rcartogram uses a solution provided by Gastner and Newman (2004), while

---

[1]Many thanks to Sebastian Jeworutzki for tips for this section.

cartogram uses the solution provided by Dougenik et al. (1985) but also does not require any external program libraries. In most cases, the end results are comparable. After importing the shapefile, which we extract Germany from, we import the gross domestic product data for the federal states and assign these data to the shapefile. We select one Brewer colour palette respectively for the western and for the eastern federal states (the values 2–5 from two five-colour palettes, as otherwise the first value would be too bright) and categorise the data into four classes. We then create two map objects DEWest and DEOst, which we draw one after the other. In a third step, we draw Berlin in separately in grey. After drawing in the legends, we repeat the entire process with the distorted representation by using the cartogram() function to create the polygon DEC from DE. The usual header and footer complete the script.

Chapter 11
Illustrative Examples

The final chapter includes a few, rather "illustrative" figures. With these, the effort put into creating them is potentially higher, and the epistemological value potentially lower than in the preceding chapters. After a table chart we visualise data of the "Better Life Index" and from GAPMINDER. Finally we show a variant on the "probably best statistical graphic of all times" (Edward Tufte), a map of Napoleon's Russian campain in 1812/13, by Charles Joseph Minard from 1869.

11.1 Table with Symbols of the "Symbol Signs" Type Face

The 'Leaky Pipeline' 2005

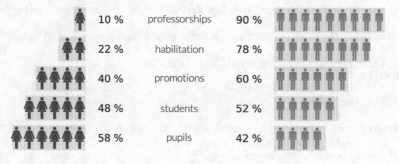

Source: Wissenschaftsrat, Drucksache Drs. 8036-07.

© Springer Nature Switzerland AG 2019
T. Rahlf, *Data Visualisation with R*, https://doi.org/10.1007/978-3-030-28444-2_11

The figure shows the downright striking "leaky pipeline", i.e., the decreasing proportion of women in science with increasing status. While women still make up the majority of pupils, their proportion continuously decreases with each additional step. In 2005, women only made up 10% of C4/W3 professorships. To visualise this fact, a back-to-back bar chart is a good choice, with women being shown on the left and men on the right. In line with Otto Neurath's idea, the percentages in the individual steps are shown as isotypes, red for women and blue for men. Every symbol stands for 10%, and we round either up or down. Data derive from a Wissenschaftsrat (Science Council) publication (Drs. 8036-07). Sander Baumann's typeface "Symbol Signs" is freely available for download from http://www.fontsquirrel.com/fonts/Symbol-Signs.

```
pdf_file<-"pdf/tablecharts_symbol_signs.pdf"
cairo_pdf(bg="grey98",pdf_file,width=9,height=4)

par(omi=c(0.5,0.25,0.5,0.25),mai=c(0,0,0,0),family="Lato Light",\
    cex=1.2)

# Import data

library(gdata)
myData<-read.xls("myData/leaking_pipeline.xlsx", encoding="\
    latin1")
attach(myData)

# Create graphics

b1<-barplot(Men+75,horiz=T,xlim=c(-175,175),border=NA,col="\
    gainsboro",axes=F)
barplot(-Women-75,horiz=T,border=NA,add=T,col="gainsboro",axes=F\
    )
barplot(rep(75,5),horiz=T,border=par("bg"),add=T,col=par("bg"),\
    axes=F)
barplot(rep(-75,5),horiz=T,border=par("bg"),add=T,col=par("bg"),\
    axes=F)
abline(v=seq(-175,195,by=10),col=par("bg"))
text(0,b1,Level)

# Titling

mtext("The 'Leaky Pipeline' 2005",3,line=0.25,adj=0,cex=1.75,\
    family="Lato Black",outer=T)
mtext("Source: Wissenschaftsrat, Drucksache Drs. 8036-07.",1,\
    line=0.25,adj=1.0,cex=0.65,outer=T,font=3)

# Symbols

par(family="Symbol Signs")
for (i in 1:5)
{
MyMen_Number<-Men[i]
```

```
text(seq(10,10*round(MyMen_Number/10),by=10)+73.5,rep(b1[i],5),
      rep("M",MyMen_Number),
  cex=2.75,col="cornflowerblue")
MyWomen_Number<-Women[i]
text(-seq(10,10*round(MyWomen_Number/10),by=10)-68,rep(b1[i],5),
      rep("F",MyWomen_Number),
  cex=2.75,col="deeppink")
}

par(family="Lato Bold")
text(55,b1,paste(Men, "%", sep=" "))
text(-55,b1,paste(Women, "%", sep=" "))
dev.off()
```

In the script, we superimpose four bar charts for our illustration, and print the text in the centre. To this end, the frequencies are shifted by 75 to either the left or right. This principle matches the approach for population pyramids from Sect. 7.2. The first barplot() call is saved in a variable b1, as we still require the bars' y-positions. We then draw the symbols for women and men in the "Symbol Signs" font over the bars. For this purpose, the values rounded to tens are used in increments of ten. As the glyphs are not centred, we have to add 73.5 for the men, and subtract 68 for the women. Both numbers are determined by trial and error. At the end, the percentages are added in "Lato Bold" font to the left or right of the labels.

11.2   Polar Area Charts with Labels (Panel)

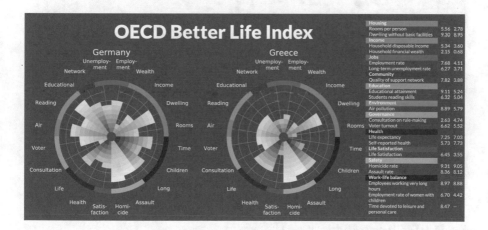

The presented polar area charts are a variant of the radial polygons used in previous examples. As in those, only the segments' radius is varied here, but not the angle. This means they are rotated column charts (just like the original Nightingale figures). It is generally possible to stack the segments. However, I do not recommend that practice, as it is easy to lose track that way. In the current and the next two examples, we present a variant with three design features:

- There is a hole in the centre. It does not have to be there, but it looks better than having the tips of all segments converge to a point.
- There is a rim on the outside. It does not have to be there either, but it is useful for orientation. The rim represents the border that cannot be crossed by the columns.
- The columns show a fading-out effect. This is only for visual reasons and causes a headlight effect with radial representation.

We use polar area charts created in this way to first present a panel figure with 2 charts, then a reduced variant with 16 charts, and finally a "poster" with a chart and explanatory text. In all three examples, we base the visualisation on data that the OECD calls the "Better Life Index". This comprises the indicators "Housing", "Income", "Jobs", "Community", "Education", "Environment", "Civic Engagement", "Health", "Life Satisfaction", "Safety", and "Work-Life Balance". The motivation behind this combination is an attempt to find the most comprehensive measure of life standard. Moritz Stefaner developed an award-winning interactive visualisation that presents the data in the shape of flower petals. This is what the data look like (Fig. 11.1).

YOUR BETTER LIFE INDEX - ORIGINAL DATA

TOPIC		Housing		Income		Jobs		Community	Education		Environment	Gov
INDICATOR		Rooms per person	Dwelling without basic facilities	Household disposable income	Household financial wealth	Employment rate	Long-term unemployment rate	Quality of support network	Educational attainment	Students reading skills	Air pollution	Consultation on rule-making
ISO3	COUNTRY											
AUS	Australia	2.4	n.a.	27.039	28745	72.30	1.00	95.4	69.72	515	14.28	10.
AUT	Austria	1.7	1.30	27.670	43734	71.73	1.13	94.6	81.04	470	29.03	7.
BEL	Belgium	2.3	0.60	26.008	69487	62.01	4.07	92.6	69.58	506	21.27	4.
CAN	Canada	2.5	n.a.	27.015	59479	71.68	0.97	95.3	87.07	524	15.00	10.
CHL	Chile	1.3	9.36	8.712	n.a.	59.32	n.a.	85.2	67.97	449	61.55	2.
CZE	Czech Republic	1.3	0.70	16.690	12685	65.00	3.19	88.9	90.90	478	18.50	6.
DNK	Denmark	1.9	0.00	22.929	27180	73.44	1.44	96.8	74.56	495	16.26	7.
EST	Estonia	1.2	12.20	13.486	11202	61.02	7.84	84.6	88.48	501	12.62	3.
FIN	Finland	1.9	0.80	24.246	18616	68.15	2.01	93.4	81.07	536	14.87	9.
FRA	France	1.8	0.80	27.508	42253	63.99	3.75	93.9	69.96	496	12.94	3.
DEU	Germany	1.7	1.20	27.665	45113	71.10	3.40	93.5	85.33	497	16.21	4.
GRC	Greece	1.2	1.80	21.499	15856	59.55	5.73	86.1	61.07	483	32.00	6.
HUN	Hungary	1.0	7.10	13.858	11426	55.40	5.68	88.4	79.70	494	15.60	7.
ISL	Iceland	1.6	0.30	n.a.	n.a.	78.17	1.35	97.6	64.13	500	14.47	5.
IRL	Ireland	2.1	0.30	24.313	23072	59.96	6.74	97.3	69.45	496	12.54	9.*
ISR	Israel	1.1	n.a.	n.a.	62684	59.21	1.85	93	81.23	474	27.57	2.*

Fig. 11.1 OECD-provided data "Better Life Index"

About the figure: In this example, 20 indicators in 11 groups are presented for
Germany and Greece. Colours match the colours that the OECD use for the groups
in their XLS table. In this example, indicators are used as labels and attached to
the outer radius in abbreviated form; the longer form is used for the group titles in
the legend on the right. Additionally, the values are output in two columns. In my
opinion, column titles can be omitted at this point. As stated earlier, our polar area
charts feature a hole in the centre, an outer rim that specifies the maximal possible
values, and a fading-out effect. For ease of orientation, gridlines are also added.
As the figure clearly shows, Germany has higher values than Greece for a series of
indicators. In other areas, such as security and the low proportion of apartments with
sanitation (= high index value), differences are moderate.

Data: see annex A, BetterLifeIndex_Data_2011V6.xls.

```
pdf_file<-"pdf/radial_columncharts_1x2_inc.pdf"
cairo_pdf(bg=rgb(0.3137,0.3137,0.3137),pdf_file,width=14,height\
     =7)

par(omi=c(0,0,0,0),family="Lato Regular",mfcol=c(1,2),cex.axis\
     =1.25, col.lab=par("bg"))
library(plotrix)
library(gdata)

mySelection<-c("DEU","GRC")
source("scripts/inc_data_radial_columncharts.r")
source("scripts/inc_colours_radial_columncharts.r")
myGridColour<-"grey"
myRadial_mar<-c(4,0,6,0)
myMLine<-4
myNames<-c(
  "Rooms",
  "Dwelling",
  "Income",
  "Wealth",
  "    Employ-\nment",
  "Unemploy-    \nment",
  "Network    ",
  "Educational",
  "Reading",
  "Air",
  "Voter",
  "Consultation",
  "Life",
  "Health",
  "Satis-\nfaction",
  "Homi-\ncide",
  "Assault",
  "Long",
  "Children","Time")
source("scripts/inc_plot_radial_columncharts.r")
dev.off()
```

The following code is integrated:

```
# inc_data_radial_columncharts.r
myData<-read.xls("myData/BetterLifeIndex_Data_2011V6.xls",\
      pattern="ISO3",sheet=3, encoding="latin1")
n<-nrow(myData)-2
myData<-myData[1:n, ]
myData<-myData[,c(1:12,14,13,15:length(myData))]

row.names(myData)<-myData$ISO3
myDataMargin<-NULL
myDataCenter<-NULL
for (i in 3:length(myData))
{
myDataMargin<-c(myDataMargin,list(11.5:12.5))
myDataCenter<-c(myDataCenter,list(0:1))
}
```

and

```
# inc_colours_radial_columncharts.r
myC1<-rgb(0.2392,0.6480,0.5804)
myC2<-rgb(0.1725,0.6392,0.8784)
myC3<-rgb(0.1451,0.4980,0.7412)
myC4<-rgb(0.8078,0.2824,0.3647)
myC5<-rgb(0.4941,0.6627,0.2627)
myC6<-rgb(0.1882, 0.6431, 0.3412)
myC7<-rgb(0.8627, 0.6627, 0.1333)
myC8<-rgb(0.4863,0.2275,0.4510)
myC9<-rgb(0.8863,0.3843,0.2157)
myC10<-rgb(0.6765,0.6765,0.6765)
myC11<-rgb(0.5882,0.1569,0.1569)

myRColours<-c(myC1,myC1,myC2,myC2,myC3,myC3,myC4,myC5,myC5,myC6,\
      myC7,myC7,myC8,myC8,myC9,myC10,myC10,myC11,myC11,myC11)
myMColours<-rep(par("bg"),length(myRColours))
```

as well as

```
# inc_plot_radial_columncharts.r

# go through all data in the selection

for (i in 1:length(mySelection)) {
myCountry<-myData[mySelection[i],3:length(myData)]+1
myCountry[is.na(myCountry)]<--1     # replacing missing values

# ... produce a list for every country

myCountryData<-NULL
for (j in 1:length(myCountry))
{
myCountryData<-c(myCountryData,list(0:as.numeric(myCountry[j])))
}
radial.pie(myDataMargin,label.prop=1.35,show.grid.labels=F,boxed\
      .radial=F,show.radial.grid=F,grid.bg=par("bg"),grid.col=\
```

```
       myGridColour,labels=myNames,sector.colors=myRColours,mar=↘
       myRadial_mar,radial.lim=c(0,11),xpd=T)

radial.pie(myCountryData,label.prop=1.35,show.grid.labels=F,↘
       boxed.radial=F,show.radial.grid=T,grid.bg=par("bg"),grid.↘
       col=myGridColour,labels=myNames,sector.colors=myRColours,↘
       mar=myRadial_mar,radial.lim=c(0,11),xpd=T,add=T)

radial.pie(myDataCenter,label.prop=1.35,show.grid.labels=F,boxed↘
       .radial=F,show.radial.grid=F,grid.bg=par("bg"),grid.col=↘
       myGridColour,labels=myNames,sector.colors=myMColours,mar=↘
       myRadial_mar,radial.lim=c(0,11),xpd=T,add=T)

myTitle<-myData[mySelection[i], "COUNTRY"]
mtext(myTitle,3,line=myMLine,adj=0.5,cex=2,col="white")
}
```

We also require a LaTeX file:

```
\documentclass{article}
\usepackage[paperheight=14cm, paperwidth=29.7cm, top=0.75cm, ↘
       left=0cm, right=0cm, bottom=0cm]{geometry}
\usepackage{colortbl}
\usepackage[abs]{overpic}
\usepackage{fontspec}
\setmainfont[Mapping=text-tex]{Lato Black}
\definecolor{hintergrund}{rgb}{0.3137,0.3137,0.3137}
\pagecolor{hintergrund}
\begin{document}
\pagestyle{empty}
\textcolor{white}{
\fontsize{36pt}{11pt}\selectfont
\vspace{0.35cm}\hspace{5.5cm}OECD Better Life Index}
\begin{center}
\setmainfont[Mapping=text-tex]{Lato Regular} \fontsize{9pt}{11pt↘
       }\selectfont
\hspace{-6.5cm}
\begin{overpic}[scale=0.65,unit=1mm]{../pdf/radial_columncharts_↘
       1x2_inc.pdf}
\put(229,67){\begin{minipage}[t]{16cm}
\textcolor{white}{
\begin{tabular}{p{4.5cm}p{0.41cm}p{0.41cm}}
\cellcolor[rgb]{0.2392,0.6480,0.5804}\textbf{Housing} &   \\
\raggedright Rooms per person &   5.56 & 2.78   \\
\raggedright Dwelling without basic facilities & 9.30 & 8.95 \\
\cellcolor[rgb]{0.1725,0.6392,0.8784}\textbf{Income} &   \\
\raggedright Household disposable income & 5.34 & 3.60 \\
\raggedright Household financial wealth & 2.15 & 0.68 \\
\cellcolor[rgb]{0.1451,0.4980,0.7412}\textbf{Jobs} &   \\
\raggedright Employment rate &   7.68 & 4.11 \\
\raggedright Long-term unemployment rate & 6.27 & 3.71 \\
\cellcolor[rgb]{0.8078,0.2824,0.3647}\textbf{Community} &   \\
\raggedright Quality of support network & 7.82 & 3.88\\
\cellcolor[rgb]{0.4941,0.6627,0.2627}\textbf{Education} & \\
\raggedright Educational attainment &   9.11 & 5.24\\
```

```
\raggedright Students reading skills & 6.32 & 5.04 \\
\cellcolor[rgb]{0.1882, 0.6431, 0.3412}\textbf{Environment} & \\
\raggedright Air pollution &   8.89 & 5.79 \\
\cellcolor[rgb]{0.8627, 0.6627, 0.1333}\textbf{Governance} & \\
\raggedright Consultation on rule-making & 2.63 & 4.74 \\
\raggedright Voter turnout &    6.62 & 5.52\\
\cellcolor[rgb]{0.4863,0.2275,0.4510}\textbf{Health} & \\
\raggedright Life expectancy &  7.25 & 7.03\\
\raggedright Self-reported health &   5.73 & 7.73\\
\cellcolor[rgb]{0.8863,0.3843,0.2157}\textbf{Life Satisfaction}  \
     & \\
\raggedright Life Satisfaction & 6.45 & 3.55\\
\cellcolor[rgb]{0.6765,0.6765,0.6765} \textbf{Safety} & \\
\raggedright Homicide rate &    9.31 & 9.05\\
\raggedright Assault rate &  8.36 & 8.12\\
\cellcolor[rgb]{0.5882,0.1569,0.1569} \textbf{Work-life balance}\
     & \\
\raggedright Employees working very long hours &  8.97 & 8.88\\
\raggedright Employment rate of women with children &   6.70 & \
     4.42\\
\raggedright Time devoted to leisure and personal care & 8.47 & \
     -\\
\end{tabular}
}
\end{minipage}}
\end{overpic}
\end{center}
\end{document}
```

In the script, we make use of the option for embedding code from an external file with the source() function, as we previously did in other examples. This also means that we can use a large proportion of the script in our next example. We use the radial.pie() function from Jim Lemon's plotrix package for our polar area charts. We begin by specifying a vector selection with the values GER and GRC. The data frame is then read via the inc_data_polar_area_chart.r file. Normalised values can be found in the third sheet. We only need the data part, and therefore only read data from the row that has ISO3 in its first column. We also do not need the last two rows, as they contain the minmum and maximum. Next, columns 13 and 14 are swapped because the labels would otherwise overlap in our figure. As both items belong to the same group, there are no further consequences. We create row labels from the ISO3 code and then create two variables myDatamargin and myDatacentre that are filled with two values, respectively, as a list. The first list will later set the outer border, the second the centre point. We then embed the inc_colours_polar_area_chart.r file, which specifies the colours. Every one of the 11 groups contains a colour. Using the group colours, we set the individual column colours via myRColours; the centre is set to match the background colour. This means we require the same number of elements as for the border colours, even though all individual segments have the same colour. Back in the calling script, myGridColour is specified. Since we will be using the same code in our next example but will not be adding a grid then, we define the colours as a variable in both examples. The same goes for myRadial_mar, which is a variable we use for specifying the distances within the radial.pie function, and for myMline, which sets the vertical position for figure titles. The abbreviated labels for the segments are specified next, before calling the inc_plot_polar_area_chart.r file, which plots the figures. We start with a loop that passes all elements of selection, here, the elements GER and GRC. All values of the country's index variables are first written into a variable myCountry; each value is increased by 1, since we start with a circle with the radius of 1 in the centre (the "hole"). Missing values have to be replaced with 1 (or 0). In the next step, a list myCountrydata containing the index values is generated for each country. We now have three lists (myDataMargin, myCountryData, and myDataCenter) that are handed over to the radial.pie() function. The second and third call are done with the add=T parameter, so that all three are superimposed over each other: first the border, then the actual data, and last the central "hole". At the end, the name of the country, which is saved in the COUNTRY column, is output as title. We have now created a figure with two polar area charts. What is left is to embed the result into LaTeX, which is where we generate our legend. The procedure is described in Sect. 4.1. We begin by loading the colortable package for colouring the cell backgrounds. The other commands match the previous examples. However, this time, we make use of the option for shifting text vertically or horizontally using \vspace{} or \hspace{} on multiple occasions. After integration of the figure with overpic, we use \put(229,67){\begin{minipage}[t]{16cm}...} to specify the position at which we can generate a table-shaped legend using the tabular environment. The \cellcolor command gives the option to use the same RGB values as specified in the R script.

11.3   Polar Area Charts Without Labels (Panel)

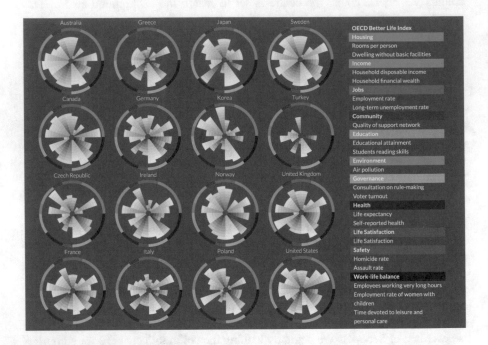

About the figure: In this case, we display 16 countries. With so many charts, the representation should be honed down. This is why we omit grid lines and individual graphic labels; the legend is limited to the labels. In this representation form, significant differences between countries stand out. Especially Turkey shows comparatively low values for a series of indicators.

Data: see annex A, BetterLifeIndex_Data_2011V6.xls

```
pdf_file<-"pdf/radial_columncharts_4x4_inc.pdf"
cairo_pdf(bg=rgb(0.3137,0.3137,0.3137), pdf_file,width=18,height⌄
    =18)

par(omi=c(0,0,0,0),family="Lato Light",mfcol=c(4,4))
library(plotrix)
mySelection<-c("AUS","CAN","CZE","FRA","GRC","DEU","IRL","ITA","⌄
    JPN","KOR", "NOR","POL","SWE","TUR","GBR","USA")
source("scripts/inc_data_radial_columncharts.r")
source("scripts/inc_colours_radial_columncharts.r")
myGridColour<-par("bg")
myRadial_mar<-c(0,0,2,0)
myMLine<-0
myNames<-rep("", (length(myData)-2))
source("scripts/inc_plot_radial_columncharts.r")
dev.off()
```

The LaTeX file is as follows:

```
\documentclass{article}
\usepackage[paperheight=21cm, paperwidth=29.7cm, top=0.25cm, \
    left=0cm, right=0cm, bottom=0cm]{geometry}
\usepackage{graphicx,color}
\usepackage{fontspec}
\usepackage{colortbl}
\usepackage[abs]{overpic}
\setmainfont[Mapping=text-tex]{Lato Regular}
\definecolor{hintergrund}{rgb}{0.3137,0.3137,0.3137}
\definecolor{text}{gray}{.95}
\pagecolor{hintergrund}
\begin{document}
\pagestyle{empty}
\begin{center}
\fontsize{11pt}{17pt}\selectfont
\hspace{-8cm}
\begin{overpic}[scale=0.45,unit=1mm]{../pdf/radial_columncharts_\
    4x4_inc.pdf}
\put(213,100){\begin{minipage}[t]{16cm}
\textcolor{text}{
\begin{tabular}{p{6.0cm}p{0.01cm}}
\textbf{OECD Better Life Index}\\
\cellcolor[rgb]{0.2392,0.6480,0.5804}\textbf{Housing} &  \\
\raggedright Rooms per person &    \\
\raggedright Dwelling without basic facilities &  \\
\cellcolor[rgb]{0.1725,0.6392,0.8784}\textbf{Income} &  \\
\raggedright Household disposable income &   \\
\raggedright Household financial wealth &  \\
\cellcolor[rgb]{0.1451,0.4980,0.7412}\textbf{Jobs} &  \\
\raggedright Employment rate &   \\
\raggedright Long-term unemployment rate &  \\
\cellcolor[rgb]{0.8078,0.2824,0.3647}\textbf{Community} &  \\
\raggedright Quality of support network &  \\
\cellcolor[rgb]{0.4941,0.6627,0.2627}\textbf{Education} &  \\
\raggedright Educational attainment &   \\
\raggedright Students reading skills &  \\
\cellcolor[rgb]{0.1882, 0.6431, 0.3412}\textbf{Environment} &  \\
\raggedright Air pollution &   \\
\cellcolor[rgb]{0.8627, 0.6627, 0.1333}\textbf{Governance} &  \\
\raggedright Consultation on rule-making &  \\
\raggedright Voter turnout &   \\
\cellcolor[rgb]{0.4863,0.2275,0.4510}\textbf{Health} &  \\
\raggedright Life expectancy &  \\
\raggedright Self-reported health &   \\
\cellcolor[rgb]{0.8863,0.3843,0.2157}\textbf{Life Satisfaction} \
    & \\
\raggedright Life Satisfaction &  \\
\cellcolor[rgb]{0.3765,0.3765,0.3765} \textbf{Safety} &  \\
\raggedright Homicide rate &   \\
\raggedright Assault rate &  \\
\cellcolor[rgb]{0.6882,0.1560,0.1560} \textbf{Work-life balance}\
    & \\
\raggedright Employees working very long hours &   \\
```

```
\raggedright Employment rate of women with children &   \\
\raggedright Time devoted to leisure and personal care & \\
\end{tabular}
}
\end{minipage}}
\end{overpic}
\end{center}
\end{document}
```

In the script, we essentially reuse the last example's elements. We begin by specifying a 4×4 figure layout using mfcol, and then select 16 ISO3 abbreviations as country selection. That done, the files for data import and colour specification that were described in the previous example are read. In this example, myGridColour is set to the background colour so that grid lines are invisible, and borders are set narrower than in the previous example. The title's vertical position is 0, the baseline, in this example. Labels are suppressed by leaving the names vector empty. With these specifications, the file that plots the polar area charts (16 in this case) is called as in the previous example. Finally, the generated figure is again embedded into a LaTeX file that largely matches the one from the last example; however, we do not specify numbers in the table legend.

```
# inc_plot_radial_columncharts.r

# go through all data in the selection

for (i in 1:length(mySelection)) {
myCountry<-myData[mySelection[i],3:length(myData)]+1
myCountry[is.na(myCountry)]<-1    # replacing missing values

# ... produce a list for every country

myCountryData<-NULL
for (j in 1:length(myCountry))
{
myCountryData<-c(myCountryData,list(0:as.numeric(myCountry[j])))
}
radial.pie(myDataMargin,label.prop=1.35,show.grid.labels=F,boxed\
      .radial=F,show.radial.grid=F,grid.bg=par("bg"),grid.col=\
      myGridColour,labels=myNames,sector.colors=myRColours,mar=\
      myRadial_mar,radial.lim=c(0,11),xpd=T)

radial.pie(myCountryData,label.prop=1.35,show.grid.labels=F,\
      boxed.radial=F,show.radial.grid=T,grid.bg=par("bg"),grid.\
      col=myGridColour,labels=myNames,sector.colors=myRColours,\
      mar=myRadial_mar,radial.lim=c(0,11),xpd=T,add=T)

radial.pie(myDataCenter,label.prop=1.35,show.grid.labels=F,boxed\
      .radial=F,show.radial.grid=F,grid.bg=par("bg"),grid.col=\
      myGridColour,labels=myNames,sector.colors=myMColours,mar=\
      myRadial_mar,radial.lim=c(0,11),xpd=T,add=T)

myTitle<-myData[mySelection[i], "COUNTRY"]
```

```
mtext(myTitle,3,line=myMLine,adj=0.5,cex=2,col="white")
}
```

11.4   Polar Area Chart (Poster)

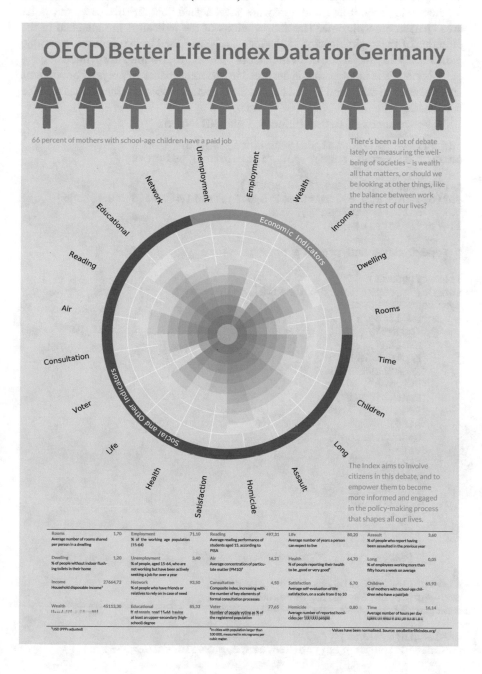

The figure is based on a template by DensityDesign, a "Research Lab" in the Design Department of the Polytechnical University of Milan. Gregor Aisch has raised awareness of a series of errors in the details of the figure; overall though, it is a successful visualisation. In this instance, we limit ourselves to one country and supplement the graphic with a person bar indicating the percentage of working mothers with school-age children who have a paid job. In this case, the polar area chart is monochrome. The abbreviations of the indicators are attached in the direction of the angle bisector. Indicators can be assigned to two groups: "conomic indicators" and "social and other indicators". These group names are printed into the outer border. The outer border is designed in different shades of grey. Abbreviations are explained in a detailed legend below; the non-normalised original values are also listed here.

Data: see annex A, BetterLifeIndex_Data_2011V6.xls.

```
pdf_file<-"pdf/radial_columncharts_inc.pdf"
cairo_pdf(bg=par("bg"), pdf_file,width=9,height=9)

par(bg=rgb(0.8627,0.8627,0.8627))
par(omi=c(0,0,0,0),mai=c(0,2,0,0.5),family="Droid Sans Mono", ⌍
    col.lab=par("bg"))
library(plotrix)

# Import data and prepare chart

library(gdata)
source("scripts/inc_data_radial_columncharts.r")
myNames<-c("Rooms",
    "Dwelling",
    "Income",
    "Wealth",
    "Employment",
    "Unemployment",
    "Network",
    "Educational",
    "Reading",
    "Air",
    "Consultation",
    "Voter",
    "Life",
    "Health",
    "Satisfaction  ",
    "  Homicide",
    "Assault",
    "Long",
    "Children",
    "Time")

g1<-rgb(150,150,150,maxColorValue=255)
g2<-rgb(80,80,80,maxColorValue=255)
m<-rgb(180,180,180,maxColorValue=255)

n<-length(myNames)
myRadialColours<-rep(rgb(255,128,0,maxColorValue=255),n)
```

```
myMarginColours<-c(rep(g1,6),rep(g2,n-6))
myCenterColours<-rep(m,n)

radial_mar<-c(7,7,7,7)

myCountry<-myData["DEU",3:length(myData)]+1
myCountry[is.na(myCountry)]<-1

myCountryData<-NULL

for (j in 1:length(myCountry)) myCountryData<-c(myCountryData,↘
    list(0:as.numeric(myCountry[j])))

# Create chart

radial.pie(myDataMargin,label.prop=1.35,show.grid.labels=F,↘
      radlab=T,boxed.radial=F,show.radial.grid=F,grid.bg=par("bg↘
      "),grid.col="white",labels=myNames,sector.colors=↘
      myMarginColours,mar=radial_mar,radial.lim=c(0,11),xpd=T)

radial.pie(myCountryData,label.prop=1.35,show.grid.labels=F,↘
      radlab=T,boxed.radial=F,show.radial.grid=T,grid.bg=par("bg↘
      "),grid.col="white",labels=myNames,sector.colors=↘
      myRadialColours,mar=radial_mar,radial.lim=c(0,11),xpd=T,↘
      add=T)

radial.pie(myDataCenter,label.prop=1.35,show.grid.labels=F,↘
      radlab=T,boxed.radial=F,show.radial.grid=F,grid.bg=par("bg↘
      "),grid.col="white",labels=myNames,sector.colors=↘
      myCenterColours,mar=radial_mar,radial.lim=c(0,11),xpd=T,↘
      add=T)

par(family="Droid Sans Mono")
arctext("Economic Indicators",col="white",radius=12,start=1.3)
arctext("Social and Other Indicators",col="white",radius=12,↘
      start=4.3)
dev.off()
```

The integration into LaTeX is as follows:

```
\documentclass{article}
\usepackage[paperheight=29.7cm, paperwidth=21cm, top=1cm, left↘
      =0.5cm, right=1cm, bottom=0cm]{geometry}
\usepackage{graphicx,color}
\usepackage{fontspec}
\usepackage[abs]{overpic}
\setmainfont[Mapping=text-tex]{Lato Regular}
\definecolor{hintergrund}{rgb}{0.8627,0.8627,0.8627}
\definecolor{text}{gray}{.55}
\newcommand{\titel}[1]{{\fontspec[Color=808080, Scale=3]{Lato ↘
      Bold} #1}}
\newcommand{\maennerg}[1]{{\fontspec[Color=808080, Scale=1.5]{↘
      Symbol Signs} #1}}
\newcommand{\maennero}[1]{{\fontspec[Color=FF8000, Scale=1.5]{↘
      Symbol Signs} #1}}
```

```
\pagecolor{hintergrund}
\linespread{1.2}
\begin{document}
\pagestyle{empty}
\vspace*{-0.5cm}
 \hspace*{0cm}\titel{OECD Better Life Index Data for Germany↘
      }\\[0.2cm]
\fontsize{72pt}{20pt}\selectfont\maennerg{FF}\maennero{FFFFFF}\\↘
      maennerg{F}\\
\vspace*{-0.8cm}
\begin{figure}[h!]
\begin{overpic}[scale=0.82,unit=1mm]{../pdf/radial_columncharts_↘
      inc.pdf}
\put(0,185)
{\begin{minipage}[t]{10cm}
\raggedright \textcolor{text}{66 percent of mothers with school-↘
      age children have a paid job}
\end{minipage}}
\put(152,185)
{\begin{minipage}[t]{4.6cm}
\raggedright \textcolor{text}{There's been a lot of debate ↘
      lately on measuring the well- being of societies - is ↘
      wealth all that matters, or should we be looking at other ↘
      things, like the balance between work and the rest of our ↘
      lives?}
\end{minipage}}
\put(152,30)
{\begin{minipage}[t]{4.6cm}
\raggedright \textcolor{text}{The Index aims to involve citizens↘
      in this debate, and to empower them to become more ↘
      informed and engaged in the policy-making process that ↘
      shapes all our lives.}
\end{minipage}}
\put(5,-55)
{\begin{overpic}[scale=0.65,unit=1mm]{tables_multivariate_↘
      manually.pdf}
\end{overpic}}
\end{overpic}
\end{figure}
\end{document}
```

The script's structure essentially matches the previous examples. First, data are read and abbreviations are specified. In contrast to the previous examples, we only need two colours for the border and one for the centre. As before, borders are individually set and then a vector myCountry is generated; this vector is converted into a list for use in the radial.pie() function. This function is called three times: The first call plots the outer border; here, we use radlab=T to attach the labels in the direction of the angle bisector of the sectors. The second call generates the actual polar area charts, the third one the "hole" in the centre. Last, labels for the two groups "conomic indicators" and "social and other indicators" are output into the outer border; we use the non-proportional font "Droid Sans Mono", which is freely available from http://www.fontsquirrel.com/fonts/droid-sans-mono. The completed figure is integrated into a LaTeX file. In this case, the legend is a little more extensive, so we forgo detailed explanations and refer to the source text at https://link.springer.com/chapter/10.1007/978-3-030-28444-2_1 instead. The legend is also a file (separately) created in LaTeX, which is integrated using the overpic environment like the PDF generated by R.

11.5   Nighttime Map of Germany as Scatter Plot

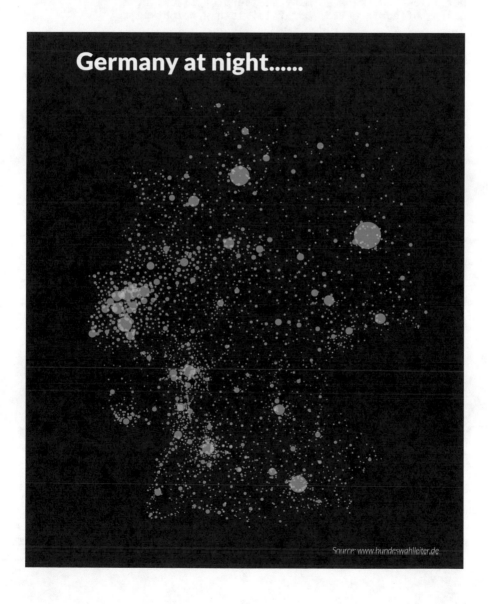

About the figure: Obviously, this is not really a night map, but rather an effect illustration. We choose black for the background. The radius is the square root of the population magnitude of the approximately 3000 German municipalities with population of more than 5000. A transparent yellow was selected as point colour to mimic luminance. As this is meant to be a "pseudo night image", we also omit a legend or title. Given the large number and spread of points, Germany as map is recognisable without an outline, just as it was in Sect. 9.2.3.

Data: see annex A, v_frauen_maenner.

```
pdf_file<-"pdf/maps_germany_scatterplot_black.pdf"
cairo_pdf(bg="grey98", pdf_file,width=9,height=11)

par(bg="black",mai=c(0,0,0,0),omi=c(0.5,1,1,0.5),family="Lato ↘
     Light",las=1)
library(RColorBrewer)

# Import data

source("scripts/inc_datadesign_dbconnect.r")
sql<-"select * from v_women_men"
myDataset<-dbGetQuery(con,sql)
attach(myDataset)

# Create chart

plot(lng,lat,type="n",axes=F)
myColour<-rgb(255,255,0,100,maxColorValue=255)
points(lng,lat,pch=19,col=myColour,cex=4*sqrt(bevinsg),lwd=0)

# Titling

mtext("Germany at night......",side=3,adj=0,line=1,cex=3,family=↘
     "Lato Black",col="white")
mtext("Source: www.bundeswahlleiter.de",side=1,adj=1,line=0,cex↘
     =1,font=3,col="white")
dev.off()
```

The script can be limited to a few lines in this instance. We colour the background black. Data are read from the MySQL table, a plot() is defined, and a transparent yellow hue is set. Points are plotted as a scatter plot without axes using points(); the point size is the square root of the total population, multiplied by a factor of 4. We use the square root, as doubling the radius of a circle causes quadrupling of its area, and we are therefore left with proportional areas.

11.6   Scatter Plot Gapminder

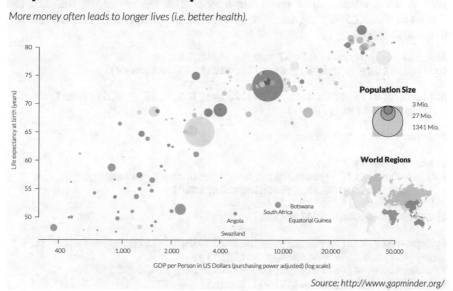

Source: http://www.gapminder.org/

About the figure: Hardly any statistics visualisations of the past years have gained as much attention as the animated figures by Hans Rosling on http://www.gapminder.org. In these, macrostatistical parameters for multiple countries are correlated from a historical perspective. Animated scatter plots for two selected variables can be watched on the website, with the locations of the points varying as a function of time. This allows impressive visualisations of, e.g., the development of the relationship between national income and life expectancy between 1800 and the present day. Aside from the animated versions, the website also offers a more elaborate static figure as a PDF version. This figure serves as a guide for our example, even though we present a slightly "slimmer" version. Its most important elements are:

- The income on the x-axis is represented logarithmically.
- The population size is taken in as a third variable and defines point size.
- The point size is explained in a legend.
- The point colour indicates the world region in which the country is located.
- The world region is explained in a second legend, in the form of a world map.
- Some individual points (in the original: all points) are labelled.

Data: The XLS files for the scatter plot can be downloaded from http://www.
gapminder.org/data/. The allocation of countries to continental regions is not
provided as an XLS file, but as a list that can be found by clicking on "Gapminder
Geographic Regions" in the figure's legend at the top right. This opens a table in
your browser, which can be saved into a local XLS file via copy/paste. You can
download the XLS file regions.xlsx at https://link.springer.com/chapter/10.1007/
978-3-030-28444-2_1.

```
pdf_file<-"pdf/scatterplots_gapminder.pdf"
cairo_pdf(bg="grey98", pdf_file,width=13,height=9)

par(omi=c(0.25,0.25,1.25,0.25),mai=c(1.5,0.85,0,0.5),family="\
     Lato Light",las=1)

# Import data and prepare chart

library(gdata)
myGdp<-read.xls("myData/gapminder/indicatorgapmindergdp_per_\
     capita_ppp.xls", encoding="latin1")
mySelection<-c("X","X2010")
myGdp2010<-myGdp[mySelection]

myExp<-read.xls("myData/gapminder/indicatorlife_expectancy_at_\
     birth.xls", encoding="latin1")
mySelection<-c("Life.expectancy.at.birth","X2010")
myExp2010<-myExp[mySelection]

myGdpExp2010<-merge(myGdp2010,myExp2010,by.x="X",by.y="Life.\
     expectancy.at.birth",all =T)

myPop<-read.xls("myData/gapminder/indicatorgapminderpopulation.\
     xls",dec=".", encoding="latin1")
mySelection<-c("Total.population","X2010")
myPop2010<-myPop[mySelection]

myGdpExpPop2010<-merge(myGdpExp2010,myPop2010,by.x="X",by.y="\
     Total.population",all =T)

myRegions<-read.xls("myData/gapminder/regions.xlsx", encoding="\
     latin1")

myData<-merge(myGdpExpPop2010,myRegions,by.x="X",by.y="Entity",\
     all =T)
myData<-na.omit(myData)

attach(myData)
X2010<-as.numeric(gsub(",","",X2010))/10000000

xmax<-round(max((X2010)),1)
x75<-round(quantile((X2010),probs=0.75),1)
x25<-round(quantile((X2010),probs=0.25),1)

xmax_leg<-round(max((X2010)^0.5)/3,1)
```

```
x75_leg<-round(quantile((X2010)^0.5,probs=0.75)/3,1)
x25_leg<-round(quantile((X2010)^0.5,probs=0.25)/3,1)

mySize<-(X2010)^0.5
myData$mySize<-mySize

myOld<-c("Sub-Saharan Africa","South Asia","Middle East & North ⬎
     Africa",
   "America","Europe & Central Asia","East Asia & Pacific")
myNew<-c(rgb(0,115,157,150,maxColorValue=255),
     rgb(158,202,229,150,maxColorValue=255),
     rgb(84,196,153,150,maxColorValue=255),
     rgb(255,255,0,150,maxColorValue=255),
     rgb(246,161,82,150,maxColorValue=255),
     rgb(255,0,0,150,maxColorValue=255))
myColours<-as.character(Group)
for (i in 1:length(myOld)) {myColours[myColours == myOld[i]]<-⬎
     myNew[i]}

# Define chart and other elements

plot(log10(X2010.x),X2010.y,type="n",axes=F,xlab="",ylab="")
points(log10(X2010.x),X2010.y,cex=mySize,pch=19,col=myColours,⬎
     lwd=0)
axis(1,at=log10(c(200,400,1000,2000,4000,10000,20000,50000)),⬎
     label=format(c(200,400,1000,2000,4000,10000,20000,50000),⬎
     big.mark=".")) 
axis(2)
title(xlab="GDP per Person in US Dollars (purchasing power ⬎
     adjusted) (log scale)",ylab="Life expectancy at birth (⬎
     years)",font=3)

myFit<-lm(X2010.y ~ log10(X2010.x))
myData$resid<-residuals(myFit)
myData$myFit<-fitted(myFit)

myData.sort<-myData[order(-abs(myData$resid)) ,]
myData.sort_begin<-myData.sort[1:5,]

attach(myData.sort_begin)
text(log10(X2010.x),X2010.y,X,cex=0.95,pos=1,offset=0.8)

# Titling

mtext("Gapminder World Map 2010",3,line=3,adj=0,cex=3,family="⬎
     Lato Black",outer=T)
mtext("More money often leads to longer lives (i.e. better ⬎
     health). ",3,line=0,adj=0,cex=1.75,font=3,outer=T)
mtext("Source: http://www.gapminder.org/",1,line=5.5,adj=1.0,cex⬎
     =1.55,font=3)

text(log10(30000),72.5,"Population Size",family="Lato Black",cex⬎
     =1.35,adj=0)
text(log10(65000),70,paste(10*x25," Mio.",sep=""),adj=0)
```

```
text(log10(65000),68,paste(10*x75," Mio.",sep=""),adj=0)
text(log10(65000),66,paste(10*xmax," Mio.",sep=""),adj=0)

# Map Legend

library(mapplots)
legend.bubble(log10(45000),67,z=c(x25_leg,x75_leg,xmax_leg*0.7),↘
      maxradius=xmax_leg*0.7,bg=NA,txt.cex=0.01,txt.col=NA,pch↘
      =21,pt.bg="#00000020",bty="n",round=1)

# Integration of the chart

par(new=T, mai=c(1,9,3.5,0.75))
library(maptools) # contains  wrld_simpl
library(rgdal) # for spTransform

data(wrld_simpl)
myW<-wrld_simpl[wrld_simpl@data[,"NAME"] != "Antarctica",]
m<-spTransform(myW,CRS=CRS("+proj=merc"))

myCountries<-m@data$ISO2
n<-length(myCountries)
myMapColours<-numeric(n)

myR1<-"Sub-Saharan Africa"
myR2<-"South Asia"
myR3<-"Middle East &amp; North Africa"
myR4<-"America"
myR5<-"Europe & Central Asia"
myR6<-"East Asia & Pacific"

myF1<-rgb(0,115,157,150,maxColorValue=255)
myF2<-rgb(158,202,229,150,maxColorValue=255)
myF3<-rgb(84,196,153,150,maxColorValue=255)
myF4<-rgb(255,255,0,150,maxColorValue=255)
myF5<-rgb(246,161,82,150,maxColorValue=255)
myF6<-rgb(255,0,0,150,maxColorValue=255)

myRegion<-c(myR1,myR2,myR3,myR4,myR5,myR6)
myColour<-c(myF1,myF2,myF3,myF4,myF5,myF6)

myRegions<-read.xls("myData/gapminder/regions.xlsx", encoding="↘
      latin1")

for (i in 1:length(myRegion))
{
myRegionSelection<-subset(myRegions$ID,myRegions$Group==myRegion↘
      [i])
myCountrySelection<-NULL
for (j in 1:length(myRegionSelection)) myCountrySelection<-c(↘
      myCountrySelection, trim(as.character(myRegionSelection[j↘
      ])))
for (j in 1:length(myCountrySelection))
{
```

```
myMapColours[grep(paste("^",myCountrySelection[j],"$",sep=""),↘
    myCountries)]<-myColour[i]
}
}

plot(m,col=myMapColours,border=F, bg=NA)
mtext("World Regions",3,line=-2,adj=0.5,cex=1.25,family="Lato ↘
    Black")

dev.off()
```

In the script, we start by reading the necessary data from four different XLS tables; data are linked using the merge() function. To do this, similar to an SQL command, we have to specify the variables whose values are to be used to link data. all=T causes data without a link counterpart to be included in the result as well. The data frames structure arranges countries by rows, and years by columns. The year of interest for us is therefore located in a column with the header "2010". The header of the country row is the respective file name: for, e.g., indicatorlife_expectancy_at_birth.xls, the header reads "Life expectancy at birth". As variables in R are not allowed to contain spaces and have to start with a letter, R generates a variable in the form of Life.expectancy.at.birth and X2010. Since we are linking three data frames each of which contains a variable called X2010, and since variable names have to be clearly differentiable, the variables X2010 are renamed to x2010.x and X2010.y during the linking process of the first two files. When this new dataframe is then linked to the third one, the X2010 file name from the third data frame can be maintained, as the first two have been renamed. The variables are then

```
X2010.x
GDP per person in $US (purchasing power adjusted)
X2010.y
Life Expectancy at birth (years)
X2010
Total Population
```

For the legend, we specify three values, twice: the maximal value of the data and the first and third quartile, once for the number outputs as label and a second time as square root for the radii of the circles, respectively. The point size is specified as the square root of the population (X2010). Now we have to convert the world regions into suitable colours. To this end, we generate a vector myColours from the Group variable, and the region name is replaced with the colour definitions using myColours[myColours == myOld[i]]<-myNew[i]. As R has automatically made Group a factor variable, we have to first convert it using as.character(). Then, the logarithmised GDP and the life expectancy are plotted using plot() and points(). x-Axis labels have thousands separators; x- and y-axis titles are added using title(), and the font is set to italics (font=3). Once this is done, we still require the five points with the greatest deviation from the subordinate (log-)linear correlation for their labels. That is done as described in Sect. 9.1.2. Then come the usual heading and sources, and the legend. The respective labels and parameters for this had been

specified earlier. After the labelling, we can use the legend.bubble() function from
the mapplots package to plot the legend. At the end, the map with the world regions
is added in. To do this, we use the previously explained code from the script in
Sect. 10.3.5. The map is placed in the lower right corner of the figure using new=T
and suitable margin settings.

11.7 Map of Napoleon's Russian Campain in 1812/13, by Charles Joseph Minard, 1869

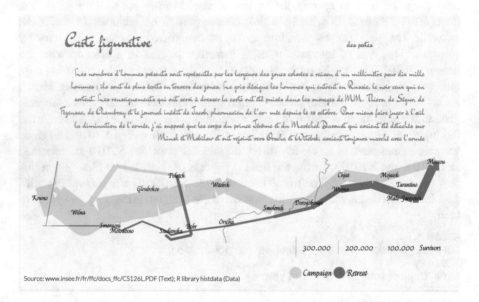

About the figure: In our last example, we would like to elaborate on the "probably
best statistical graphic of all times" (Edward Tufte). By now, there are a series
of versions of this graphic, that have been collated by Michael Friendly. These
also include a version created in R, but using the ggplot function and significantly
differing from the original. The variant described here does not attempt to be a
most exact copy of the original either, but tries to reflect its look and feel. For this
reason, we choose the "Dr Sugiyama" font from the Google font repertoire for the
explanatory text, and the font "Felipa" for the labels of the map. Both are available
under the SIL Open Font License at Google Fonts. We right-align the explanatory
text. Our map of troop movements is largely based on the original, but we add a
modern legend. Data derive from the data frame "HistData".

```
pdf_file<-"pdf/maps_minard_napoleon_inc.pdf"
cairo_pdf(bg="grey98", pdf_file,width=11.69,height=7.26)
```

```
par(omi=c(1,0.1,0.75,0.1),mai=c(0.25,0,3,0),lend=1,family="↘
     Felipa",las=1)
library(HistData); library(mapdata)

# Import data

#load("myData/Minard.troops_de"); data(Minard.cities)
data(Minard.troops); data(Minard.cities)
towards_all<-subset(Minard.troops,direction=="A")
backwards_all<-subset(Minard.troops,direction=="R")

# Define chart and other elements

plot(Minard.troops$long,Minard.troops$lat,type="n",axes=F,xlab="↘
     ",ylab="")
for (i in 1:3)
{
towards<-towards_all[towards_all$group==i,]
n<-nrow(towards)-1
for (j in 1:n)
{
z<-j+1
x<-towards[j:z,]
x
lines(x$long,x$lat,type="l",col="bisque2",lwd=x$survivors[1]/↘
     4500)
}
backwards<-backwards_all[backwards_all$group==i,]
n<-nrow(backwards)-1
for (j in 1:n)
{
backwards<-backwards_all[backwards_all$group==i,]
z<-j+1
x<-backwards[j:z,]
x
lines(x$long,x$lat,type="l",col="azure4",lwd=x$survivors[1]/↘
     4500)
}
}
attach(Minard.cities)
text(long,lat,city)
rivers<-map("rivers",plot=F,add=T)
points(rivers$x,rivers$y,col=rgb(0,0,255,120,maxColorValue=255),↘
     type="l")
par(xpd=T)
legend(32.5,54.15,c("300,000","200,000","100,000"),border=F,pch=↘
     "|",col=c("bisque2","bisque2","bisque2"),bty="n",cex=1.3,↘
     pt.cex=c(3,2,1),xpd=NA,ncol=3)
legend(32.5,53.6,c("Campaign","Retreat"),border=F,pch=19,pt.cex↘
     =4,col=c("bisque2","azure4"),bty="n",cex=1.3,xpd=NA,ncol↘
     =3)
text(37.4,53.93,"Survivors",cex=1.1,xpd=T)
par(family="Dr Sugiyama")
```

```
# Titling

mtext("Carte figurative ",line=-0.5,adj=0.12,cex=3,col="azure4",↘
      outer=T)
mtext("des pertes ",line=-0.5,adj=0.80,cex=1.5,col="azure4",↘
      outer=T)
par(family="Lato Light")
mtext(side=1,adj=0,line=3,"Source: www.insee.fr/fr/ffc/docs_ffc/↘
      CS126L.PDF (Text); R library histdata (Data)",outer=T)
dev.off()
```

The integration into LaTeX is as follows:

```
\documentclass{article}
\usepackage[paperheight=18.5cm, paperwidth=29.7cm, top=0cm, left↘
      =0cm, right=0cm, bottom=0cm]{geometry}
\usepackage{graphicx,color}
\usepackage{fontspec}
\usepackage[abs]{overpic}
\setmainfont[Mapping=text-tex]{Dr Sugiyama}
\definecolor{hintergrund}{rgb}{0.9412,0.9412,0.9412}
\definecolor{text}{gray}{.55}
\pagecolor{hintergrund}
\linespread{1.2}
\begin{document}
\pagestyle{empty}
\begin{center}
\fontsize{18pt}{20pt}\selectfont
\begin{overpic}[scale=1,unit=1mm]{../pdf/maps_minard_napoleon_↘
      inc.pdf}
\put(23.5,144){\begin{minipage}[t]{25cm}
\raggedleft\textcolor{text}{Les nombres d'hommes présents sont ↘
      représenté s par les largeurs des zones colorées a raison ↘
      d'un millimètre pour dix mille hommes ; ils sont de plus é↘
      crits en travers des zones. Le gris désigne les hommes qui↘
       entrent en Russie, le noir ceux qui en sortent. Les ↘
      renseignements qui ont servi à dresser la carte ont été ↘
      puisés dans les ouvrages de MM. Thiers, de Ségur, de ↘
      Fezensac, de Chambray et le journal inédit de Jacob, ↘
      pharmacien de l'ar- mée depuis le 28 octobre. Pour mieux ↘
      faire juger à '÷lil la diminution de l'armée, j'ai supposé↘
       que les corps du prince Jérôme et du Maréchal Davoust qui↘
      avaient été  détachés sur Minsk et Mobilow et ont rejoint↘
      vers Orscha et Witebsk, avaient toujours marché avec l'↘
      armée}
\end{minipage}}
\end{overpic}
\end{center}
\end{document}
```

About the script: The data frame Minard.troops contains 51 rows and is structured like this:

```
> Minard.troops[1:7,]
  long  lat Survivors Direction group
```

```
1 24.0 54.9 340000    campaign    1
2 24.5 55.0 340000    campaign    1
3 25.5 54.5 340000    campaign    1
4 26.0 54.7 320000    campaign    1
5 27.0 54.8 300000    campaign    1
6 28.0 54.9 280000    campaign    1
7 28.5 55.0 240000    campaign    1
```

The data frame Minard.cities contains the longitude and latitude for 20 cities, as well as their names. We first set the figure margins so that there is enough space above it for the text that we will later add with LaTeX. We then build the two partial data frames: one data frame towards_all for the campaign, and one data frame backwards_all for the retreat. Overall, there are three groups for which the movements are plotted. We use a loop to draw a line in the colour bisque2 from each position j defined in the data frame to the next position j+1. Line width corresponds to the number of survivors; in consideration of the figure dimensions, the number is divided by 4500. We do the same for the retreat, but in colour azure4. Now, the cities and the river have to be plotted, and a legend has to be created. For the legend, we use the legend() function twice and the text() function once. The "Dr Sugiyama" font is selected before the title output at the end, and then switched back to "Lato Light" for the sources. The text above the figure is added with LaTeX. The structure of the LaTeX file matches the previous examples, with the only differences being that the text is coloured grey with \definecolor{text} {gray} {.55} and right-aligned using \raggedleft.

Chapter 12
Interactive Visualisation with JavaScript: Highcharts and Mapael

For a long time, there was no alternative to Adobe's Flash if you wanted to create animated, dynamic or interactive visualisations. However, things have now changed: with the unforeseen dynamic that JavaScript has experienced in recent years, and with the option of representing and animating pixel graphics directly in a browser with Canvas, or even vector graphics with SVG, very powerful alternative tools are now available. Wikipedia lists almost 40 JavaScript modules for visualisation. D3 is one of the most prominent of those. Anyone who has seen the spectacular examples from Mike Bostok knows that there is a sheer inexhaustible potential for future data visualisations slumbering here. No wonder that there are now around a dozen books on D3 alone. However, in contrast to static visualisations, the level of programming required here is significantly higher. There are four different languages/formats involved (HTML, CSS, SVG, JavaScript) and the number of programming lines is much higher than with R, for example. R now provides multiple concepts and packages that can be used to create JavaScript visualisations more or less directly. Ultimately, such packages form a type of container in R. In each case, a specific syntax developed by the authors of these packages translates the scripts written in this form into the notation required for the underlying JavaScript library. This means that we are dependent on the scope of the language of the R package and on the quality and flexibility of the translation routines. We want to recommend a different path here: with Mapael and Highcharts, you have access to pre-engineered solutions for a JavaScript library that is easy to use and whose graphics look outstanding. Highcharts is developed and supplied by the Norwegian company Highsoft. Other software they provide includes Highstock (specially developed for representing share prices), Highmaps (for maps), Highcharts Cloud (an online platform for providing Highcharts graphics) and Highslide JS (a type of film viewer). Highcharts is free of charge for personal use and for non-profit organisations. For other entities, licence costs are charged for use. The pre-engineered graphics currently provided by version 4 are 9 types of line diagram, 10 types of area diagram, 13 types of bar/column charts, 7 types of circle diagrams, and 3 types of scatter plot. There are also 3 dynamically updated diagram types, 6 combination diagrams

© Springer Nature Switzerland AG 2019
T. Rahlf, *Data Visualisation with R*, https://doi.org/10.1007/978-3-030-28444-2_12

made up from various elements, 6 (pseudo) 3D diagrams, 5 tachometer diagrams (apparently popular in the BI departments of some companies), 2 heat maps, and 3 tree maps, as well as 10 more exotic types of diagrams such as various radial diagrams or box plots. For all graphics, you can choose between four different CSS topics. The simplicity of operation and the successful appearance has helped Highcharts to become widespread very quickly. The effort involved in programming JavaScript choropleth maps is somewhat greater. However, these maps do offer a real added value, as the "Election time machine: results of all Bundestag elections from SPIEGEL Online"[1] or zensus-unzensiert[2] show. For solutions that are less technically challenging, there are now a number of "toolkits" that simplify the creation of animated visualisations significantly. Examples include Datawrapper, which is developed in Germany, or Highmaps, which we have already mentioned. However, with both of these, the maps are already specified, which means that using your own maps is not easy. To do so, you need a tool that you can use to define the maps freely. One such option is Kartograph from Gregor Aisch, but this is unfortunately no longer being developed any further. An alternative is Mapael from Vincent Brouté. Mapael is a jQuery plug-in that is based on raphael.js. Mapael also has the great advantage that it can be used to create choropleth maps in such cases where statistics data are not available for all polygons. In the following, we will use R to prepare data such that they can be integrated in the JavaScript code for creating the visualisations. For viewing, either an HTML page that contains the JavaScript code for integration and visualisation of the map must be created, or alternatively, a new contribution must be created in a content management system (CMS). If WordPress is used as the CMS, we also need a plug-in because JavaScript code cannot be executed directly in a WordPress contribution. For this purpose, we can install the plug-in "Header and Footer Scripts" from Anand Kumar. This plug-in provides a separate field "Insert Script to" for each contribution. The definition of CSS classes and the integration of the required JavaScript sources are entered in this field: In the contribution, the illustration is then integrated with the following:

```
<div id="container" style="min-width: 310px; height: 400px; ↘
    margin: 0 auto"></div>
```

In the following, we will use three examples to illustrate the use of interactive visualisations with JavaScript: one scatter plot, one time series with Highcharts, and choropleth maps with Mapael.

12.1   Scatter Plot in Highcharts

To recap (see Chap. 9): up to four variables can be portrayed in a scatter plot: two numerical variables on the x-axis and y-axis, one numeric or ordinal variable can define the size of the points, and one nominal variable can define the colour.

---

[1] http://www.spiegel.de/politik/deutschland/wahl-zeitmaschine-ergebnisse-aller-bundestagswahlen-a-918701.html.

[2] http://zensus-unzensiert.de.

Supplementary elements can include: a smoothing, e.g. a regression line, labelling of individual data points, an average value intersection, an area or line (ellipse) that identifies the bivariate distribution, and a line that connects the individual points to one another. Using an example for the correlation between the number of employees in industry and commerce in 1907 and the number of votes for the SPD in the electoral districts of the German Empire in 1912, here we show how to present a scatter plot with two variables with JavaScript. For the third variable, which defines the size of the points, we select the number of eligible voters, and for the colour of the points, the electoral district province or federal state. For demonstration purposes, we select three electoral district provinces/federal states.

The data record can be downloaded from GESIS (http://dx.doi.org/10.4232/1.8145).

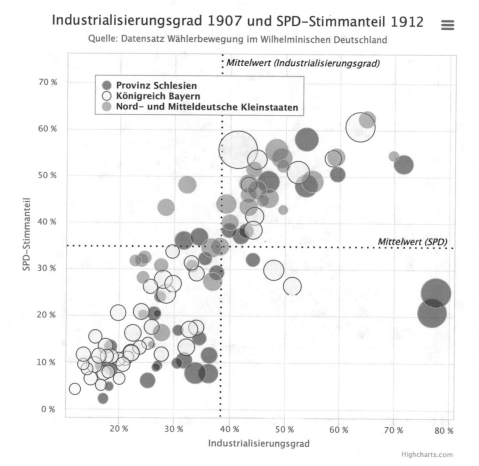

Let us look at the data with R first. To do this, we use the package sjmisc, which enables an easy import of SPSS data with variable labels and value labels.

```
library(sjmisc)
ZA8145 <- read_spss("myData/ZA8145_wdk_1912.sav")
attach(ZA8145)
plot(kbe07igp, spd12p, cex=sqrt(bev10abs/100000), col=prov, pch↘
    =19)
```

The resulting image is a scatter plot which—unsurprisingly—shows an increasing number of votes for the SPD as the number of employees in industry and commerce increases (Fig. 12.1).

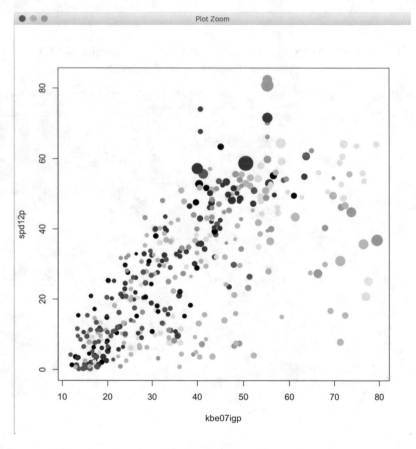

Fig. 12.1 Number of votes for the SPD and number of employees in industry and commerce

Using the function frq(ZA8145$wkrprov), we can output a frequency count for the electoral district provinces/federal states:

```
> frq(ZA8145$wkrprov)
 val                                  label frq raw.prc valid.prc╲
        cum.prc
   1             Provinz Ostpreußen  17    4.28     4.28╲
            4.28
   2             Provinz Westpreußen 13    3.27     3.27╲
            7.56
   3                    Stadt Berlin  6    1.51     1.51╲
            9.07
   4          Potsdam/Frankfurt a.d. O. 20  5.04     5.04╲
           14.11
   5                 Provinz Pommern 14    3.53     3.53╲
           17.63
   6                  Provinz Posen  15    3.78     3.78╲
           21.41
   7              Provinz Schlesien  35    8.82     8.82╲
           30.23
   8                Provinz Sachsen  20    5.04     5.04╲
           35.26
   9      Provinz Schleswig-Holstein 10    2.52     2.52╲
           37.78
  10                Provinz Hannover 19    4.79     4.79╲
           42.57
  11                Provinz Westfalen 17   4.28     4.28╲
           46.85
  12           Provinz Hessen-Nassau 14    3.53     3.53╲
           50.38
  13                Provinz Rheinland 36    9.07     9.07╲
           59.45
  14               Königreich Bayern 48   12.09    12.09╲
           71.54
  15              Königreich Sachsen 23    5.79     5.79╲
           77.33
  16          Königreich Württemberg 17    4.28     4.28╲
           81.61
  17             Grossherzogtum Baden 14    3.53     3.53╲
           85.14
  18            Grossherzogtum Hessen  9    2.27     2.27╲
           87.41
  19 Nord- und Mitteldeutsche Kleinstaaten 30  7.56   7.56╲
           94.96
  20                    Hansestädte   5    1.26     1.26╲
           96.22
  21       Reichsland Elsas-Lothringen 15    3.78    3.78╲
          100.00
  22                            NA    0    0.00      NA╲
           NA
```

For the scatter plot with JavaScript, we select "Provinz Schlesien", "Königreich Bayern", and "Nord- und Mitteldeutschen Kleinstaaten". When we move the mouse over a point, the name of the electoral district and the number of eligible voters is displayed as the name. The individual groups can be displayed or hidden simply by clicking them in the legend. The scaling on the x-axis and y-axis then adapts to match the available data automatically.

The illustration is created according to the principle that we will also apply in the following examples: we use a pre-engineered schema from those available from Highcharts and adapt it slightly. To create the data in the required structure we use R, replacing the example data in the template with our own data. The following steps are required:

1. (One-time) installation of the Highcharts JavaScript libraries; this step can be omitted if you refer to the script versions available on the Highcharts website
2. One-time, if WordPress is used) installation of the plug-in "Header and Footer Scripts"
3. Download of the data record
4. Creation of the contribution with the code (via "Header and Footer Scripts")
5. Manual editing of the legend
6. Creation of the code to be integrated with R
7. Insertion of the code created by R in the header field of the contribution in the JavaScript section "areas"

For the scatter plot, we select a bubbleplot template from Highcharts. From Version 4, Highcharts offers a corresponding template in which the size of the points is scaled correctly automatically. Therefore, unlike the scatter plot in R, we do not have to transform the root first.

The finished script looks as follows:

```
<script type="text/javascript" src="https://ajax.googleapis.com/↘
    ajax/libs/jquery/1.8.2/jquery.min.js"></script>
<style type="text/css">
${demo.css}
</style>
<script type="text/javascript">
$(function () {
    Highcharts.chart('container', {
        chart: {
            type: 'bubble',
            plotBorderWidth: 1,
            zoomType: 'xy'
        },

        legend: {
            enabled: false
        },

        title: {
            text: 'Industrialisierungsgrad 1907 und SPD-↘
                Stimmanteil 1912'
        },
```

```
subtitle: {
    text: 'Source: Datensatz Wählerbewegung im ⬎
          Wilhelminischen Deutschland'
},

xAxis: {
    gridLineWidth: 1,
    title: {
        text: 'Industrialisierungsgrad'
    },
    labels: {
        format: '{value} %'
    },
    plotLines: [{
        color: 'black',
        dashStyle: 'dot',
        width: 2,
        value: 38.4,
        label: {
            rotation: 0,
            y: 15,
            style: {
                fontStyle: 'italic'
            },
            text: 'Mittelwert (Industrialisierungsgrad)'
        },
        zIndex: 3
    }]
},

yAxis: {
    startOnTick: false,
    endOnTick: false,
    title: {
        text: 'SPD-Stimmanteil'
    },
    labels: {
        format: '{value} %'
    },
    maxPadding: 0.2,
    plotLines: [{
        color: 'black',
        dashStyle: 'dot',
        width: 2,
        value: 34.8,
        label: {
            align: 'right',
            style: {
                fontStyle: 'italic'
            },
            text: 'Mittelwert (SPD)',
            x: -10
        },
        zIndex: 3
```

```
        }]
    },

    tooltip: {
        useHTML: false,
        headerFormat: '<table>',
        pointFormat: '<tr><th colspan="2"><h3>{point.name}</
            h3></th></tr>' +
                      '<tr><th> - Wahlber.: </th><td>{point.
                          bev}</td></tr>',
        footerFormat: '</table>',
        followPointer: true
    },
legend: {
        layout: 'vertical',
        align: 'left',
        verticalAlign: 'top',
        x: 100,
        y: 80,
        floating: true,
        backgroundColor: (Highcharts.theme && Highcharts.
            theme.legendBackgroundColor) || '#FFFFFF',
        borderWidth: 1
    },
    plotOptions: {
      bubble: {
            minSize: 1,
            maxSize: 50
        }, series: {
            dataLabels: {
                enabled: false,
                format: '{point.name}'
            }
        }
    },
    series: [{            name: 'Provinz Schlesien',
        data: [

            { x: 18.6, y: 8.5, z: 22300, name: 'Memel
                Heydekrug', bev: '22.300' },
            { x: 16.6, y: 9.2, z: 21443, name: 'Labiau
                Wehlau', bev: '21.443' },


            { x: 46.8, y: 48.8, z: 48593, name: 'Berlin Ost'
                , bev: '48.593' },
            { x: 37.6, y: 29.2, z: 25343, name: 'Berlin
                Innen-Nord', bev: '25.343' }
        ],
        marker: {
            fillColor: '#d7191c'
        }
    }, {
        name: 'Königreich Bayern',
```

```
        data: [

                { x: 44.3, y: 41.4, z: 34639, name: 'Memel ⟍
                        Heydekrug', bev: '34.639' },
                { x: 41.3, y: 55.7, z: 128912, name: 'Labiau ⟍
                        Wehlau', bev: '128.912' },


                { x: 20.7, y: 9.5, z: 25834, name: 'Arnswalde', ⟍
                        bev: '25.834' },
                { x: 32.2, y: 13.2, z: 31372, name: 'Landsberg –⟍
                        Soldin', bev: '31.372' }
        ],
        marker: {
            fillColor: '#ffffbf'
        }
    }, {
        name: 'Nord- und Mitteldeutsche Kleinstaaten',
        data: [

                { x: 24.1, y: 31.8, z: 21848, name: 'Memel ⟍
                        Heydekrug', bev: '21.848' },
                { x: 28.4, y: 43.2, z: 30486, name: 'Labiau ⟍
                        Wehlau', bev: '30.486' },


                { x: 46.3, y: 36, z: 10709, name: 'Flatow', bev:⟍
                        '10.709' },
                { x: 36.9, y: 27.3, z: 34648, name: '⟍
                        Deutschkrone', bev: '34.648' }
        ],
        marker: {
            fillColor: '#2c7bb6'
        }
    }]
  });
});
    </script>
  </head>
  <body>
<script src="https://code.highcharts.com/highcharts.js"></script⟍
    >
<script src="https://code.highcharts.com/highcharts-more.js"></⟍
    script>
<script src="https://code.highcharts.com/modules/exporting.js"><⟍
    /script>
```

We can generally adopt the default settings. Bubble was used only to add the size of the points: minSize: 1, maxSize: 50, to insert a legend, and to make the labelling of the points somewhat leaner. The average values of the number of employees in industry and commerce in 1907 and the number of votes for the SPD are shown as dotted lines.

To create the data portion, we now use R. Firstly, we create a temporary file. The data should then be output for the provinces/federal states 7, 14 and 19. Furthermore, colours are defined for these three subsets (from a ColorBrewer colour palette). This is followed by two loops: the outer loop runs through the three subsets (provinces/federal states), the inner loop runs through the data for each of these subsets. As the designations of the provinces/federal states, just like the electoral districts, are stored in the SPSS file as value labels, they are initially saved with the function get_labels() of the sjmisc package as ZA8145.val.

```
ZA8145.val <- get_labels(ZA8145)
file.create("temp.txt")
d<-c(7, 14, 19)

farbe<-1:19
farbe[7]<-"#d7191c"
farbe[14]<-"#ffffbf"
farbe[19]<-"#2c7bb6"

for(i in d)
{
auswahl<-ZA8145[ZA8145$wkrprov==i, ]
attach(auswahl)
n<-nrow(auswahl)

output<-paste("                  name: '", ZA8145.val$prov[i], "',\n ⬎
                data: [\n", sep="")
write(output, file="temp.txt", append=T)
for(j in 1:n)
{
output<-paste("                    { x: ", kbe07igp[j], ", y: ", ⬎
      spd12p[j], ", z: ", wbr12abs[j], ", name: '", ZA8145.val$⬎
      wkr_nr[j], "', bev: '", format(wbr12abs[j], big.mark=".", ⬎
      scientific=FALSE), "' },", sep="")
write(output, file="temp.txt", append=T)
}

output<-paste("                ],\n              marker: {\n ⬎
                fillColor: '", farbe[i], "'\n                    }⬎
      n          }, {", sep="")
write(output, file="temp.txt", append=T)
}
```

Finally, the contents of the file temp.text must be transferred to the JavaScript code via copy and paste: we thus replace the existing data section with the section created by R.[3]

12.2   Time Series in Highcharts

With time series, you sometimes have to expect missing values. For representations of time series as line diagrams, we must differentiate between two cases here: if individual or sequential values within a series are missing, these missing values are shown as interruptions in the line. From the point of view of technical representation, this is not a major problem. What is more difficult is when individual values that are available are surrounded by gaps. In these cases, a combination of points and lines should be selected for representation (see Sect. 8.4.2).

In the following, we use the illustration of a time series on life expectancy in Germany as an example. The example covers four time series for the German Empire (Deutsches Reich), the Federal Republic of Germany (Bundesrepublik), the Deutsche Demokratische Republik (DDR), and Germany (Deutschland) since 1990. The data can be downloaded from GESIS (http://dx.doi.org/10.4232/1.12202).

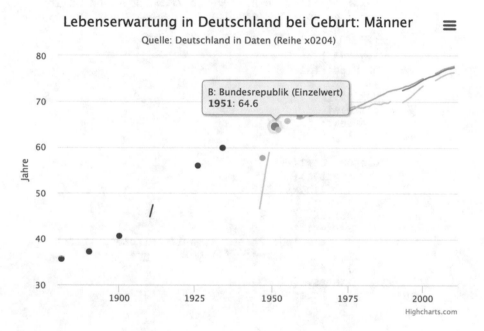

For the representation, just like for the scatter plot, we use the JavaScript library Highcharts. Again, we use an existing template, which we adapt.

```
<script type="text/javascript" src="https://ajax.googleapis.com/
    ajax/libs/jquery/1.8.2/jquery.min.js"></script>
<style type="text/css">
${demo.css}
</style>
<script type="text/javascript">
$(function () {
```

```javascript
$('#container').highcharts({
    chart:{
        type:'scatter'
    },
    tooltip: {
    backgroundColor: '#F0F0F0',
    borderColor: 'grey',
    borderRadius: 10,
    borderWidth: 1,
    formatter: function() {
        return this.series.name + '<br /><b>' + this.x + ↘
            '</b>: ' + this.y ;
    }
    },
    title: {
    text: 'Lebenserwartung bei Geburt: Männer'
    },
    subtitle: {
    text: 'Quelle: <a href="http://www.deutschland-in-↘
        daten.de">Deutschland in Daten (Reihe x0204)</↘
        a>'
    },
    legend: {
    enabled: false
    },
    yAxis: {
    title: {
        text: 'Jahre'
        }
    },
    plotOptions:{
        scatter:{
            lineWidth:2
        },
        series: {
        marker: {
            enabled: false, symbol: 'circle', states: {↘
                hover: {fillColor: 'grey'}}
        },
        states: {
            hover: {
                enabled: true
            },
        }
    }

    },
    series: [{
    name: ' A: Zollverein / Deutsches Reich ',
    data: [
    [1881,35.6],[1882,null],[1883,null],[1884,null],[1885,null],[↘
        1886,null],[1887,null],[1888,null],[1889,null],[1890,37.2]↘
        ,[1891,null],[1892,null],[1893,null],[1894,null],[↘
        1895,null],[1896,null],[1897,null],[1898,null],[1899,null]↘
```

```
      ,[1900,40.6],[1901,null],[1902,null],[1903,null],[↖
      1904,null],[1905,null],[1906,null],[1907,null],[1908,null]↘
      ,[1909,null],[1910,44.8],[1911,47.4],[1912,null],[↖
      1913,null],[1914,null],[1915,null],[1916,null],[1917,null]↘
      ,[1918,null],[1919,null],[1920,null],[1921,null],[↖
      1922,null],[1923,null],[1924,null],[1925,null],[1926,56],[↖
      1927,null],[1928,null],[1929,null],[1930,null],[1931,null]↘
      ,[1932,null],[1933,null],[1934,59.9]
], lineColor: 'rgba(215,25,28, .9)'},
{
   name: 'A: Zollverein / Deutsches Reich (Einzelwert)',
   color: 'rgba(215,25,28, .9)',
   data: [
[1881,35.6],
[1890,37.2],
[1900,40.6],
[1926,56],
[1934,59.9]
], marker: {enabled: true}, lineWidth: 0},
{
   name: ' B: Bundesrepublik ',
   data: [
[1947,57.7],[1948,null],[1949,null],[1950,null],[1951,64.6],[↖
      1952,null],[1953,null],[1954,null],[1955,null],[1956,null]↘
      ,[1957,null],[1958,null],[1959,66.8],[1960,null],[↖
      1961,null],[1962,66.9],[1963,67.1],[1964,67.3],[1965,67.4]↘
      ,[1966,67.6],[1967,67.6],[1968,67.6],[1969,67.4],[↖
      1970,67.2],[1971,67.3],[1972,67.4],[1973,67.6],[1974,67.9]↘
      ,[1975,68],[1976,68.3],[1977,68.6],[1978,69],[1979,69.3],[↖
      1980,69.6],[1981,69.9],[1982,70.2],[1983,70.5],[1984,70.8]↘
      ,[1985,71.2],[1986,71.5],[1987,71.8],[1988,72.2],[↖
      1989,72.4],[1990,72.6],[1991,72.7],[1992,72.9],[1993,73.1]↘
      ,[1994,73.4],[1995,73.5],[1996,73.8],[1997,74.1],[↖
      1998,74.4],[1999,74.8],[2000,75.1],[2001,null],[2002,null]↘
      ,[2003,null],[2004,76.2],[2005,76.5],[2006,76.9],[↖
      2007,77.2],[2008,77.4],[2009,77.6],[2010,77.8]
], lineColor: 'rgba(253,174,97, .9)'},
{
   name: 'B: Bundesrepublik (Einzelwert)',
   color: 'rgba(253,174,97, .9)',
   data: [
[1947,57.7],
[1951,64.6],
[1959,66.8],
], marker: {enabled: true}, lineWidth: 0},
{
   name: ' C: Deutsche Demokratische Republik ',
   data: [
[1946,46.6],[1947,50.8],[1948,55.3],[1949,58.9],[1950,null],[↖
      1951,null],[1952,63.9],[1953,null],[1954,null],[1955,65.8]↘
      ,[1956,null],[1957,null],[1958,66.4],[1959,66.3],[↖
      1960,66.5],[1961,67.1],[1962,67.3],[1963,67.9],[1964,67.7]↘
      ,[1965,68],[1966,68.2],[1967,68.4],[1968,68],[1969,67.8],[↖
      1970,68.1],[1971,68.5],[1972,68.5],[1973,68.8],[1974,68.9]↘
```

```
     ,[1975,68.5],[1976,68.8],[1977,69],[1978,68.8],[1979,68.7]\
     ,[1980,68.7],[1981,69],[1982,69.1],[1983,69.5],[1984,69.6]\
     ,[1985,69.5],[1986,69.5],[1987,69.8],[1988,69.7],[\
     1989,70.1],[1990,null],[1991,null],[1992,null],[1993,69.9]\
     ,[1994,70.3],[1995,70.7],[1996,71.2],[1997,71.8],[\
     1998,72.4],[1999,73],[2000,73.5],[2001,null],[2002,null],[\
     2003,null],[2004,74.7],[2005,75.1],[2006,75.5],[2007,75.8]\
     ,[2008,76.1],[2009,76.3],[2010,76.4]
], lineColor: 'rgba(171,221,164, .9)'},
{
   name: 'C: Deutsche Demokratische Republik (Einzelwert)',
   color: 'rgba(171,221,164, .9)',
   data: [
[1952,63.9],
[1955,65.8],
], marker: {enabled: true}, lineWidth: 0},
{
   name: ' D: Deutschland ',
   data: [
[1993,72.5],[1994,72.8],[1995,73],[1996,73.3],[1997,73.6],[\
     1998,74],[1999,74.4],[2000,74.8],[2001,75.1],[2002,75.4],[\
     2003,75.6],[2004,75.9],[2005,76.2],[2006,76.6],[2007,76.9]\
     ,[2008,77.2],[2009,77.3],[2010,77.5]
], lineColor: 'rgba(43,131,186, .9)'},
{
   name: 'D: Deutschland (Einzelwert)',
   color: 'rgba(43,131,186, .9)',
   data: [
], marker: {enabled: true}, lineWidth: 0}]
        });
     });
</script>
</head>
<body>
<script src="https://code.highcharts.com/highcharts.js"></script\
     >
<script src="https://code.highcharts.com/modules/exporting.js"><\
     /script>
```

We select "scatter" as the chart type and thus we can show both points and lines in one illustration. We then define a specific tool tip that encompasses the name of the series as well as the year and the specific value for the year. As we are dealing with four series, each with the same content which is specified in the title, we do not need to show a legend. Most of the elements that influence the appearance of the illustration can be specified at various points: depending on where they are inserted, they then relate either to an individual data point, to an individual series, to all series, or to the entire illustration. We first define a general point symbol and a line thickness for all series, and then adapt these for the individual series. The series are integrated such that initially, a value pair is specified for each year value. If the corresponding series value (y-value) is missing, "zero" is entered. The series is represented as a line. This creates a line which is interrupted for the missing values. If we had not entered these "zero" values, an interpolated line would have been drawn between the available values. In a second step, wherever individual values are available, these are drawn as points. To do this, we use a procedure that looks for such "individual values" automatically and which we have already become familiar with in Sect. 8.4.2. The values for the series are taken from the XLS table "K05_1_Gesundheitswesen_-_Lebenserwartungen_nach_Geschlecht_für_ausgewählte_Perioden.xls", which can be downloaded from GESIS (dx.doi.org/10.4232/1.12202).

```
library(gdata)
daten<-read.xls("myData/K05_1_Gesundheitswesen_-_\
    Lebenserwartungen_nach_Geschlecht_für_ausgewählte_Perioden\
    .xls", skip=10)

file.create("temp.txt")

A<-daten[1:7, c(1,2)]
jahre<-1881:1934
reihenname<-"A: Zollverein / Deutsches Reich"
names(A)<-c("x", "y")
A<-merge(jahre, A, by=1, all.x=T)
attach(A)
farbe<-"rgba(215,25,28, .9)"
source("r-Skripte/inc_highcharts_zeitreihen.r")


B<-daten[9:67, c(1,8)]
jahre<-1947:2010
reihenname<-"B: Bundesrepublik"
names(B)<-c("x", "y")
B<-merge(jahre, B, by=1, all.x=T)
attach(B)
farbe<-"rgba(253,174,97, .9)"
source("r-Skripte/inc_highcharts_zeitreihen.r")


C<-daten[8:67, c(1,14)]
jahre<-1946:2010
reihenname<-"C: Deutsche Demokratische Republik"
names(C)<-c("x", "y")
C<-merge(jahre, C, by=1, all.x=T)
attach(C)
farbe<-"rgba(171,221,164, .9)"
source("r-Skripte/inc_highcharts_zeitreihen.r")


D<-daten[50:67, c(1,20)]
jahre<-1993:2010
reihenname<-"D: Deutschland"
names(D)<-c("x", "y")
D<-merge(jahre, D, by=1, all.x=T)
attach(D)
farbe<-"rgba(43,131,186, .9)"
source("r-Skripte/inc_highcharts_zeitreihen.r")
```

The data here are essentially imported from four different columns of the XLS table and then a subroutine is called that writes the data out in the required structure: first in the form for the line representation, and then the isolated values as points.

```
# first the lines ---------------------------------

output<-paste("{
   name: '", reihenname, "',
   data: [")
write(output, file="temp.txt", append=T)

output<-paste("[", as.character(x), ",", as.character(y), "]", ↘
      collapse=",", sep="")
write(output, file="temp.txt", append=T)
write(paste("], lineColor: '",farbe,"'},", sep=""), file="temp.↘
      txt", append=T)

# then the points ------------------------------------

output<-paste("{
   name: '", reihenname, " (Einzelwert)',
   color: '",farbe,"',
   data: [", sep="")
write(output, file="temp.txt", append=T)

# erst vorne ...

if (!is.na(y[1]) & is.na(y[2]))
{
output<-paste("[", x[1], ",", y[1], "],", sep="")
write(output, file="temp.txt", append=T)
}

# ... then the middle ...

von<-2; bis<-length(y)-1
for (i in von:bis)
{
if (is.na(y[i-1]) & !is.na(y[i]) & is.na(y[i+1]))
{
output<-paste("[", x[i], ",", y[i], "],", sep="")
write(output, file="temp.txt", append=T)
}
}

# ... and the end

if (!is.na(y[length(y)]) & is.na(y[length(y)-1]))
{
output<-paste("[", x[length(y)], ",", y[length(y)], "]", sep="")
write(output, file="temp.txt", append=T)
}
write("], marker: {enabled: true}, lineWidth: 0},", file="temp.↘
      txt", append=T)
```

Finally, the contents of the file temp.text must be transferred to the JavaScript code via copy and paste: just like in the previous example, we thus replace the existing data section with the section created by R.

In the blog entry, the illustration is again integrated with the following:

```
<div id="container" style="min-width: 310px; height: 400px; 
      margin: 0 auto"></div>
```

12.3 Choropleth Maps with Mapael

To understand the basic principle of linking statistical data with map geometries, it is helpful to first understand the creation of "static" choropleth maps with R, as explained in Sect. 10.3 (Fig. 12.2).

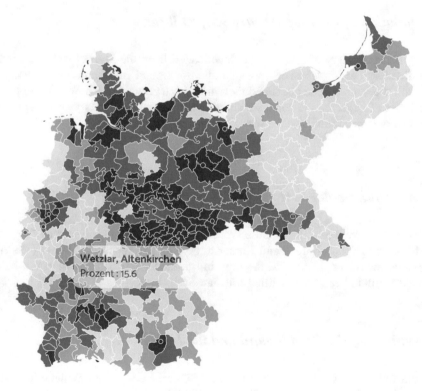

Fig. 12.2 Choropleth map with Mapael (1)

Interactive choropleth maps are created in the following steps:

1. (One-time) installation of the Mapael JavaScript libraries
2. (One-time, if WordPress is used) installation of the plug-in "Header and Footer Scripts"
3. Download of the data record
4. Download of the map data in SHP format
5. Conversion of the SHP file into an SVG file with Mapshaper
6. Conversion of the SVG File into JS
7. Creation of the contribution with the code (via "Header and Footer Scripts")
8. Manual editing of the legend
9. Creation of the code to be integrated with R
10. Insertion of the code created by R in the header field of the contribution in the JavaScript section "areas"

Installation of the Mapael JavaScript Libraries

The package jQuery-Mapael can be downloaded from the Mapael website. From the directory "examples/basic", we take the file legend_areas.html as a template. We edit the source code contained therein and integrate it in the WordPress page used here. To execute the code, we simply have to store the files jquery.mapael.js and jquery.mapael.min.js on the server and in a subdirectory "maps", store the maps in .js format.

Header and Footer Scripts

If, as in the current example, we use WordPress, we need a plug-in such as "Header and Footer Scripts" from Anand Kumar to be able to execute JavaScript (see the previous two examples). With the plug-in, in each contribution a separate field "Insert Script to head" can be filled with JavaScript code.

Download of the Data Record and the Map Data

In the following, we use the data record "Wählerbewegung im Wilhelminischen Deutschland. Die Reichstagswahlen von 1890 bis 1912" (voter movements in Wilhelmine Germany. The Reichstag elections of 1890–1912) as an example, as well as the corresponding map, which we used in Sect. 10.3.2.

Conversion of the SHP File into an SVG File with Mapshaper

The map that we use here is a shapefile created by Daniel Zieblatt and Jeffrey C. Blossom, which first has to be converted into the SVG format. Mapshaper is a suitable tool for this, and it offers a variety of options (Fig. 12.3).[4]

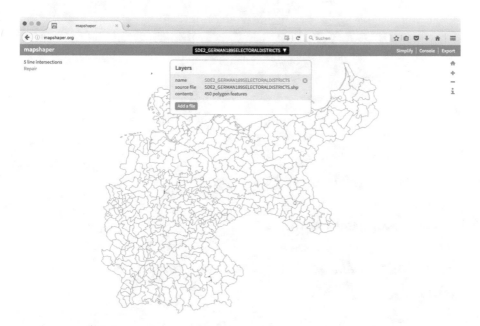

Fig. 12.3 Mapshaper: display of the shapefile

We have to specify which ID field is to be used (in this example, ID field=OBJECTID). The map can also be "simplified" so that the polygons do not have lots of unnecessary corners and edges (Fig. 12.4).

The result of the export is an SVG file (that is, an XML file) that looks as follows (abbreviated):

```
<?xml version="1.0"?>
<svg xmlns="http://www.w3.org/2000/svg" version="1.2" ⤵
     baseProfile="tiny" width="800" height="728" viewBox="0 0 ⤵
     800 728" stroke-linecap="round" stroke-linejoin="round">
<g id="SDE2_GERMAN1895ELECTORALDISTRICTS">
<path d="M 57.3058 667.944 ... 57.3058 667.944 Z" id="6"/>
<path d="M 159.7408 655.7696 ... 655.7696 Z" id="7"/>
...
```

---

[4]http://mapshaper.org.

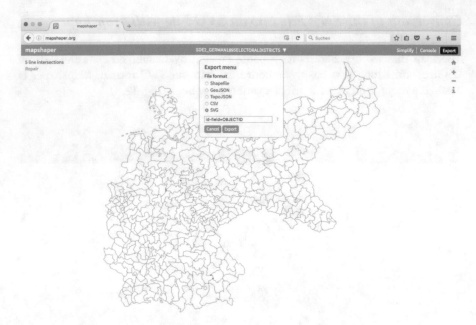

Fig. 12.4 Mapshaper: Export als SVG-Datei

```
<path d="M 767.4287 85.1217 ... 767.4287 85.1217 Z" id="435"/>
</g>
</svg>
```

Conversion of the SVG File into JS

For Mapael, the map must be available in JS format. We can use the online tool SVG To Mapael from Vincent Brouté for this conversion (Fig. 12.5).

The exported map looks as follows (abbreviated):

```
/*!
 *
 * Jquery Mapael - Dynamic maps jQuery plugin (based on raphael.\
     js)
 * Requires jQuery and Mapael >=2.0.0
 *
 * Map of Germany 1895
 *
 * @author Daniel Zieblatt / Jeffrey C. Blossom (Shapefile)
 */
(function (factory) {
    if (typeof exports === 'object') {
        // CommonJS
```

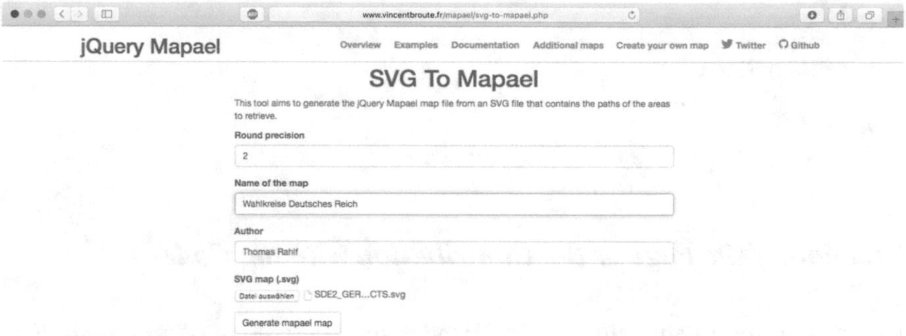

Fig. 12.5 SVGToMapael

```
        module.exports = factory(require('jquery'), require('↘
            mapael'));
    } else if (typeof define === 'function' && define.amd) {
        // AMD. Register as an anonymous module.
        define(['jquery', 'mapael'], factory);
    } else {
        // Browser globals
        factory(jQuery, jQuery.mapael);
    }
}(function ($, Mapael) {

    "use strict";

    $.extend(true, Mapael,
        {
            maps :   {
                germany_1895 : {
                    width : 800,
                    height : 728,
                    getCoords : function (lat, lon) {
                        // todo
                        return {"x" : lat, "y" : lon};
                    },
                    'elems': {
                        "6" : "M 57.31 667.94 57.35 669.24 .... ↘
                            Z",
                        "7" : "M 159.74 655.77 160.37 656.05 ↘
                            .... Z",
                        "9" : "M 80.74 653.86 81.28 654.17 81.61↘
                            .... Z",
                        ...
                        "435" : "M 767.43 85.12 767.67 85.62 ↘
                            .... Z"
                    }
                }
            }
        }
```

```
    );

    return Mapael;

}));
```

Creation of the Page or the Contribution with the Code

We had already installed the plug-in "Header and Footer Scripts" to do this. This plug-in provides a separate "Insert Script to" field for every contribution. The definition of CSS classes and the integration of the required JavaScript sources are initially entered in this field:

```css
<style type="text/css">
        .container {
            max-width: 800px;
            margin: auto;
        }

        /* Specific mapael css class are below
         * 'mapael' class is added by plugin
         */

        .mapael .map {
            position: relative;
        }

        .mapael .mapTooltip {
            position: absolute;
            background-color: #fff;
            moz-opacity: 0.70;
            opacity: 0.70;
            filter: alpha(opacity=70);
            border-radius: 10px;
            padding: 10px;
            z-index: 1000;
            max-width: 200px;
            display: none;
            color: #343434;
        }
 </style>

<script src="https://cdnjs.cloudflare.com/ajax/libs/jquery/3.0.0↘
      /jquery.min.js"charset="utf-8"></script>
<script src="https://cdnjs.cloudflare.com/ajax/libs/jquery-↘
      mousewheel/3.1.13/jquery.mousewheel.min.js" charset="utf-8↘
      "</script>
<script src="https://cdnjs.cloudflare.com/ajax/libs/raphael/↘
      2.2.0/raphael-min.js"charset="utf-8"></script>
<script src="../../js/jquery.mapael.js"charset="utf-8"></script>
```

```
<script src="../../js/maps/wahlkreise_deutsches_reich.js"charset\
    ="utf-8"></script>
```

Manual Editing of the Legend

The definition of the CSS classes and the integration of the JavaScript sources are
followed immediately by the main portion. Here, the first step is the definition of the
legend. This also involves the classification into classes: using the values defined
here, the polygons are classified and coloured. The legend can initially be taken
from the example file and then adapted through various options where necessary.

```
<script type="text/javascript">
    $(function () {
        $(".mapcontainer").mapael({
            map: {
                name:"wahlkreise_deutsches_reich",
                defaultArea: {
                    attrs: {
                        stroke:"#fff",
                        "stroke-width": 1
                    },
                    attrsHover: {
                        "stroke-width": 2
                    }
                }
            },
            legend: {
                area: {
                    title:"Legende",
                    titleAttrs: {"font-family":"Oxygen"},
                    labelAttrs: {"font-family":"Oxygen"},
                    slices: [
                        {
                            max: 10,
                            attrs: {
                                fill:"#fee5d9"
                            },
                            label:"bis 10 Prozent"
                        },
                        {
                            min: 10,
                            max: 20,
                            attrs: {
                                fill:"#fcae91"
                            },
                            label:"10 bis 20 Prozent"
                        },
                        {
                            min: 20,
                            max: 40,
```

```
                          attrs: {
                              fill:"#fb6a4a"
                          },
                          label:"20 bis 40 Prozent"
                      },
                      {
                          min: 40,
                          attrs: {
                              fill:"#cb181d"
                          },
                          label:"über 40 Prozent"
                      }
                  ]
              }
          },
```

Next comes the data portion. As our example, we take the number of votes for the SPD in the Reichstag election of 1912. The end result must look as follows: the individual paragraphs begin with the respective electoral district ID in quotation marks, followed by the actual value (the number of votes for the SPD, as a whole number) and a tool tip that shows the labelling (name and percentage value) when the mouse is moved over the corresponding area.

```
              areas: {

  "6": {
                          value:"24",
                          href:"#",
                          tooltip: {content:"<span style=\"font-\
                              weight:bold;\"> Altkirch – Thann <\
                              /span><br />Prozent :  24.2"}
                      },
  "7": {
                          value:"11",
                          href:"#",
                          tooltip: {content:"<span style=\"font-\
                              weight:bold;\"> Überlingen–Meß\
                              kirch–Stockach–Konstanz </span><br\
                              />Prozent :  10.8"}
                      },
  "9": {
                          value:"33",
                          href:"#",
                          tooltip: {content:"<span style=\"font-\
                              weight:bold;\"> Colmar Elsass </\
                              span><br />Prozent :  33.2"}
                      },
  ...
  "434": {
                          value:"3",
                          href:"#",
```

```
                              tooltip: {content:"<span style=\"font-⤸
                                  weight:bold;\"> Goldap - Darkehmen⤸
                                  </span><br />Prozent :   3.4"}
                        },
"435": {
                              value:"15",
                              href:"#",
                              tooltip: {content:"<span style=\"font-⤸
                                  weight:bold;\"> Gumbinnen - ⤸
                                  Insterburg </span><br />Prozent : ⤸
                                  14.8"}
                        }
                    }
                });
            });
        </script>
```

Creation of the Code to be Integrated with R

To create the data portion, we need the statistics information from the data record. The easiest option is to create the required code directly from R.

```
library(Hmisc)
library(maptools)
library(plyr)

ZA8145<-spss.get("myData/ZA8145_wdk_1912.sav", use.value.labels=⤸
    F)

vars <- c("wkr.nr", "spd12p")
auswahl<-ZA8145[vars]

ZA8145<-spss.get("myData/ZA8145_wdk_1912.sav", use.value.labels=⤸
    T)
auswahl$wkr.nr2<-ZA8145$wkr.nr

x<-readShapeSpatial("shp/SDE2_GERMAN1895ELECTORALDISTRICTS.shp")

xdata<-x@data

auswahl$DISTRICT<-auswahl$wkr.nr
xdata$DISTRICT<-xdata$DISTRICT_N
z<-join(xdata, auswahl, type="left")

file.create("temp.txt")

attach(z)
n<-nrow(z)
for(i in 1:n)
{
```

```
output<-paste('"',OBJECTID[i],'": {\n
      value: "',round(spd12p[i], digits=0),'",\n
                        href: "#",\n
                        tooltip: {content: "<span style=\"
      font-weight:bold;\">',wkr.nr2[i],'</span><br />Prozent : '
      ,spd12p[i],'"}\n                        },')
write(output, file="temp.txt", append=T)
}
```

We import the data record twice—once to get the electoral district numbers and once to get the electoral district designations (the electoral district designations are saved in the data record as "labels"). Because we have more polygons than electoral districts, we must first link the shapefile with the data record using "left join". Otherwise, we would only have had to import the data record. Then, in a loop, the required details for each polygon are written to a temporary file in the form required in the JavaScript code. Finally, this part must be inserted into the JavaScript block in the area "areas:".

Choropleth Map for the Population Census of 1925: Employees in Industry and Trade

Using a second example (Fig. 12.6), we want to show how we can use maps in which the statistical data have to be assigned to multiple areas. This is always the case, for example, if an administrative district consists of multiple smaller subareas. Based on the population census of 1925, the following map shows, for around one thousand small administrative districts, the proportion of the overall workforce that is employed in industry and trade. The map is based on the data record of the election and social data of the districts and communities of the German Empire from 1920 to 1933, as well as on a map with the administrative boundaries created by the Max Planck Institute for Demographic Research in Rostock. The data record can be downloaded from GESIS (http://dx.doi.org/10.4232/1.8013), and the map can be downloaded from the website censusmosaic.org.

In 1925, more than forty percent of all workers were employed in industry and trade. The distribution over the regions was very different. Alongside the Ruhr region, Saxony was a second high-density industrial area.

Creation of the Map

The basic procedure corresponds to that in the previous section. In this case, one additional step is necessary because the IDs of the region units in the data record do not match those in the shapefile.

Arbeiter in Industrie und Handwerk 1925

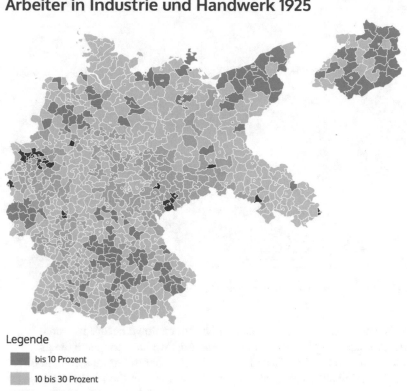

Legende

■ bis 10 Prozent

▢ 10 bis 30 Prozent

▢ 30 bis 50 Prozent

■ über 50 Prozent

Fig. 12.6 Choroplethenkarte mit Mapael (2)

Firstly, the map German_Empire_1925_v.1.0.shp is converted into an SVG file with Mapshaper as described above. In doing so, for our purposes, we can simplify the complexity of the polygons to 10 percent (using a slider in Mapshaper) without losing any information. This reduces the size of the file significantly and accelerates the loading time of the page on which the map is presented. We specify the column "ID" as the ID.

In contrast to the map of the electoral districts of the German Empire, however, here, the different region units that belong to an electoral district do not have the same (electoral district) ID. Instead, they are grouped as one, as we can see in QGIS from the example of Bernburg (Fig. 12.7).

In the SVG export, this produces a grouping with the tag "g":

```
<g id="11005">
<path d="M 290.0323 266.5992 ..... 266.5992 Z"/>
<path d="M 294.9746 279.2339 ..... 279.2339 Z"/>
</g>
```

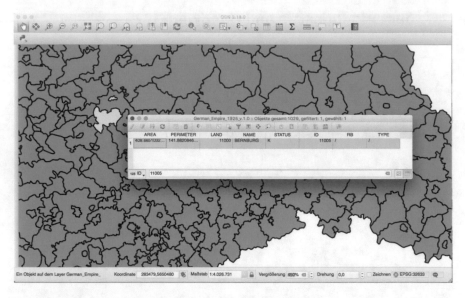

Fig. 12.7 Example for two separate regions with the same ID (Bernburg)

This type of region group must first be broken down before we can display it in Mapael for use in a choropleth map. To do this, we have to delete the group tags (" <g id="11005" >... </g >") and assign the IDs that the groups contain directly to the individual paths. In each case, the first path receives the ID that was assigned to the group. All subsequent paths receive the same ID but followed by a hyphen and a number (2, 3, etc.). In our specific example, this looks as follows:

```
<path id="11005" d="M 290.0323 266.5992 ..... 266.5992 Z"/>
<path id="11005-2" d="M 294.9746 279.2339 ..... 279.2339 Z"/>
```

The edited SVG file can now be converted into a JS file using the tool SVG To Mapael. This JS file is then the basis for the representation in Mapael. The result is a JS version of the map in which the polygons have been correctly adopted:

```
/*!
 *
 * Jquery Mapael - Dynamic maps jQuery plugin (based on raphael.↘
       js)
 * Requires jQuery and Mapael >=2.0.0
 *
 * Map of Kreise Deutsches Reich 1925
 *
 * @author MPIDR Population History GIS Collection
 */
(function (factory) {
    if (typeof exports === 'object') {
        // CommonJS
        module.exports = factory(require('jquery'), require('↘
            mapael'));
```

```
    } else if (typeof define === 'function' && define.amd) {
        // AMD. Register as an anonymous module.
        define(['jquery', 'mapael'], factory);
    } else {
        // Browser globals
        factory(jQuery, jQuery.mapael);
    }
}(function ($, Mapael) {

    "use strict";

    $.extend(true, Mapael,
        {
            maps :  {
                kreise_deutsches_reich_1925 : {
                    width : 800,
                    height : 626,
                    getCoords : function (lat, lon) {
                        // todo
                        return {"x" : lat, "y" : lon};
                    },
                    'elems': {
                        "11001" : "M 330.61 268.17 ... 268.17 Z"\
                            ,
                        "11003" : "M 326.57 276.86 ... 276.86 Z"\
                            ,
                        "11005" : "M 290.03 266.6 ... 266.6 Z",
                        "11005-2" : "M 294.97 279.23 ... 279.23 \
                            Z",
                        "11004" : "M 303.26 279.78 ... 279.78 Z"\
                            ,
                                .
                                .
                                .
```

The beginning of the JavaScript code for the display of the map corresponds largely to that from the previous example, but here we integrate a different map (kreise_deutsches_reich_1925.js):

```
<style type="text/css">

    .container {
        max-width: 800px;
        margin: auto;
    }


    /* Specific mapael css class are below
     * 'mapael' class is added by plugin
     */

    .mapael .map {
        position: relative;
    }
```

```css
    .mapael .mapTooltip {
        position: absolute;
        background-color: #fff;
        moz-opacity: 0.70;
        opacity: 0.70;
        filter: alpha(opacity=70);
        border-radius: 10px;
        padding: 10px;
        z-index: 1000;
        max-width: 200px;
        display: none;
        color: #343434;
    }
</style>
```

```html
<script src="https://cdnjs.cloudflare.com/ajax/libs/jquery/3.0.0\
    /jquery.min.js"charset="utf-8"></script>
<script src="https://cdnjs.cloudflare.com/ajax/libs/jquery-\
    mousewheel/3.1.13/jquery.mousewheel.min.js" charset="utf-8\
    "></script>
<script src="https://cdnjs.cloudflare.com/ajax/libs/raphael/\
    2.2.0/raphael-min.js"charset="utf-8"></script>
<script src="../../js/jquery.mapael.js"charset="utf-8"></script>
<script src="../../js/maps/kreise_deutsches_reich_1925.js"\
    charset="utf-8"></script>

    <script type="text/javascript">
        $(function () {
            $(".mapcontainer").mapael({
                map: {
                    name:"kreise_deutsches_reich_1925",
                    defaultArea: {
                        attrs: {
                            stroke:"#fff",
                            "stroke-width": 1
                        },
                        attrsHover: {
                            "stroke-width": 2
                        }
                    }
                },
```

For the classification into classes, which is in turn defined via the legend (Fig. 12.8), let us look at the distribution of the data in R using the function hist().

```r
hist(daten$p25iarb)
```

The result is shown in Fig. 12.8.

Accordingly, a total of four classes makes sense: "Up to 10 percent", "10 to 30 percent", "30 to 50 percent", and "More than 50 percent". For the colour palette, we select "4-class RdYlGn" from the page colorbrewer2.org, where the colour

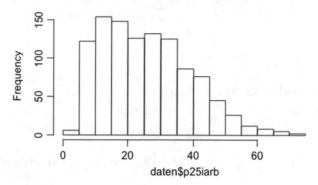

Fig. 12.8 Histogram der Variable p25iarb

designations are available in hexadecimal form. We therefore enter the class limits
and the colours (in hexadecimal form):

```
legend: {
    area: {
        title:"Legende",
        titleAttrs: {"font-family":"Oxygen"},
        labelAttrs: {"font-family":"Oxygen"},
        slices: [
            {
                max: 10,
                attrs: {
                    fill:"#bae4bc"
                },
                label:"bis 10 Prozent"
            },
            {
                min: 10,
                max: 30,
                attrs: {
                    fill:"#7bccc4"
                },
                label:"10 bis 30 Prozent"
            },
            {
                min: 30,
                max: 50,
                attrs: {
                    fill:"#43a2ca"
                },
                label:"30 bis 50 Prozent"
            },
            {
                min: 50,
                attrs: {
```

```
                                    fill:"#0868ac"
                         },
                         label:"über 50 Prozent"
                 }
             ]
         }
     }
```

This is followed by the "data portion":

```
                 areas: {
"2": {
                     value: "16",
                     href: "#",
                     tooltip: {content: "<span style=\"font-↴
                         weight:bold;\">FISCHHAUSEN</span><↴
                         br />Prozent : 16"}
             },
"4": {
                     value: "21",
                     href: "#",
                     tooltip: {content: "<span style=\"font-↴
                         weight:bold;\">KOENIGSBERG (PR) S<↴
                         /span><br />Prozent : 21"}
             },
"5": {
                     value: "11",
                     href: "#",
                     tooltip: {content: "<span style=\"font-↴
                         weight:bold;\">KOENIGSBERG (PR) L<↴
                         /span><br />Prozent : 11"}
             },
   .
   .
   .
   .
"19085": {
                     value: "21",
                     href: "#",
                     tooltip: {content: "<span style=\"font-↴
                         weight:bold;\">DONAUESCHINGEN</↴
                         span><br />Prozent : 21"}
             },
"19086": {
                     value: "37",
                     href: "#",
                     tooltip: {content: "<span style=\"font-↴
                         weight:bold;\">MANNHEIM</span><br ↴
                         />Prozent : 37"}
             }
         }
     });
    });
  </script>
```

We import the data from the SPSS data record with R and use these data in turn to create the required information. The map and the data record must be linked using the name designations of the administrative units in the two files, which are not identical. One option for the linking is to use "fuzzy matching" to connect the data automatically, as proposed by Mark Heckmann for a current choropleth map, for example. However, for our purposes, the result would not be satisfactory as the writing variations in the two files are not insignificant. Here, the better (although more time-consuming) variant is to compare the approximately one thousand entries manually and create a concordance table that assigns the region units of the data record uniquely to those of the map in each case (Konkordanz.xlsx).

```r
# ZA8013_to_mapael.r

library(gdata)
library(memisc)
library(plyr)

konkordanz<-read.xls("konkordanz.xlsx")

ZA8013<-spss.system.file("myData/ZA8013_Sozialdaten.sav")
daten<-subset(ZA8013, select=c(lfnr, krnr, agglvl, name, wkr, \
        regbez, c25iarb, c25wohn))
daten<-daten[daten$agglvl %in% c(4, 5, 6) & daten$c25iarb>0, ]
daten$p25iarb<-100*daten$c25iarb/daten$c25wohn

dfdaten<-as.data.frame(daten)

z<-join(konkordanz, dfdaten, by="krnr", type="left")

file.create("temp.txt")

attach(z)
n<-nrow(z)
for(i in 1:n)
{
for(j in 1:Anzahl[i])
{
mapael_id<-ID[i]
if (j>1) mapael_id<-paste(toString(mapael_id), "-", toString(j))
output<-paste('"',toString(mapael_id),'": {\n \
                                value: "',round(p25iarb[i], digits\
        =0),'",\n                       href: "#",\n \
                                tooltip: {content: "<span style\
        =\\\"font-weight:bold;\\\">',name[i],'</span><br />Prozent\
        : ',round(p25iarb[i], digits=0),'"}\n                            \
        },', sep="")
write(output, file="temp.txt", append=T)
}
}
```

Once the required variables of the SPSS data record have been imported (here, using the package memisc), we restrict the data to the district level (agglvl %in% c(4, 5, 6)). Then, as shown in the example of the number of votes for the SPD in 1912, in a loop, the required details for each polygon are written to a temporary file in the form required in the JavaScript code. The result must then be inserted into the JavaScript block in the area "areas:".

Appendix

A Data

Here are the sources of all data, which are often used in this book.

ZA2391: Partial Cumulation of Politbarometers

See http://dx.doi.org/10.4232/1.12512

ZA4753: European Values Study 2008: Germany (EVS 2008)

See http://dx.doi.org/10.4232/1.10151

ZA4804: European Values Study Longitudinal Data File

See http://dx.doi.org/10.4232/1.12253

ZA4972: Eurobarometer 71.2 (May-Jun 2009)

See http://dx.doi.org/10.4232/1.10990
The view is created with the following SQL:

```
--------------------------------
CREATE VIEW v_za4972_laender AS
(
select
(
case when (za4972.v7 in (4,14)) then 4
     when (za4972.v7 in (9,10)) then 10
     else za4972.v7 end) AS v7,
avg(za4972.v84) AS v84,
avg(za4972.v85) AS v85,
avg((case when (za4972.v100 = 1) then 1
```

© Springer Nature Switzerland AG 2019
T. Rahlf, *Data Visualisation with R*, https://doi.org/10.1007/978-3-030-28444-2

```
            when (za4972.v100 = 2) then -(1)
            when (za4972.v100 = 3) then 0 end)) AS v100,
avg((case when (za4972.v114 = 1) then 1
            when (za4972.v114 = 2) then -(1)
            when (za4972.v114 = 3) then 0 end)) AS v114
from za4972
group by
(
case when (za4972.v7 in (4,14)) then 4
     when (za4972.v7 in (9,10)) then 10
     else za4972.v7 end));
---------------------------------
```

BetterLifeIndex_Data_2011V6.xls

See http://www.oecdbetterlifeindex.org.

weltenergiemix.xlsx

See http://www.bpb.de. Data were entered into an XLSX spreadsheet.

personen.xlsx

See http://www.celebheights.com. I'm not kidding. Data were entered into an XLSX spreadsheet.

v_women_men

This view combines four tables: "gemeinden", "zipcode", "city", and "state". The tables "zipcode", "city", and "state" are from Daniel Lichtblau, see http://www. lichtblau-it.de/downloads. The table "gemeinden" can be downloaded from GV-ISys, see http://destatis.de. Data were entered into an SQL database. The view is created with the following SQL:

```
------------------------------------
CREATE VIEW v_women_men AS
(
select
d.name AS bundesland,
c.state_id AS state_id,
a.gemeinde AS gemeinde,
(a.bevinsg / 1000000) AS bevinsg,
(8 * sqrt((a.bevinsg / 1000000))) AS transbev,
c.lat AS lat,c.lng AS lng,
(a.bevw / a.bevm) AS wm
from
(((gemeinden a join zipcode b)
join city c)
```

```
join state d)
where ((a.plz = b.zipcode)
and (b.city_id = c.id)
and (c.state_id = d.id)));
--------------------------------
```

gadm.org

The homepage says: "GADM is a spatial database of the location of the world's administrative areas (or adminstrative boundaries) for use in GIS and similar software. Administrative areas in this database are countries and lower level subdivisions such as provinces, departments, bibhag, bundeslander, (...). GADM describes where these administrative areas are (the "spatial features"), and for each area it provides some attributes, such as the name and variant names." For Germany we use the files "DEU_adm0.RData", "DEU_adm1.RData", "DEU_adm2.RData", and "DEU_adm3.RData".

NUTS-Maps

See http://ec.europa.eu/eurostat/web/gisco/geodata.

B Bibliography

Adler, Joseph (2012): R in a Nutshell (2nd edition). Sebastopol, CA: O'Reilly.

Annink, Ed / Bruinsma, Max (Ed.) (2010): Gerd Arntz, Graphic Designer. Rotterdam: NAi 010 Publishers.

Barker, Tom (2013): Pro Data Visualization Using R and JavaScript. Berkeley, CA: Apress.

Bertin, Jacques (2011): Semiology of Graphics: Diagrams, Networks, Maps. Redlands, CA: Esri Press.

Bivand, Roger S. / Pebesma, Edzer J. / Gómez-Rubio, Virgilio (2008): Applied Spatial Data Analysis with R. New York: Springer.

Brath, Richard / Jonker, David (2015): Graph analysis and visualization: discovering business opportunity in linked data, Indianapolis, Ind.: Wiley.

Brewer Cynthia A. (1999), Color Use Guidelines for Data Representation, in: Proceedings of the Section on Statistical Graphics, American Statistical Association, pp. 55–60.

Brewer, Cynthia (2005): Designing Better Maps: A Guide for GIS Users. Redlands, CA: ESRI Press.

Brunsdon, Chris / Comber, Lex (2015): An Introduction to R for Spatial Analysis and Mapping. Los Angeles u.a.: Sage Publications.

Burlingame, Noreen (2012): The Little Book of Data Science. Wickford, RI: New Street Communications LLC.

Cairo, Alberto (2012): The Functional Art: An Introduction to Information Graphics and Visualization. San Francisco: New Riders Publishers.

Cairo, Alberto (2016): The Truthful Art: Data, Charts, and Maps for Communication. San Francisco: New Riders Publishers.

Camões, Jorge (2016): Data at Work: Best practices for creating effective charts and information graphics in Microsoft Excel, San Francisco: New Riders.

Chang, Winston (2013): R Graphics Cookbook. Sebastopol, CA: O'Reilly.

Chen, Chun-houh / Härdle, Wolfgang / Unwin, Antony (Eds.) (2008): Handbook of Data Visualization (Springer Handbooks of Computational Statistics). Berlin u.a.: Springer.

Cleveland, William S. (1994): The Elements of Graphing Data. 2nd Ed., Murray Hill, NJ: AT & T Bell Laboratories.

Cleveland, William S. (1993): Visualizing Data. Murray Hill, NJ: AT & T Bell Laboratories.

Crawley, Michael J.(2012): The R Book. 2nd Ed., Chichester: John Wiley & Sons.

de Vries, Andrie / Meys, Joris (2012): R For Dummies. 2nd Ed., Chichester: John Wiley & Sons.

Dougenik, J. A. / Chrisman, N. R. / Niemeyer D.R. (1985): An Algorithm to construct continuous area cartograms, in: Professional Geographer, 37(1), S. 75-81.

Eve, Matthew / Burke, Christopher (Ed.) / Otto Neurath (2010): From Hieroglyphics to Isotype: A Visual Autobiography. London: Hyphen Press.

Few, Stephen (2009): Now You See It: Simple Visualization Techniques for Quantitative Analysis. Burlingame, CA: Analytics Press.

Few, Stephen (2012): Show Me the Numbers: Designing Tables and Graphs to Enlighten. 2nd Ed., Burlingame, CA: Analytics Press.

Few, Stephen (2013): Information Dashboard Design: Displaying Data for At-A-Glance Monitoring. 2nd Ed., Burlingame, CA: Analytics Press.

Friendly, Michael (2008): The Golden Age of Statistical Graphics. In: Statistical Science 23/4 (2008), pp. 502–535.
   URL: http://www.datavis.ca/papers/golden-STS268.pdf.

Friendly, Michael / Meyer, David (2015): Discrete Data Analysis with R: Visualization and Modeling Techniques for Categorical and Count Data. Boca Raton: Chapman & Hall/CRC.

Gastner, M. T. / Newman, M. E. J. (2004): Diffusion-based method for producing density equalizing maps, in: Proceedings of the NAS, 101(20), S. 7499-7504.

Gelman, Andrew / Unwin, Antony (2013): Infovis and Statistical Graphics: Different Goals, Different Looks. In: Journal of Computational and Graphical Statistics 22/1, pp. 2–28. Discussion by Robert Kosara: InfoVis Is So Much More: A Comment on Gelman and Urwin and an Invitation to Cosider the Opportunities; Paul Murrell: Comment; Hadley Wickham: Graphical criticism: some historical notes, pp. 29–44. Answer pp. 45–49.

Gray, Jonathan / Bounegru, Liliana / Chambers, Lucy (Ed.) (2011): The Data Journalism Handbook. How Journalists Can Use Data to Improve the News.
   URL: http://datajournalismhandbook.org/1.0/en/index.html.

Harmon, Katharine A. (2009): The Map as Art: Contemporary Artists Explore Cartography. New York, NY: Princeton Architectural Press.

Hornik, Kurt / Zeileis, Achim / Meyer, David (2006): The Strucplot Framework: Visualizing Multiway Contingency Tables with vcd, in: Journal of Statistical Software 17/3, pp. 1–48.

Kabacoff, Robert I. (2011): R in Action: Data Analysis and Graphics with R. Shelter Island: Manning.

Kenthirapalan, Sanketha et al. (2016): Functional profiles of orphan membrane transporters in the life cycle of the malaria parasite, in: Nature Communications 7, http://dx.doi.org/10.1038/ncomms10519

Loukides, Mike (2012): What Is Data Science? Sebastopol, CA: O'Reilly.

Mittal, Hrishi V. (2011): R Graphs Cookbook. Birmingham: Packt Publishing.

Murrell, Paul (2011): R Graphics. 2nd Ed., Boca Raton: Chapman & Hall / CRC Press.

Neurath, Otto (1930): Gesellschaft und Wirtschaft: Bildstatistisches Elementarwerk. Leipzig: Bibliografisches Institut.

Neurath, Otto (1939): Modern Man in the Making. 1st edition. New York u.a.: Alfred A. Knopf, Random House LLC.

Parker, Heidi G. et al (2017): Genomic Analyses Reveal the Influence of Geographic Origin, Migration, and Hybridization on Modern Dog Breed Development, in: Cell Reports 19/4 , S. 697 - 708

Perpinan Lamigueiro, Oscar (2014): Displaying Time Series, Spatial, and Space-Time Data with R. Boca Raton: Chapman and Hall/CRC, Press, Taylor & Francis Group.

Phillips, David / Barker, Gwendolyn E. / Brewer, Kimberly M. (2010): Christmas and New Year as Risk Factors for Death. Social Science & Medicine 71, pp. 1463–1471.

Rogowitz, Bernice E. / Treinish, Lloyd A., Why Should Engineers and Scientists Be Worried About Color?
http://www.research.ibm.com/people/l/lloydt/color/color.HTM

Segel, Edward / Heer, Jeffrey (2010): Narrative Visualization: Telling Stories with Data. In: IEEE Trans. Visualization & Comp. Graphics (Proc. InfoVis).
URL: http://vis.stanford.edu/papers/narrative

Slocum, Terry A. / McMaster, Robert B. / Kessler, Fritz / Howard, H. H. (Ed.) (2010): Thematic Cartography and Geovisualization, 3rd Ed., Upper Saddle River, NJ: Pearson Prentice Hall.

Steele, Julia / Iliinsky, Noah P. N. (Ed.) (2010): Beautiful Visualization: Looking at Data through the Eyes of Experts (Theory in Practice). Beijing [a.o.]: O'Reilly.

Symons, John / Pombo, Olga / Torres, Juan Manuel (2011): Logic, Epistemology, and the Unity of Science Volume 18. Otto Neurath and the Unity of Science. Dordrecht u.a.: Springer Science+Business Media.

Tufte, Edward (1990): Envisioning Information. Cheshire, CT: Graphics Press.

Tufte, Edward (1997): Visual Explanations. Images and Quantities, Evidence and Narrative. Cheshire, CT: Graphics Press.

Tufte, Edward (2000): The Visual Display of Quantitative Information, 2nd Ed., Cheshire, CT: Graphics Press.

Unwin, Antony (2015): Graphical Data Analysis with R. Boca Raton, FL: CRC Press.

van der Loo, Mark, P. J. / de Jonge, Eswin (2012): Learning RStudio for R Statistical Computing. Birmingham: Packt Publishing.

Wallace, Timothy R. / Huffman, Daniel P. (Ed.) (2012): Atlas of Design, Bd.1. Milwaukee, WI: nacis.

Wilkinson, Leland (2005): The Grammar of Graphics. 2nd Ed., New York: Springer.

Wong, Dona M. (2010): The Wall Street Journal Guide to Information Graphics: The Dos and Don'ts of Presenting Data, Facts, and Figures. New York u.a.: W.W. Norton & Co.

Yau, Nathan (2011): Visualize This: The FlowingData Guide to Design, Visualization, and Statistics. Indianapolis, Ind.: John Wiley & Sons.

Yau, Nathan (2013): Data Points: Visualization That Means Something. Indianapolis, Ind.: John Wiley & Sons.

Zeileis, Achim / Hornik, Kurt / Murrell, Paul (2009): Escaping RGBland: Selecting Colors for Statistical Graphics. In: Computational Statistics & Data Analysis 53/9, pp. 3259–3270.

Websites

http://flowingdata.com
http://driven-by-data.net
http://visualizingeconomics.com
http://rgraphgallery.blogspot.de

Printed in the United States
By Bookmasters